# Biogeology: Evolution in a Changing Landscape

# CRC Biogeography Series

## Malte C. Ebach

*School of Biological, Earth and Environmental Sciences, Australia*
*University of New South Wales*

**Biogeography and Evolution in New Zealand**
*By Michael Heads*

**Handbook of Australasian Biogeography**
*Edited by Malte C. Ebach*

**Neotropical Biogeography**: Regionalization and Evolution
*By Juan J. Morrone*

**Evolutionary Biogeography of the Andean Region**
*By Juan J. Morrone*

**Biogeology:** Evolution in a Changing Landscape
*By Bernard Michaux*

For more information about this series, please visit: https://www.crcpress.com/CRC-Biogeography-Series/book-series/CRCHANOFBIO

# Biogeology: Evolution in a Changing Landscape

## A Journey through Space, Time and Form

Bernard Michaux

CRC Press
Taylor & Francis Group
Boca Raton London New York

CRC Press is an imprint of the
Taylor & Francis Group, an **informa** business

CRC Press
Taylor & Francis Group
6000 Broken Sound Parkway NW, Suite 300
Boca Raton, FL 33487-2742



Printed on acid-free paper

International Standard Book Number-13: 978-0-367-14723-5 (Paperback)
International Standard Book Number-13: 978-0-367-14793-8 (Hardback)



**Library of Congress Cataloging-in-Publication Data**

Names: Michaux, Bernard, author.
Title: Biogeology : evolution on a changing landscape.
Description: Boca Raton : CRC Press, [2020] | Series: CRC biogeography series | Includes bibliographical references and index.
Identifiers: LCCN 2019011898| ISBN 9780367147938 (hardback) | ISBN 9780367147235 (pbk.)
Subjects: LCSH: Environmental geology. | Biogeography. | Biodiversity.
Classification: LCC QE38 .M53 2020 | DDC 550--dc23
LC record available at https://lccn.loc.gov/2019011898

**Visit the Taylor & Francis Web site at**
**http://www.taylorandfrancis.com**

**and the CRC Press Web site at**
**http://www.crcpress.com**

# Contents

# chapter one

# Setting the scene

## What is biogeology?

Biogeology is the name I use for the study of animal and plant distribution patterns resulting from an interplay between geology – specifically tectonics – and evolution. I prefer this term over 'biogeography' because it makes explicit the relationship between speciation and tectonically mediated changes of the earth's surface. Also, the term biogeography has become hopelessly fragmented and ill-defined even with qualifiers such as 'historical' or 'ecological'. Historical biogeography, for instance, refers to a number of disparate and philosophically divergent ideas such as dispersalist biogeography, panbiogeography and vicariance biogeography. Each of these approaches to historical biogeography is built on different 'ways of seeing' the world (Berger 1973) and is based on assumptions that are either explicitly stated (earth and life evolve together) or are implicitly held (distributional patterns are essentially modern, i.e. Neogene). Being explicit about underlying assumptions is, in my view, better than leaving them unstated and hidden, because then they are open to scrutiny and modification, and openess allows readers to better evaluate ideas and arguments. The causal relationship between tectonics and evolution is central to biogeology. The assumptions that underpin this causality are:

- *Evolutionary change is episodic and produces new species*

Evidence from the fossil record demonstrates that new species evolve suddenly and then remain unchanged (Eldredge and Gould 1972; Gould and Eldredge 1977; Michaux 1988, 1989). While it is common to refer to any change as 'evolutionary' (e.g. the evolution of stars or ties), all such examples are qualitatively different from evolutionary change that produces something that is biologically novel and disjunct (new species), rather than being a development within a given system (such as the growth of an organism throughout its life). The 'evolution' of stars or ties is, therefore, analogous to an individual's development rather than speciation. Small and cyclic changes in species through time ('microevolution') are not cumulative and do not lead to new species ('macroevolution').

- *Speciation is allopatric and requires disruption of an ancestral species' range*

Tectonic changes can cause a species' range to fragment, for example, on a regional scale when back-arc basins open or continents fragment, or on a more local scale, when land movements cause fluctuations in lake levels resulting in isolated embayments (Williamson 1981). There have been a number of suggestions as to how range fragmentation can promote speciation. Eldredge and Gould (1972) argued that significant genetic change could be achieved through chance events in small isolated populations leading to rapid evolution, but in truth, nobody knows. In my view, the formation of isolated, often peripheral populations promotes speciation through the disruption of individuals' developmental systems in response to increased physiological stress leading to heritable epigenetic changes (Michaux 2014). However, the pattern of speciation is what is important and there is, in my view, little or no convincing evidence for sympatric speciation, notwithstanding theoretical models and proposed examples that have been aired in the literature (e.g. Barluenga et al. 2006; Bolnick and Fitzpatrick 2007). The logic underpinning biogeology is that because tectonic events promote fragmentation of species' ranges, it also promotes speciation and thus leaves a biological imprint.

- *Tectonic processes can result in evolutionary changes in multiple lineages*

Tectonic changes can be large-scale processes that can act over a wide geographical area and on many species simultaneously. While species will react individually to tectonically mediated change, speciation in multiple lineages is to be expected. The repetition of distributional patterns in diverse and unrelated organisms, termed generalised tracks by Croizat (1964), is a consequence of speciation in multiple lineages following regional tectonic processes.

- *Range expansion versus chance dispersal*

Modern distributions are often explained by a series of individual, chance-dispersal events. The term 'chance' was first used by Darwin to describe *extraordinary* means of dispersal that enabled organisms to cross dispersal barriers such as oceans. The means have to be extraordinary; otherwise, by definition, there is no barrier to dispersal to cross. However, an organism's distribution in space not only requires movement but also survival. Consider the problems faced by a small group of organisms – or even an individual gravid female – that unexpectedly finds itself in a novel environment to which it is completely naïve. Survival and establishment depend on factors such as initial population size, sex ratio and ability to cope with a new biological context, all of which may be difficult for individual immigrants.

An example illustrating the phenomenon of range expansion was reported in the *Guardian* newspaper (29 September 2017) and concerned some 300 Japanese marine species that 'invaded' the west coast of North America following the 2011 Tōhoku tsunami off the east coast of Japan. An estimated one million creatures – including crustaceans, sea slugs and sea worms – made the 7725 km journey on a flotilla of tsunami debris. At first glance, this appears to be an example of chance dispersal in the sense that a random event (an earthquake) resulted in Japanese species crossing an oceanic barrier (the North Pacific). Indeed, Crisp et al. (2011) explicitly invoked rare events such as tsunami as potential causes of trans-oceanic disjunctions. However, things are often not what they seem at first glance. The 2011 tsunami was just the most recent of *many* such events that have occurred historically, such as the tsunami commemorated in the famous woodcut – *Kanagawa Oki Uranami* – by Katsushika Hokusai [Plate 1.1]. This is because Japan is adjacent to an active subduction zone just offshore that frequently generates large and shallow enough earthquakes to produce tsunami. So, why aren't there many Japanese marine species naturalised on the west coast of North America? One might argue that previous immigrant species did not establish themselves, but with 300 species recently making a successful trip, and with at least 29 tsunami recorded in historical times, you'd expect some to have survived and naturalised. Yet apart from Japanese species brought over in ballast water, as hull fouling, or from the aquarium trade, there are no examples of Japanese (or East Asian) species self-introducing on the Pacific coast of North America (Ruiz et al. 2000). So, what was different in 2011? Well, the answer is surprisingly

***Plate 1.1*** Kanagawa Oki Uranami by Katsushika Hokusai. The most celebrated of historical tsunamis to have occurred off the east coast of Japan. Mount Fuji in the background.

prosaic – plastic. Natural debris washed far out to sea after a tsunami will always carry attached organisms, but this 'natural' debris doesn't survive the long journey across the Pacific, becoming waterlogged or physically degraded and sinking to the ocean bottom. But plastic doesn't do either, so this time the Japanese species made it all the way. If plastic had always been part of the environment, then the North Pacific would not have been a barrier to dispersal for some Japanese marine invertebrates, and the west coast of North America would, in all probability, have been part of their normal range.

Whether these species will ultimately survive and become naturalised is not predictable. Only time will tell how this fascinating natural experiment will turn out. What is germane to the argument is that large-scale, tectonically mediated events interacting within a physical and biological context can affect many species. Tectonic events may be of sufficient scale to create new environments or disrupt existing environments, making it possible for whole biotas (or parts thereof) to expand their range. Because tectonically mediated range expansion affects many taxa and large numbers of organisms, survival probabilities will usually be higher than for individuals who disperse into new environments by chance (e.g. rare vagrants). As an explanatory mechanism for modern distributions, range expansion does not rely on a series of very low probability events such as chance dispersals or survival of individuals, and the events that promote range expansion are potentially discoverable because they may be significant enough to leave their mark in the geological record.

## *Dynamic earth*

The Geological Society of London marked the 50th anniversary of the publication of McKenzie and Parker's (1967) groundbreaking paper introducing plate tectonic theory to the world with a special conference in 2017. Jonathan Amos, the *Guardian*'s science correspondent, suggested that plate tectonics should appear on any list of great scientific achievements of the twentieth century because of the tremendous explanatory power that the theory had throughout the entire field of geology (Amos 2017). I couldn't agree more and would liken the revolution wrought in geology to similar effects in physics and chemistry brought about by quantum theory and in astronomy by general relativity. Like all truly revolutionary ideas, the concept of a tectonically active planetary surface also had profound implications for subjects beyond geology, and in particular for biological disciplines such as biogeography.

While McKenzie and Parker's (1967) paper was the first to be published on the subject, the central idea behind plate tectonics – that the earth's surface was in a constant flux when viewed from a geological perspective – has a much longer history that can be traced to a small book

*Die Entstehung der Kontinente und Ozeane* (*The Origin of Continents and Oceans*) published in 1915. Its author was a 35-year-old German meteorologist named Alfred Wegener who suggested that continents moved over time. As early as 1910, Wegener had noticed a fit between coastlines on either side of the south Atlantic and late in 1911 read, by chance it is said, a summary of fossil evidence for a former land bridge between Brazil and Africa. Wegener dismissed the idea of sunken land bridges on geophysical grounds and suggested instead that the fit between continental margins and the palaeontological and geological similarities between Brazil and Africa were a result of them being joined in the past. He called his theory *continental drift*.

The Achilles' heel of continental drift was the lack of a plausible mechanism of continental movement, leading most geologists of the time to dismiss Wegener's ideas outright. However, he was not without some influential allies such as Arthur Holmes in England. Holmes' early support for continental drift was probably based as much on his intellectual commitment to 'big picture' thinking as it was on his acceptance of Wegener's geological and palaeontological evidence. His knowledge of geophysical processes led Holmes to propose that slow-moving convection currents within the mantle were the driving force behind continental drift (Holmes 1931). Although this explanation provided a credible mechanism for Wegener's theory, it was largely ignored. Holmes was to become a very influential figure in British geological circles, and during his long teaching career, he continued to promote continental drift to his students. It's interesting to speculate what effect this influence might have had on British geologists such as McKenzie, Parker, Oxburgh and Dewey, who were important contributors to the early development of plate tectonic theory.

Wegener's reputation was eventually rehabilitated when technological advances provided irrefutable evidence for horizontal continental movement in the 1960s. Sonar, originally developed in World War 1, had allowed increasingly accurate maps of the ocean floor to be made. To everyone's surprise, the ocean floor was not smooth. The Atlantic Ocean had a major mountain range running down the centre – the prosaically named Mid-Atlantic Ridge. Then seismic arrays, established around the globe in the early 1960s to monitor nuclear testing as part of the Nuclear Non-Proliferation Treaty, showed that earthquakes were largely aligned along narrow zones that defined the edges of plate-like structures on the earth's surface, a pattern that was also mirrored by volcanic activity. The final piece of evidence was the discovery that the earth's magnetic field reverses frequently and leaves an imprint on volcanic rocks when magnetic iron oxide minerals are fixed in their magnetic orientation as the molten rock solidifies. It was found that the reversal patterns in the rocks of the ocean floor ran parallel to the mid-ocean ridges and were symmetrical about them. The implication was that new ocean crust was being made at the mid-ocean ridges and thus the idea of sea-floor spreading gained general acceptance. If new crust was being generated at mid-ocean ridges, then older oceanic crust had to be destroyed (unless the earth was expanding) and it was quickly understood that this occurred at ocean trenches. So active is this process that while continental rocks as old as four billion years can be found (Bowring and Williams 1991), the oldest oceanic crust is only 200 million years old. Thus, the theory of plate tectonics was born and continental drift proved to be correct after all.

Plate tectonic theory has come a long way in past 50 years, and while the general idea of the earth's surface being composed of a dozen or so large lithospheric plates still holds, the theory has advanced beyond this classical stage. Biogeology was developed in relation to the biogeographic regions I've been interested in, namely Wallacea, New Guinea, Melanesia and New Zealand. All of these regions are tectonically complex, and their development is not explicable in terms of rigid plate interactions but rather in terms of exotic terranes (island arcs, continental fragments and marginal basins), their movement and interaction with other terranes or adjacent continental regions and continent–continent collision zones. Here, the tectonics occur at the margins of great plates – the Indo-Australian and Pacific in particular – where the crust is fragmented into small basins, arcs, rift zones and continental fragments that are (over geological time) continuously forming, amalgamating and rearranging their configurations. Oxburgh (1972) descriptively termed the tectonics at the margins of plates as flake tectonics, but even this picture doesn't quite capture the *plasticity* of the tectonic processes at plate margins in regions of prolonged high-heat flow.

Biogeography was born at a time when biogeographers thought that the earth's surface was permanent in the sense that the present configuration of continents and islands was unchanging. Sure, new islands could come into being through such processes as volcanic eruption or coral growth, and all land was subject to erosion and periods of mountain building, but their relative positions remained unchanged. While geologists have altered their 'ways of seeing' and pre-plate tectonic geological concepts simply aren't relevant any more, biogeographers don't seem to have embraced this change in perspective with quite the same enthusiasm. Neo-dispersalism (Ebach 2017) is a nineteenth-century idea, derived directly from Darwin (and later Wallace) via Darlington, Simpson and other adherents of the 'modern' synthesis with a virtually unchanged conceptual schema. Decorating old ideas with the trappings of modernity in the form of modern techniques cannot disguise this fact. The earth's surface is dynamic and because individual species' longevity is in the order of millions of years, the history of clades can be expected to have been influenced by processes operating over geological time scales.

## *Organisation of the book*

Each chapter of the book is introduced by a publication that traces the chronological development of biogeology. Individual chapters cover regions that form a broad swathe of islands and archipelagos stretching from Wallacea (approximately modern Indonesia) through New Guinea and Melanesia to New Zealand, the New Zealand Subantarctic Islands and Antarctica. The tectonic development of this region was dominated by the breakup of the Australian sector of East Gondwana during the Late Cretaceous, Australia's subsequent northward drift in the Eocene and its on-going collision with South East Asia/Sundaland, and a long-lived and complex tectonic boundary between the Pacific and Australian plates. How tectonic events help explain the highly diverse and endemic biotas now inhabiting individual islands and archipelagos is central to each publication.

Chapter 2 provides an overview of biogeographic patterns within the southwest Pacific and discusses Leon Croizat's seminal contribution to biogeographic thought. This overview underlines the connection, both geological and phylogenetic, between these diverse islands and serves to unite the individual treatments that follow. Leon Croizat was an important early influence in the development of my ideas, as was Alfred Russel Wallace (Chapter 5), two authors who are not normally associated with each other. Indeed, Croizat was scathing about Wallace and his supposed support for 'centres of origin', but to my mind, Wallace was the first person to systematically think about modern distributions in terms of past geographies. I revisit and expand my thoughts about the relationship between biology and geology and use an analysis of the evolutionary history of the birds-of-paradise as an illustration. Chapter 3 critiques neo-Dispersalism and explores ways in which historical research can be interpreted without recourse to ad hoc explanations. Chapter 4 revisits New Guinea and updates the biogeology of this most fascinating island, placed at the leading edge of the Australian craton. An evaluation of the use of single taxon studies in New Guinean biogeology in particular and biogeographical studies in general concludes the chapter.

Chapter 5 is one of two devoted to Wallacea. It begins with a short biographical essay of my great hero A.R. Wallace, reassesses the biogeology of Sulawesi and concludes with a schema designed to plan an ideal biogeological investigation. Chapter 6 discusses the history of the New Zealand Subantarctic Islands and evaluates the contribution of Charles Fleming – who was one of the earliest scientists to work in the New Zealand Subantarctic Islands – to the development of New Zealand biogeography. The chapter also includes an update on Campbell Plateau geology, the phenomenon of old taxa on young islands, and concludes with an analysis of the biota of the Chatham Islands that are – erroneously, in my opinion – usually included as part of the New Zealand Subantarctic Islands.

Chapter 7 is concerned with the reconstruction of polar Gondwana at 100 Ma and discusses the hypothesis that a proto-alpine fault was already an active tectonic boundary at this time. Chapter 8 is the second of the chapters devoted to Wallacea. It starts with an essay exploring the development of the concept of natural groups and how this might apply to areas of endemism. An evaluation of different sorts of areas of endemism is illustrated with reference to the biogeology of the Banda Arcs. Chapter 9 is concerned with the biogeology of New Zealand, which is illustrated by examining the changes in its flora and molluscan fauna during the Late Cretaceous and Cenozoic. Chapter 10 is a synthesis of the biogeology of the Indonesian-west Pacific region.

# Points of View

*Syst. Zool.*, 38(4):390–398, 1989

## Generalized Tracks and Geology

B. Michaux

*Evolutionary Genetics Laboratory, Department of Zoology, University of Auckland, Private Bag, Auckland, New Zealand*

> Main massings and major "tracks" . . . are the biological expression of basic geological underlying realities. *Flesh and rocks evolve together.* There is reason of the strongest why the biogeography of the New World (and all the earth) does not necessarily agree with its present geography. This biogeography is primarily answerable as a matter of fact indeed to former geographies, not the current one. (Croizat, 1961:77)

Croizat's important insight into distribution patterns is that they are highly repetitious. He called repetitive patterns "generalized tracks" (Croizat et al., 1974), lines on a map which summarize distributions of many diverse individual taxa. He rejected chance dispersal as the cause of generalized tracks because, as he frequently remarked, the pattern is independent of the individual taxon's dispersal abilities. He argued against a series of special explanations to explain a single pattern.

He envisaged that during periods of "mobilism" the range of an ancestral taxon expands—to an extent consistent with both its ecological requirements and dispersal ability ("means of survival")—until a new range is stabilized and a period of "immobilism" begins. If the expanded range subsequently becomes fragmented, "form making" differentiates these vicariant populations. In Croizat's terminology, "dispersal" is the history of "form making." The term dispersal, as generally used, can be equated with his term "mobilism." If the process causing fragmentation operates on a sufficiently large scale (such as the opening of a marine basin, or the uplift of a mountain chain), then ranges of many individual taxa would be similarly fragmented and a "generalized track" would result.

Croizat noted that some major tracks are associated with geographically extensive mountain belts. An example, much discussed by Croizat (Croizat, 1958, 1961, 1964), is the Tethyan track running from Europe via northern India to southeast Asia. Extensions of the track run west to the Caribbean and east into the Australasian region. The track runs along major Tertiary orogenic belts formed during the destruction of the Tethys ocean.

A relationship between tracks and geosynclines was regarded by Croizat as fundamentally important. In discussing effects of mountain building associated with destruction of geosynclines (Croizat, 1964: 59–60), he noted not only the negative effect of climatic change on the biota of the two approaching "shields" (geologically stable continental areas) leading to extinction, but also the positive effects promoting "form making," which occurs in a number of ways. As new living space is created and made available to the original biotas a period of mobilism would occur resulting in a mixed biota. As this (biogeographically) composite area becomes uplifted the biota would respond to changing conditions by changing in form. "Form making" is also promoted as mountain chains become both higher and more extensive. Rising mountain chains promote periods of mobilism for those groups living in various altitudinal zones, which in turn provides potential for further "form making" if these new extended ranges become vicariated. Lastly, Croizat noted the major effects that erosion have on the physical geography of the areas adjacent to the mountain belts, and the promotion of form making associated with these physical changes.

It should be clear from the discussion above that the Tethyan track resulted, in Croizat's view, from the combination of two distinct biotas, with their own separate biogeographical histories, and subsequent in situ evolution. Croizat also recognized another type of "dispersal," which he termed "horstian." He defined this as "Dispersal along insular chains . . . or interplays between islands "in front" and land "in back . . ." (Croizat, 1964:62). "Horstian dispersal," associated with island arc and foreland systems, is thus distinct from geosynclinal-type dispersal in the sense that fragmentation of an existing biota occurs, rather than the combination of biotas. Both merging and fragmentation of biotas are, as noted by Craw (1982), of equal importance in understanding how geographic distributions have arisen, and, in conjunction with "form making" associated with these events, give insight into Croizat's dictum that "earth and life evolve together."

## RELATIONSHIP BETWEEN BIOGEOGRAPHY AND GEOLOGY

. . . it is basically around the concept of geosyncline . . . that biogeography and geology do come together. (Croizat, 1964:62)

Craw and Weston (1984) have recently developed one of the corollaries of Croizat's view of the natural relationship between biogeography and geology—that of novel geological prediction arising from track analysis. Seberg (1986) criticised this claim on two counts, namely that these authors concentrated only on corroborative evidence, and that alternative geological histories for many areas conflict. Seberg (1986) concluded that the predictive power ". . . with respect to historical geologic events is a fiction."

While Craw and Weston (1984) are certainly correct in drawing attention to the implications for biogeography of advances in understanding the geological history of southeast Asia, the evidence presented portrays a rather simplified picture of what is a complex of contentious issues. A number of different geological models discussed in the literature place parts of southeast Asia adjacent to Gondwana. These include the expanding earth hypothesis (e.g., Carey, 1975; Kremp, 1982), the Pacifica hypothesis (e.g., Nur and Ben-Avraham, 1982), as well as more 'conventional' Gondwanic reconstruction models such as those of Ridd (1980), Parker and Gealey (1983), Wongwanich and Burrett (1983) and Burrett and Stait (1985). Biogeographic predictions, based solely on spatial relationships between areas, cannot distinguish between these hypotheses (Platnick and Nelson, 1988).

The geology of the southeast Asian and western Pacific areas is complicated, and there is a danger that simplification of the evidence leads not to illumination but to confusion. Take the example of the Pacifica hypothesis. Craw and Weston (1984:9) seemed to imply that western North America, parts of Asia and other Pacific fragments may have formed a Pacifica continent. Craw (1982:fig. 1C) also invoked a Pacifica continent sited west of New Zealand during Jurassic/Triassic times. At its most basic level, Nur and Ben-Avraham's (1982) argument concerned the role of micro-continental fragments in orogeny. They argued that orogenic events are initiated by the attempted subduction and consequent accretion of terranes before continent–continent collision. In support of their hypothesis they pointed out that fragmentation commonly accompanied continental rifting (e.g., Africa), and that there are numerous oceanic plateaus that may be continental fragments. Their reconstruction of Pacifica (Nur and Ben-Avraham, 1982:fig. 9) is one in which west Pacific plateau regions are reunited forming "an extension of Gondwana beyond New Zealand and Australia."

There appears to be little evidence that the exotic terranes forming much of the Pacific seaboard of North America (Coney et al., 1980) were derived from the west (Pacifica). Paleomagnetic evidence suggests that many of the terranes were derived from the south (Beck, 1980; Hillhouse and Grome, 1984; Champion et al., 1984; Howell, 1985). There is some indication that the oldest terranes have a distinct westerly component to their movement (Debiche and Cox, 1985). However, these terranes are at least 180 million years old, originating during the lower Jurassic or earlier. Although Croizat's biogeographic analyses clearly demonstrate that many biological relationships are ancient in origin, I find it difficult to accept that modern distribution patterns would still record such an ancient event.

The same argument applies to the history of Sibumasa block (eastern Burma and the Thai-Malay peninsula). Ridd (1980) argues that this block was adjacent to Gondwana during the Devonian. By the Cretaceous it had collided with the Indo-China block, and these as a unit with the South China block. Even accepting this argument (and see McEhinny et al. (1974), Buffetaut and Ingavat (1985), and Helmcke, (1985) for contrary evidence), what has a Devonian connection between a part of Asia and Gondwana to do with modern distributions? Much more to the point is Ridd's (1980) suggestion that northeastern Borneo and the Indian ocean sides of Sumatra and Java are younger Mesozoic and Tertiary accretions. Cameron et al. (1980) suggested that western Sumatra is a continental fragment that closed a marginal basin in the region during the Cretaceous.

Craw and Weston (1984) confounded their analysis of the relationship between biogeography and geology with their attempt to show both the superiority of panbiogeography and to distance it from vicariance biogeography. I do not believe that they have made a strong case for biogeography being able to provide geologists with novel predictions. However, if one is content with a more modest contribution of modern biogeographic data affecting geologists' "way of seeing" (Berger, 1972)—to become an accepted analytical tool with similar status to structural and stratigraphic analyses so to speak—then biogeography and geology do, very naturally, come together.

### RELATIONSHIP BETWEEN GEOLOGY AND BIOGEOGRAPHY

The other side of the biogeography/geology coin—what geology might tell us about tracks—has received little attention. The quote at the beginning of this paper sums up this relationship—"the basic . . . underlying realities" of tracks are geological. Tectonic history is explanatory. How did members of the molluscan genus *Amalda* (Olividae: Ancillinae) come to be (or to have been) distributed between the Caribbean and New Zealand? Answer; because ancilline ancestors were widespread around the Tethyan shores of Gondwana, and were rafted into position by dispersing continental fragments (Michaux, 1989a). In the following section I would like to detail a general geological history for the southwest Pacific area, and to demonstrate that biotic distributions appear to reflect this history.

*Generalized tracks and the geological history of the southwest Pacific.*—Croizat (1964:fig. 39) identified a number of tracks in eastern Australia and Polynesia (see Fig. 1E). These tracks form three "beams." One connects the northern lowlands of West Irian/Papua New Guinea to the Solomon Islands and Vanuatu, thence to Fiji and on to Polynesia in general. Two link the central Highlands of West Irian/Papua New Guinea to New Caledonia–New Zealand, and eastern Australia-Tasmania respectively. These three "beams" are linked by a track from the borders of Queensland and New South Wales to Lord Howe Island, Norfolk Island, New Caledonia, Vanuatu and onwards. To the north and west these southwest Pacific tracks link with both the Tethyan and (Indian Ocean) Gondwanic tracks.

Progress has been made in understand-

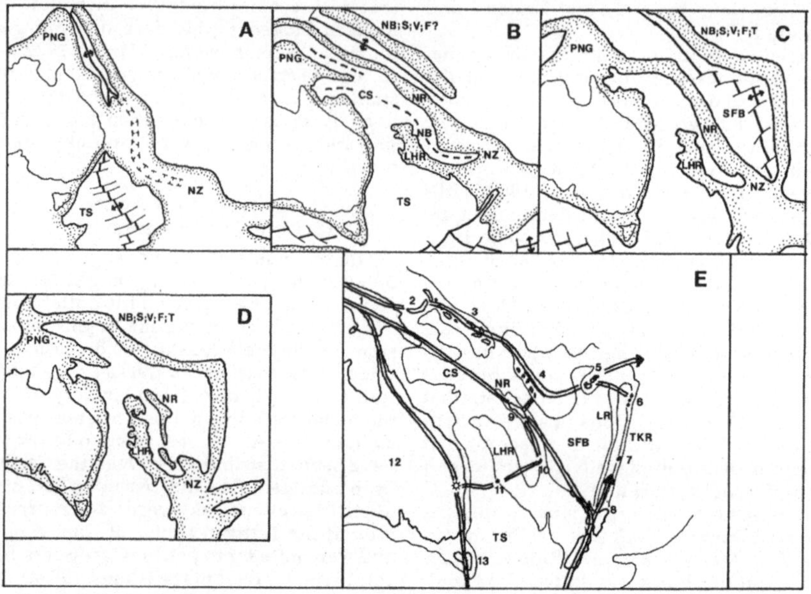


FIG. 1. Cartoons showing the tectonic development of the southwest Pacific during uppermost Mesozoic to late Miocene (A–D, from Parker and Gealey (1983)), and generalized tracks in the same area (Fig. 1E, based on Croizat 1964). Continental crust stippled margin, oceanic crust plain. A. Late Cretaceous (70MYBP). Breakup of Gondwanan east coast almost completed. Spreading in the Tasman Sea (TS) well advanced. Spreading system active in the Coral Sea to the north, with rifting into the New Zealand area (NZ). It is envisaged that the northern peninsula (incipient outer-Pacific arc) was a dynamic, restless, archipelago—similar to the Indonesian arc today. B. Middle Eocene (47MYBP). Northwestern extension of the Norfolk Ridge (NR) separated from the foreland of Papua New Guinea (PNG). Back-arc spreading separated New Britain (NB), Solomons (S), Vanuatu (V) and Fiji (F) from the later ridge system, forming the outer-Pacific arc. Norfolk Basin (NB) completed, Lord Howe Rise (LHR) largely submerged to depth. C. Upper Oligocene (25MYBP). Norfolk Ridge collided with Papuan foreland forming the Central Highlands. South Fiji Basin (SFB) opened, collision of Three Kings terrane (Fig. 2) with Northland at the start of the Miocene caused realignment of Tonga-Kermadec arc to complete the outer-Pacific arc. D. Late Miocene (MYBP). Reversal in polarity of western outer-Pacific arc caused a collision with Papua New Guinea forming the northern coastal region. E. 1 West Irian/Papua New Guinea; 2 New Britain; 3 Solomon Islands; 4 Vanuatu; 5 Fiji; 6 Tonga; 7 Kermadec Islands; 8 North Island, New Zealand; 9 New Caledonia; 10 Norfolk Island; 11 Lord Howe Island; 12 Australia; 13 Tasmania; star = MacPherson-Macleay overlap.

ing the overall tectonic development of the southwest Pacific. A general summary for the development of the entire area, with figures showing reconstructions, is given in Parker and Gealey (1983). Ballance et al. (1982) provided a finer grained resolution for New Zealand and the area to the immediate north. Figure 1A–D is based on Parker and Gealey (1983). It is clear that during the upper Cretaceous all the areas mentioned formed a single geological unit adjacent to eastern Australia. This reconstruction is obtained by closing up the basins which postdate this age.

By the end of the Cretaceous this unit was separating from eastern Australia by the opening of the Tasman Sea to the south, and the Coral Sea to the north (Fig. 1A). The pivotal region for the opening of the Tasman Sea at this time appears, on the

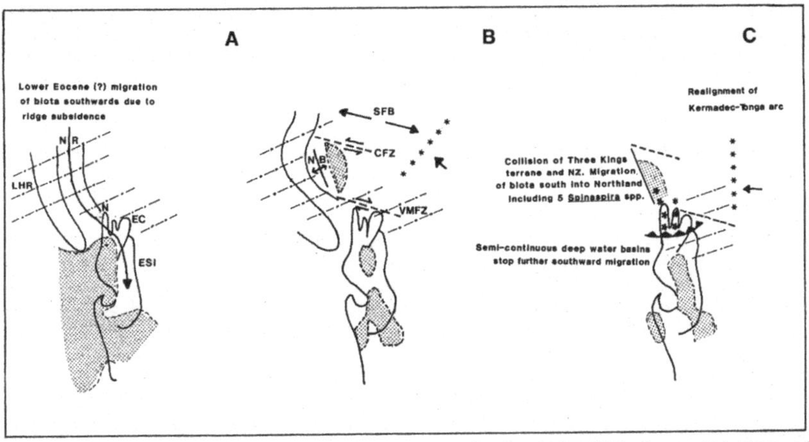


FIG. 2. Stippled = land; dash-dot = deep water; star = volcanic arc. A. Middle Eocene. Subsidence of Norfolk Ridge and/or northwards drift of New Zealand forces northern biota into the New Zealand region. EC East Cape; ESI east coast South Island; LHR Lord Howe Rise; N Northland. B. Upper Oligocene. South Fiji Basin (SFB) opens in the New Zealand region. Norfolk Basin opens, Three Kings terrane moves southeast along the Cook fracture zone (CFZ) and Vening-Meinesz fracture zone (VMFZ). Stars are active volcanoes along the Tongan trench system. C. Lowest Miocene. Collision of Three Kings terrane with New Zealand, emplacement of the Northland allochthon.

reconstruction of Parker and Gealey (1983), to be the Queensland–New South Wales border, identified by Croizat, (1964) as an important biogeographic node (the MacPherson-Macleay overlap). This separation became complete in the Palaeocene. Concurrent with the separation of the area as a whole from eastern Australia was the formation of two parallel arc systems, the inner and outer Pacific arcs (Fig. 1B, Eocene). These two arcs were initially a composite arc which divided in response to back-arc spreading. The Lord Howe Rise, existing now as a separate entity, submerged to bathyal depths.

Parker and Gealy's (1983) Eocene reconstruction shows the opening of the Solomon Sea, Vanuatu Basin and South Fiji Basin. However, Ballance et al. (1982) have suggested an Oligocene age for the opening of the South Fiji Basin based on magnetic anomalies. It is clear, however, that the Norfolk Ridge (=inner Pacific arc) was separated from the Solomon-Vanuatu ridge (=outer Pacific arc), but probably not from the Kermadec or Colville-Lau ridges, at this time. On a more local scale the middle Eocene is an interesting age—biotically—for New Zealand. As Fleming (1962) commented "The Bortonian marine shelf fauna is one of the most distinctive in New Zealand's Tertiary history." This distinctiveness is caused by the sudden influx of "Indo-Pacific" Scleractinian coral genera (Keyes, 1968), gastropods of the families Cypraeidae, Mitridae, Conidae (Fleming, 1962), and Ancillinae (Michaux, 1989b), and Foraminifera (Hornibrook, 1971). On land the vegetation was also augmented by trees of northern affinity (listed in Fleming, 1962), including *Nothofagus* spp. of the *brassi* group which today are found in New Caledonia and the highlands of West Irian/Papua New Guinea. In Figure 2A it is suggested that this biota became mobile and migrated into the region in response to New Zealands northward drift and/or subsidence of the central portions of the Norfolk Ridge. Craw's (1988) "outer eastern arc biota" may be interpretable in terms of this event.

During the Oligocene (Fig. 1C) the inner-Pacific arc reversed polarity, collided with Papua New Guinea forming the Central Highland region, and caused the emplacement of ophiolites in both Papua New Guinea and New Caledonia. The opening of the South Fiji Basin realigned Tonga (and possibly Fiji) as part of the outer Pacific arc. This outer arc, subjected to subduction and transcurrent faulting between Halmahera (Moluccas) and the Solomons, doubled up and collided with Papua New Guinea during the late Miocene forming the northern lowlands (Fig. 1D). Important local events also occurred at the start of the Miocene in New Zealand.

The lowest Miocene (Otaian Stage) of Northland is characterized by a unique biota of northern affinity. This boita is restricted to Northland during the Otaian. 17 Indo-Pacific genera of corals are recorded from localities in Northland including *Porites* and *Turbinaria*, the only recorded occurrence of reef-building coral genera in New Zealand (Keyes, 1968). Similar Indo-Pacific influences are recorded by molluscs (Fleming, 1962; Wakefield, 1977; Michaux, 1989b), larger Foraminifera (Hornibrook, 1971; Chaproniere, 1980), Ostrocoda, and a significant number of plant genera such as *Metrosideros, Dysoxylem, Freycinetia, Rhopalostylis, Elytranthe, Macropiper, Avicennia* and *Dodenaea* (Fleming, 1979). The New Zealand Miocene coconut *Cocos zeylanica* is also recorded from Northland. The majority of these newcomers were shortlived and soon became extinct. Indeed, judged by its fruit size, *Cocos zeylanica* was at the extreme of its range. Sub-tropical temperatures for the Otaian based on faunal evidence is not supported by oxygen isotope evidence (Devereux, 1967).

The second important event was the emplacement of the Northland allochthon. This very substantial body outcrops over much of Northland and parts of the East Cape area and may extend further south (Ballance et al., 1982). Its gravity slid into its present position from the north at the start of the Miocene (Ballance and Spörli, 1979). This allochthonous body is composed of Jurassic to Late Cretaceous oceanic volcanic basement, contained within a disrupted Late Cretaceous to Late Oligocene sequence, the total thickness being several kilometres. Ballance et al. (1982) argue that these sediments were derived from an area immediately to the north of Northland. The cause of the uplift required to evert the basin in which these sediments were originally deposited is unknown.

In Figure 2B and C it is suggested that these two events are linked. It is proposed that Northland's "marginally tropical fauna" was transported from the north on an exotic terrane. This explanation is preferred to postulating that the biota was derived by migration down an emergent Norfolk Ridge. The occurrence of deep water Upper Eocene sediments in the north of Eocene New Zealand and the lack of any Oligocene migrants suggest that this ridge was submerged during these periods. It is difficult to imagine what mechanism might account for its subsequent uplift at the beginning of the Miocene. It is, however, an alternative hypothesis that cannot be entirely discounted.

In this paper the terrane is identified as what is now known as the Three Kings Rise (Fig. 2C). There is no direct geological evidence to support this hypothesis (or for that matter to refute it), but to be tenable the age of the Norfolk basin should be late Oligocene. It is not envisaged that this terrane travelled far. Closing the Norfolk basin and sliding the Three King's Rise back along the Cook and Venning-Meinesz fracture zones would place the center of the Rise opposite Norfolk Island and its northern tip south of New Caledonia. The collision between this terrane and New Zealand caused basin compression which provides the mechanism for the emplacement of the Northland allochthon.

Croizat (1964), quoting Allan (1961), noted a remarkable fact about the Three Kings Islands. These islands, which are the only subaerial part of the Rise, have an area of just 8 km$^2$ but contain "175 different plant species of which about 7.5% are endemic including two monotypic genera, *Plectomirtha* (Sapindacea), and *Elingamita* (Myrsinaceae)." (Croizat, 1964:61). Cheeseman (1887), who visited the islands prior to the release of goats, noted a number of plant species identical or very similar to species found on Norfolk island. Biotic affinities to more distant places are evinced by the rail *Rallus philippensis* and the tree *Paratropis* sp., a genus known only from Fiji, Tahiti and the Philippines (Cheeseman, 1887).

Comparing the foregoing discussion on the tectonic development of the southwest Pacific with Croizat's (1964) summary of the biogeography of the area shows a remarkable concordance (Fig. 1). The MacPherson-Macleay overlap is "nodal" because it occupies a position at the northernmost extension of the Tasman Sea, connecting areas both to the south and east. The outermost "beam" is equivalent to the outer Pacific arc (northern lowlands of PNG–New Britain–Solomons–Vanuatu–Fiji–Tonga). The middle "beam" is equivalent to the inner Pacific arc (central highlands PNG–New Caledonia–New Zealand), and the inner "beam" to coastal "dispersal" along the western shore of the Tasman Sea. The track connecting the three beams was formed as ancestral biotas became vicariated through basin formation and subsequent subsidence. My only disagreement concerns that part of the central track drawn directly between New Caledonia and New Zealand, which should be drawn via Norfolk Island. This latter track (New Caledonia–Norfolk Island–New Zealand) seems to be a temporal composite resulting from separate events of "mobilism" during the Eocene and at the beginning of the Miocene.

## SUMMARY

If one accepts that "earth and life evolve together" then geology's relationship to biogeography is causal. They are "... the independent and dependent variables respectively in a cause and effect relationship ..." Rosen (1978). Craw (1988) has criticized this approach as leading to "narrative." He argues that geology and biogeography are best linked at the conceptual level. But what does this mean? We find (Craw, 1988:table 1) that the biogeographic concept of the node is equivalent to the geographic concept of "Composite geological area; new region formed by terrane collision, docking and/or suturing." Yet this is not correct in the case of the MacPherson-Macleay overlap. The equivalence of main massings with "Zones of metamorphism, granitization or sedimentation" is nonsense.

Craw (1988) further states that "The common descriptive and analytic language uniting systematics, geology and biogeography is graph theory ..." (Craw, 1988: 292). As an example of facilitating communication among these fields he provides a geological area cladogram for areas of endemism in New Zealand based on geological characters, which suggests "a historical geological relationship between these [three] areas." Similarity in geological structures has no necessary historical basis because geological structures are related primarily by process and only secondarily, if at all, by history. It would be quite feasible to link two areas with no historical connection by "geological characters" if those areas experienced similar tectonic processes. Cladograms based on such evidence would be as reliable as biological cladograms based on characters which evolved independently of genealogy.

In the example of the biogeography and geological history of the southwest Pacific I hope that I have demonstrated that the "narrative" approach makes Croizat's tracks intelligible. In other cases conflict may arise because particular geological models and biogeographic hypotheses are at odds. In this case no priority can be assigned to either set of evidence. Even though geology is ultimately causal, this does not guarantee that geologists have correctly interpreted an area's geological history. Likewise, biogeographic hypotheses may be wrong. Perhaps only time and a good deal of rethinking will resolve these cases.

Biogeography's relationship to geology is also far from clearcut. It might be argued that the hypothesis presented in this paper concerning the nature of the Three Kings Rise is a novel geological prediction but that would be elevating the scientific status of speculation—for that is what it is—to heights that it does not warrant. Speculation is not necessarily a bad thing, particularly if it focuses attention on interesting geological questions concerning the Three Kings Rise and associated basin system, or the unique biogeography of the Three Kings Islands. A knowledge of biogeographic hypotheses can only broaden a geologist's perception of tectonic problems.

## ACKNOWLEDGMENTS

I would like to thank Jack Grant-Mackie and Peter Ballance (Auckland University) for many long discussions about the geology of New Zealand and the southwest Pacific, and for comments on earlier drafts. I also acknowledge the two anonymous referees whose comments were most helpful. This work was supported by UGC grant number 394.602 and AURC grant number 449.213 and is publication number 28 from the EGL.

## REFERENCES

ALLAN, H. H. 1961. Flora of New Zealand. Volume 1. Government Printer, Wellington.

BALLANCE, P. F., AND K. B. SPÖRLI. 1979. The Northland Allochthon. J. Roy. Soc. N. Z., 9:259–275.

BALLANCE, P. F., J. P. PETTINGA, AND C. WEBB. 1982. A model of Cenozoic evolution of northern New Zealand and adjacent areas of the southwest Pacific. Tectonophysics, 87:37–48.

BECK, M. E. 1980. Paleomagnetic record of plate-margin tectonic processes along the western edge of North America. J. Geophy. Res., 85:7115–7131.

BERGER, J. 1972. Ways of seeing. Pelican, London.

BUFFETUAT, E., AND R. INGAVAT. 1985. The Mesozoic vertebrates of Thailand. Scientific American, 253: 80–87.

BURRETT, C., AND B. STAIT. 1985. Southeast Asia as part of an Ordovician Gondwanaland—A palaeobiogeographic test of a tectonic hypothesis. Earth Planet. Sci. Lett., 75:184–190.

CAMERON, N. R., M. C. G. CLARKE, D. A. ALDISS, J. ASPDEN, AND A. DJUMEDDIN. 1980. Pages 149–188 *in* The geological evolution of northern Sumatra. Proc. 9th. I.P.A. Conference, Jakarta.

CAREY, S. W. 1975. Pages 1–31 *in* Tectonic evolution of Southeast Asia. Proc. 4th. I.P.A. Conference, Jakarta.

CHAMPION, D. E., D. G. HOWELL, AND C. S. GROME. 1984. Paleomagnetic and geological data indicating 2500 km of northward displacement for the Salinian and related terranes, California. J. Geol. Res., 89:7736–7752.

CHAPRONIERE, G. C. 1980. Influence of plate tectonics on the distribution of late Palaeogene to early Neogene larger Foraminiferids in the Australasian region. Palaeogeogr., Palaeoclimatol., Palaeoecol., 31: 299–317.

CHEESEMAN, T. F. 1887. Notes on the Three Kings Islands. Trans. N.Z. Inst., 20:141–50.

CONEY, P. J., D. L. JONES, AND J. W. MONGER. 1980. Cordilleran suspect terranes. Nature, 288:329–333.

CRAW, R. C. 1982. Phylogenetics, areas, geology and the biogeography of Croizat: A radical view. Syst. Zool., 31:304–316.

CRAW, R. C. 1988. Continuing the synthesis between panbiogeography, phylogenetic systematics and geology as illustrated by empirical studies on the biogeography of New Zealand and the Chatham Islands. Syst. Zool., 37:291–310.

CRAW, R. C., AND P. WESTON. 1984. Panbiogeography: A progressive research program? Syst. Zool., 33:1–33.

CROIZAT, L. 1958. Panbiogeography. Volumes 1, 2a, and 2b. Published by the author, Caracas.

CROIZAT, L. 1961. Principia botanica. Volume 1a and 1b. Published by the author, Caracas.

CROIZAT, L. 1964. Space, time, form: The biological synthesis. Published by the author, Caracas.

CROIZAT, L., G. NELSON, AND D. E. ROSEN. 1974. Centres of origin and related concepts. Syst. Zool., 23:265–287.

DEBICHE, M., AND A. COX. 1985. Pages 58–61 *in* Terrane trajectory analysis for the northeast Pacific. Geol. Soc. Australia Abstracts 14 (3rd. cicum-Pacific Terrane Conference, Sydney).

DEVEREUX, I. 1967. Oxygen isotope paleotemperature measurements on New Zealand Tertiary fossils. N.Z.J. Sci., 10:988–1011.

FLEMING, C. A. 1962. New Zealand biogeography: A palaeontologist's approach. Tuatara, 10:53–108.

FLEMING, C. A. 1979. The geological history of New Zealand and its life. Auckland University Press, Auckland.

HELMCKE, D. 1985. The Permo-Triassic "Paleotethys" in mainland Southeast Asia and adjacent parts of China. Geol. Rdsch., 74:215–228.

HILLHOUSE, J. W., AND C. S. GROME. 1980. Paleomagnetism of the Triassic Hound Island Volcanics, Alexander terrane, southeast Alaska. J. Geophys. Res., 2594–2602.

HORNIBROOK, N. DE B. 1971. New Zealand Tertiary climate. N.Z. Geol Surv. Report 47. 19 pp.

HOWELL, D. G. (ed.). 1985. Tectonostratigraphic terranes of the circum-Pacific region. Circum-Pacific Council for Energy and Mineral Resources, Earth Science Series, number 1, Houston, Texas.

KREMP, G. O. 1982. On the continental drift of China, India and other Southeast Asian plates. Paleo Data Banks, 17:62–113.

KEYES, I. W. 1968. Cenozoic marine temperatures indicated by Scleractinian coral fauna of New Zealand. Tuatara, 16:21–25.

MICHAUX, B. 1989a. The evolution of the Ancillinae with special reference to New Zealand Tertiary and Recent species of *Amalda* and *Alocospira* (Gastropoda: Olividae: Ancillinae) (submitted for publication).

MICHAUX, B. 1989b. Cladograms can reconstruct phylogenies. An example from the fossil record. Alcheringa, 13:21–36.

McElhinny, M. W., N. S. Haile, and A. R. Crawford. 1974. Palaeomagnetic evidence shows Malay Peninsula was not a part of Gondwana. Nature, 252: 641–645.
Nur, A., and Z. Ben-Avraham. 1982. Oceanic Plateaus, the fragmentation of continents and mountain building. J. Geophys. Res., 87:3644–3661.
Parker, E. S., and W. K. Gealey. 1983. Plate tectonic evolution of the western Pacific-Indian Ocean region. Proceedings of the EAPI/ASCOPE/CCOP/IOC workshop on the geology and hydrocarbon potential of the South China Sea and possibilities of joint development, Honolulu.
Platnick, N. I., and G. Nelson. 1988. Spanning-tree biogeography: Shortcut, detour, or dead-end? Syst. Zool., 37:410–419.
Ridd, M. F. 1980. Possible Palaeozoic drift of SE Asia and Triassic collision with China. J. Geol. Soc. London, 137:635–640.
Rosen, D. E. 1978. Vicariant patterns and historical explanation in biogeography. Syst. Zool., 27:159–188.
Seberg, O. 1986. A critique of the theory and methods of Panbiogeography. Syst. Zool., 35:369–380.
Wakefield, L. L. 1977. Lower Miocene paleogeography and molluscan taxonomy of Northland, New Zealand. Unpublished Ph.D. Thesis, University of Auckland.
Wongwanich, T., and C. Burrett. 1983. The Lower Palaeozoic of Thailand. J. Geol. Soc. Thailand, 6: 21–29.
*Received 6 March 89; accepted 28 July 89*

*Syst. Zool.*, 38(4):398–406, 1989

# Higher Category Name Usage for Amphibians and Reptiles

Harold A. Dundee

*Department of Biology, Tulane University, New Orleans, Louisiana 70118*

Although biologists are to be commended for having evolved nomenclatural systems to provide orderliness in forming names for organisms and their supposed evolutionary groupings, the efforts have not succeeded in providing a universal system for categories above the family-group level. From the 1842 "Stricklandian Code," authored by Strickland and promulgated by the British Association for the Advancement of Science, and the 1901 "Règles Internationales de la Nomenclature Zoologique" (publ. 1905, edited by Blanchard), to the current 3rd edition of the "International Code of Zoological Nomenclature" no provisions are made to provide a methodical formulation of names above the family-group. At the 14th International Congress of Zoology in Copenhagen in 1953 proposals were made to extend the rules of nomenclature, with some important differences, to names above the family-group level, but at the 15th Congress in 1958 all provisions for names above the family-group were rejected, and both the 1st edition (1961) and present 3rd edition (1985) of the Code clearly indicate in Article 1 that the Code is concerned with names of taxa in family, genus, and species groups (this includes superfamily). Dubois (1984a) proposed some principles to apply in determining class-group names, but his offering is, of course, non-official.

In the absence of nomenclatural regulation the application of higher category names in herpetology has often provided excessive synonymy and at times, confusion. A reader cannot necessarily recognize the hierarchy of a name above the family-group level because no formalized endings exist (cf. *idae* for family, *inae* for subfamily). A particular journal might, e.g., carry an article title (Testudines: Emydidae), or (Squamata: Eublepharidae), or (Reptilia: Viperidae), thus an ordinal, subordinal, and class name preceding the family name, but only a specialist might necessarily recognize the group, whereas a generalist reader would be unaware of the systematic position stated. A strong case

## Attribution

**Michaux, B. 1989. Generalized tracks and geology. *Systematic Zoology*, 38:390–398.**

First published in 1989 in *Systematic Zoology*, 38(4):390–398, doi:10.2307/2992404. Reprinted with permission from Oxford University Press.

# *chapter two*

# *Flesh and rocks evolve together*

## *Leon Croizat and panbiogeography*

Despite the efforts of New Zealand and American Museum biologists including Robin Craw, Michael Heads, John Grehan, Gary Nelson and Donn Rosen to bring the work of Leon Croizat to a wider audience during the 1970s and 1980s, and South American biologists such as Juan Morrone, Jorge Crisci and colleagues in the 1990s, his contribution to the field is still largely unacknowledged (Morrone 2015). I am sure that Croizat will eventually achieve proper recognition as one of the major twentieth-century figures in the development of biogeography when future historians of science can objectively judge his life's work. Reading this paper again after so many years, the thing that struck me first was how insightful Croizat's analysis of southwest Pacific biogeology really was. In broad outline, I think his tracks correctly summarise the tectonic history of the various islands adjacent to continental Australia – a remarkable achievement considering he had no access to modern computational methods or detailed phylogenetic and molecular data. His model for southwest Pacific biogeography identified a Melanesian Arc track linking northern New Guinea to the Solomon Islands, Vanuatu and Fiji that once formed a contiguous island arc system. He also identified a Melanesian Rift track that links the Central Highlands of New Guinea with New Caledonia and New Zealand that were all continental fragments rifted from the Australian margin of Gondwana. And lastly, he identified a connection between Lord Howe Island and Norfolk Island.

Croizat's writings can be a difficult read, in part because they don't conform to a linear narrative, in part because the point he is trying to make is often difficult to discern, in part because of his iconoclastic style, but mostly, I suspect, because he was intuitive. Croizat's intuitive approach was both his greatest strength and the source of a fundamental weakness of track analysis, which is central to the panbiogeographic method. Unlike most researchers who worked on single taxonomic groups, Croizat was a generalist who collected and analysed diverse taxonomic revisions from which he discerned common patterns of distribution. He worked at a time when evolutionary taxonomy, with its emphasis on ancestor–descendent relationships, paraphyletic taxa and ill-defined, theory-driven 'methods' of determining phylogenetic relationships dominated the field. What he did was to take these data and out of his familiarity with them identify common distributional patterns that he visualised as lines drawn on a map. I think this is remarkably insightful and a testament to his unique perception.

I was never a panbiogeographer despite Croizat's significant influence on my own biogeological development. Ebach (2017) recently argued that panbiogeography was a complete evolutionary synthesis rivalling that of neo-Darwinism, and I found this wider context in which Croizat's biogeography was embedded unconvincing. Croizat was a committed gradualist who saw 'form making' rather than speciation as the product of evolutionary change. Species were simply human constructs that divided an evolutionary continuum into convenient packages. Distinct taxa emerged from the 'ground up' by the accumulation of small changes following some orthogenetic trends that eventually separated taxa into recognisable and distinct groups. Although Croizat had little time for Darwinism and considered natural selection a secondary, 'external' process operating on variation that was provided by an 'internal' orthogenetic trend, I viewed his broader evolutionary ideas as part of a class of theories – including Darwinism – that are transformational and population-based (Michaux 1988). Transformational theories of evolution (Eldredge 1979) are concerned with how taxonomic characters change and view species as emergent phenomena of this process. For example, a transformational approach to the evolution of horses (Equidae) would not be concerned so much with individual equine species but would concentrate on some aspect of morphological change such as the transformation of the pentadactyl limb structure during equine evolution from the five-toed *Eohippus* to the hoofed modern horse, interpreting these character-state changes in terms of orthogeny or adaptation as populations continuously changed through time. I thought Croizat was saying nothing new or particularly interesting about evolution.

In the final analysis, Croizat's approach of drawing tracks on a map also proved intractable to systemisation. Track analysis suffered from two major flaws – a lack of phylogenetic input and a rigorous (numerical) method for drawing generalised tracks. Cladograms order taxa in a hierarchy of relationships such as (A(B,C)) where B and C are more closely related to each other than either is to A. If A, B and C occur in areas 1, 2 and 3, then a track would be drawn between areas 2 and 3 and then this grouping connected to area 1. In other words, the phylogeny orients the track. In reality the situation is rarely as clear-cut as this example, with taxa being found in more than one area, or in overlapping areas, or different taxa supporting different area relationships. In these situations, you need a way to combine all the information in a methodologically rigorous way. Croizat's intuitive approach has proved immune to such treatment despite earlier attempts at constructing tracks using minimal spanning trees (Page 1987, 1990; Cavalcanti 2009). If this immunity is a fundamental characteristic of Croizat's intuitive way of working, then track analysis will only ever be of heuristic value, but if future researchers can find ways to translate distributional and phylogenetic data into generalised tracks that could be plotted on maps, then track analysis could become an important analytical tool (Morrone 2015).

## *Relationship between biology and geology*

A second theme developed in this paper was an exploration of the relationship between geology and biogeography, which Rosen (1978) described as 'the independent and dependent variables respectively in a cause and effect relationship'. I also viewed tectonic change as causal, but no longer see the analogy with mathematics as particularly useful because it stresses prediction at the expense of a model's heuristic value. At one extreme of a continuum are models from which testable hypotheses can be derived and which have the characteristics of accuracy and specificity (apply only to a particular situation). At the other extreme, purely heuristic models make no predictions but instead provide insights that are applicable to many situations and have the characteristic of generality (Evans 2012). Both extremes have their uses: models that make testable predictions can transform a discipline from a descriptive to an analytical stage, allowing a subject to progress beyond narrative (Ball 1975, 1990); heuristic models can stimulate debate, provide insights that are general, and guide research programmes.

Croizat's model (Michaux 1989: Figure 2.1E) lies towards the heuristic end of the spectrum because of its generality – it encompasses the whole southwest Pacific – and because it makes only weak predictions. In the original paper, I was critical of Craw and Weston (1984) – who claimed that track analysis could generate novel geological hypotheses – because tracks connecting areas could be compatible with more than one geological model. I now think that this criticism was overstated but still contend that generalised tracks, at least on a regional scale, cannot generate testable geological hypotheses because they lack analytical rigour. For example, although each track in Figure 2.1E links areas connected by a shared geological history and therefore the model predicts that they are more closely related to each other than they are to any other area, there is no detail as to the exact nature of that relatedness. Are we to assume that sister-group relationships are between geographically closest neighbours, as Page (1987) suggested? Even if we accept that as a reasonable assumption (which I don't), then which end of the track is basal? Islands such as New Caledonia that belong to multiple tracks are pivotal to understanding the regional geological history because of their relationship to areas on different tracks. Figure 2.1E provides no information about the relationship between tracks. Heuristic models at this scale are not capable of generating testable hypotheses, nor should we expect them to, but such an analysis does suggest directions for more detailed study that may, with suitable data, produce testable hypotheses.

Reciprocal illumination between independent data sets serves to constrain biogeological explanations. Geological models and general areagrams serve as checks on each other, and any incompatibility between the two could indicate problems with one or the other or both data sets, which could lead to spurious explanations. The process of refining models may lead to the generation of a number of testable hypotheses. For example, biological evidence may support one geological model over others or show that an area was geologically composite, while geological data can be used to date one or more internal nodes of a general areagram based on molecular data or throw doubt on the validity of a phylogeny.

An area that is connected by more than one track indicates the presence of a composite biota. Composite biotas have constituent parts that show sister-group relationships to biotas of two or more geographical areas (spatial composites) or are derived at different times (temporal composites). A corollary of spatially composite biotas is that these areas may be geologically composite. A phylogenetic analysis of biotas of this type can provide detailed hypotheses concerning which areas are sister areas, thus providing geologists with additional and independent data as to the source area of the constituent terranes. This is surely a valuable tool in the analysis of complex tectonic processes that supplements standard geological approaches. An example of a temporally mixed biota discussed in Michaux (1989: Figure 2) concerned the possible movement of plants and animals into northern New Zealand from the central Norfolk Ridge during the Middle Eocene and again in the Lower Miocene (Chapter 9). Temporally composite biotas are interpreted as the result of tectonic events promoting range expansion and thus provide hypotheses of, for example, initiation or intensification of subduction resulting in regional uplift. Biotas may be both spatially and temporally composite.

A general areagram represents relationships between areas based on the biological relationships deduced from their floras and faunas. Internal nodes of an areagram are usually interpreted as range-splitting events, but this interpretation is mainly a consequence of the way the data are presented. Relationships can also be presented as nested statements such as (A(B,C)) or in a Venn Diagram, where the emphasis in interpretation naturally shifts to the unification of areas. Whether nodes are interpreted as vicariant or geodispersal/range expansion events, it is possible to link nodes to tectonic processes that can be mapped onto an areagram. This allows some internal nodes of the general areagram to be dated and, if the areagram is based on molecular phylogenies, can provide absolute ages of *all* internal nodes. I agree with Head's (2014) contention that all dating methods presently used are problematic to some degree or another, and that published ages of molecular-based phylogenies consistently underestimate clade ages (Wilf and Escapa 2014). These ages are usually calculated using one of three general methods – extrapolation of known base-substitution rates, age of oldest fossils or tectonic calibration – often combined with the idea of clock-like change, although some substitution models now allow for different rate changes over different parts of the phylogeny. Both base-substitution rates and age of oldest fossils are methods built on such unrealistic assumptions that neither can be expected to yield consistently accurate estimates of divergence times. Using substitution rates derived from modern taxa and then applying them over evolutionary time, often to groups only distantly related, is not justifiable, no matter how sophisticated the analysis might be. Dating phylogenies using the age of the oldest fossil yields a minimum age for divergence that often then becomes a maximum estimate by proxy (Heads 2014). The third method of using tectonic events to calibrate phylogenies is not without its own problems – including the uncertainty of age estimates for any proposed tectonic event and the probable episodic nature of such processes (Chapter 10) – but it potentially provides a way out of this conundrum using a completely independent data set. For readers who wish to acquaint themselves more thoroughly with the techniques and assumptions of molecular dating, I recommend a clear and entertaining critique by Wilke et al. (2009).

Hipsley and Müller (2014) were critical of using geological calibration of molecular phylogenies, which according to their analysis was used in 15% of studies between 2007 and 2013, because it is tautological.

> Unfortunately many studies continue to use ages derived from geological calibrations to support biogeographic hypotheses, falling into a trap of circular reasoning by presupposing the very speciation mode they are trying to test.

Trewick and Gibb (2010) also discussed tautology:

> The assumptions made to justify use of vicariant events in clock calculations are rarely well expressed and can lead to circular reasoning; an assumption of vicariance cannot be used to date the origin of a lineage that is then used to demonstrate a role of vicariance in lineage formation.

Dating nodes using tectonic events is not designed to test modes of speciation, but to place a nested hierarchy of area relationships into a temporal framework. While there would be some justification of this criticism if you dated a node with a presumed tectonic event and then interpreted

***Plate** 2.1* Wallace's Standardwing. Woodcut based on an illustration by Keulemans, originally published in *The Malay Archipelago.*

that node in terms of the same event, this is not what is being suggested here. What such calibration does is to generate dates for the *other* internal nodes. The question then becomes whether these derived dates are congruent with other tectonic data (Ung et al. 2016).

## *Birds-of-paradise*

New Guinea is the centre of birds-of-paradise diversity where some forty species can be found; a further three species are found in Australia (Pizzey and Knight 2012) and three species – the Halmahera Paradise Crow (*Lycocorax pyrrhopterus*), the Obi Paradise Crow (*Lycocorax obiensis*) and Wallace's Standardwing (*Semioptera wallacei*) (Plate 2.1) – in Halmahera and Obi, Indonesia (Eaton et al. 2016). These localities are shown in Figure 2.1. The occurrence of birds-of-paradise on Halmahera (plus satellite islands) and Obi is something of an anomaly – if the birds got to Halmahera, what stopped them spreading to other Indonesian islands close by with suitable forest habitats? A simplified version of the Paradisaeidae phylogeny

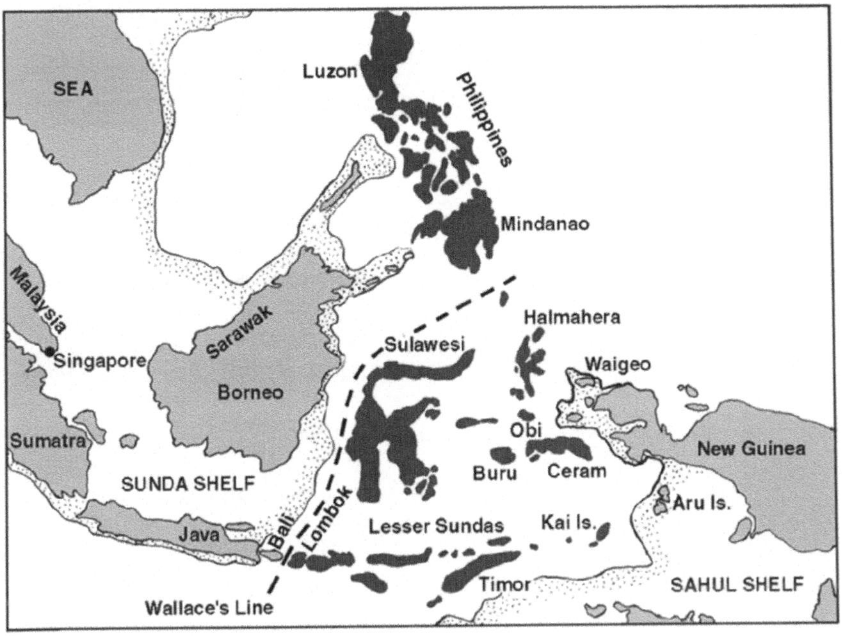


*Figure 2.1* Locality map. Dark grey areas = Wallacea.

shown in Figure 2.2 is based on mitochondrial and nuclear DNA genes published by Irestedt et al. (2009). The phylogeny identified four monophyletic groups (clades) within the Paradisaeidae. One important feature of the phylogeny is that previously identified natural groupings, such as birds-of-paradise and sicklebills, were shown to be polyphyletic as their constituent species are found in different clades. While complete accuracy of the phylogenies used in any biogeological analysis is probably more of an aspirational goal than a practical reality, the more accurate the phylogenies are the greater the chance of recovering a meaningful general areagram from all the noise. The second important feature is that the two birds-of-paradise found on Halmahera belong to different clades, one of which is more basal (and therefore older) than the other. In other words, Halmahera is temporally composite. The Paradise Crow is the oldest surviving member of the Paradisaeidae and is related to species that are otherwise restricted to cratonic areas of the lowlands and lower montane forests of New Guinea. Wallace's Standardwing evolved later and is related to species that are found on terrane areas such as the northern tip of Cape York Peninsula and eastern Australia as well as the Central Highlands and northern ranges of New Guinea.

New Guinea is geologically composite and has been formed by the amalgamation of many terranes to an autochthonous foreland region

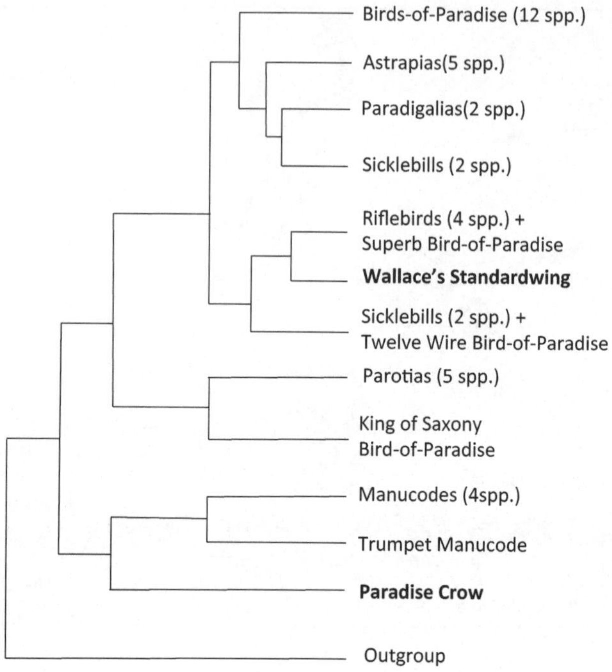


*Figure 2.2* Molecular phylogeny of the Paradisaeidae based on Irestedt et al. 2009. Maluku species in bold.

(Pigram and Davies 1987; Hill and Hall 2003; Baldwin et al. 2004, 2012; Wallace et al. 2004; Davies et al. 1997; Davies 2012). The area to the south of the Central Highlands is part of the Australian craton, as is the Bird's Head region. The southern Central Highlands are buckled cratonic rocks (Papuan Fold and Thrust Belt) deformed during collision with a variety of continental terranes, volcanic arcs and ophiolites that now form the northern part of the Central Highlands. The continental terranes were rifted from the Australian margin during the Palaeocene and, together with island arcs and associated ophiolites, were sutured to the New Guinea foreland in the Eocene. Northern New Guinea was formed from island arcs and fragments of ocean crust accreted outboard of the Central Highlands from the Oligocene to Miocene. A detailed description and discussion of the geology of New Guinea can be found in Chapter 3.

Halmahera is also geologically composite. East Halmahera is an old island arc formed during the Cretaceous to Eocene, part of a system that can be traced north into the Philippines and east into central New Guinea. West Halmahera is composed of a younger arc that formed on this older arc during the Miocene (Hakim and Hall 1991). In addition to these island arcs, there is also evidence of a collision between Halmahera and the

Australian margin 23 million years ago based on the presence of anomalous continental rocks on the islands of Bacan and Obi to the south (Ali and Hall 1995).

The interplay between evolution and tectonic events is clear in this example. A working hypothesis is that Wallace's Standardwing arrived in Halmahera following the collision of the East Halmahera Arc with the continental margin of New Guinea during the Oligocene (circa 23 Ma) as evinced by the Australian continental rocks found on Obi. This event was contemporaneous with the suturing of other arc volcanics – for example, the Bawami-Torricelli Ranges – further east (Davies 2012). Speciation within the clade is interpreted in terms of range fragmentation and biological isolation as a result of these and subsequent tectonic events. Calibrating the node linking Wallace's Standardwing with the Riflebirds at 23 million years implies that the Paradise Crow lineage evolved in the Eocene, linking the evolution of this species with the East Halmahera Arc. It also implies that the earliest member of the Paradisaeidae evolved during the Palaeocene. This proto bird-of-paradise would probably have lived in what is now southern New Guinea but which then was the northern edge of the Australasian craton. This ancestral species dispersed through range expansion onto the East Halmahera Arc some time during the Palaeocene to Eocene. In most Eocene reconstructions, this arc is presumed to be at some distance offshore from the cratonic margin, but the biological evidence does not support this hypothesis. Rather, it is more probable that the arc had either formed along the northern Australasian margin during the Eocene and was subsequently detached, or had collided with the cratonic margin by then. There is evidence of a contemporaneous Late Eocene arc-continent collision further east (Davies 2012).

What this example illustrates is that geology provides a coherent picture of the geographical context in which the evolution of this group took place, including providing key dates for major events. On the other hand, the biological evidence for different ages of the two Halmahera species suggests that following an Eocene dispersal of an ancestral bird-of-paradise onto the East Halmahera Arc, the arc became isolated again, presumably as a result of back-arc basin formation in this tectonically active region, until reorganisation of tectonic regimes resulted in its collision with continental New Guinea in the Oligocene. Continued clockwise rotation of the Pacific plate translated Halmahera westwards to its present position.

3 September 1991 *Aust. Syst. Bot.*, 4, 25–36

# Distributional Patterns and Tectonic Development in Indonesia: Wallace Reinterpreted

*B. Michaux*

Private Bag, Kaukapakapa, New Zealand.

*Abstract*

A. R. Wallace's name will always be linked with what was known as the Malay Archipelago, and thus it seems appropriate to reassess his contribution to Indonesian biogeography. Wallace is often regarded as the founder of 'dispersalist' biogeography. However, this simplified view seriously misrepresents Wallace's position, which was that animal and plant distributions in Indonesia could be understood in terms of the region's geological history. This view is contrasted with that of dispersalist biogeography, which is that present distributions are the result of unique events, that the present is a consequence of past accidents. The results of Wallace's work are outlined and are reinterpreted in the light of modern evidence about the tectonic development of this region. It is concluded that Wallace's observations are broadly intelligible in the light of what is known about Indonesia's geological history. There is much to gain by biogeographers and geologists working together more closely in Indonesia, for there are many outstanding problems that require resolution. This then is one future direction that may be taken as we stand at the crossroads.

## Introduction

There are many examples in science where one person's name has dominated a particular sphere of study. A. R. Wallace occupies that central position with respect to Indonesian biogeography through his pioneering work in the region between 1854 and 1862.

A reading of Wallace's (1880) book 'The Malay Archipelago', in which he gives an account of his travels, clearly demonstrates the best and worst sides of Victorian natural history. Descriptions of his collecting expeditions are particularly offensive to modern sensibilities. Wallace's account of collecting orangutan, in what was then Borneo (Wallace 1880, pp. 39–64), must rank as one of the worst cases of wanton destruction ever carried out in the name of science. Yet to dismiss him as a gun-happy chauvinist is to ignore the fact that he was a remarkably astute and thorough observer of nature. It is this latter characteristic that gives his work a timeless quality that makes it as relevant today as it was when published. Indeed, as a record of a world long since transformed through colonialism and the pressures of burgeoning human population, it is priceless.

Today Wallace is regarded as the founder of 'dispersalist biogeography' (e.g. Croizat 1982). However, this simplified view seriously misrepresents his position. Dispersalist biogeographers support the theory that history is, by and large, a chapter of accidents, a series of unique events determined by chance coincidences. For example, it can be observed that birds and insects are regularly blown across the Tasman Sea from Australia to New Zealand. These events are

1030–1887/91/010025$05.00

unique happenings brought about by a fortuitous interplay of events. From this observation it is concluded that these birds or insects may one day establish themselves in New Zealand and, by becoming part of the New Zealand biota, evolve independently of their parental stock. The general conclusion that follows is that *all* distributional patterns are the result of unique events.

The fallacy inherent in this type of argument is demonstrated by the example of 'Cleopatra's nose' discussed in Carr (1970) and outlined in Table 1. Carr (1970, p. 108) considers that 'the dual and reciprocal function of history . . . [is] to promote our understanding of the past in the light of the present and of the present in the light of the past. Anything which, like Anthony's infatuation with Cleopatra's nose, fails to contribute to this dual purpose is from the point of view of the historian dead and barren'. This is so because no general conclusion can be drawn, which may promote our understanding either of the past or of the present, from unique events. To do so is simply to fall into the trap of arguing that all generals lose battles because they are infatuated with beautiful women.

**Table 1. The fallacy of 'Cleopatra's Nose' or the dangers of arguing from the unique to the general**

| Example | Unique event | Specific result | General conclusion |
|---|---|---|---|
| History | Mark Antony becomes infatuated with Cleopatra | Loss of the battle of Actium | Generals lose battles because of their infatuation with beautiful women |
| Biogeography | Accidental dispersal from A to B | Establishment at B and evolution into a new species | Distribution patterns are the result of accidental dispersal |

Wallace did not try to explain distribution patterns by invoking the occurrence of unique events but rather by recourse to general principles. For Wallace that general principle was geological change. There are numerous passages in Wallace (1880) that confirm his appreciation of the importance of geological change in understanding distribution patterns in Indonesia. His discussion of the faunal relationship between the Malay Peninsula and the islands of Borneo, Sumatra, and Java is detailed and provides a clear statement of his position. Having noted the overall similarity of this area's flora and fauna to that of India, he continued by describing his views of the biotic relatedness of these large islands. He commented on the greater similarity of the mammals and birds of Borneo and Sumatra than those of Borneo and Java, and on the high degree of endemism of the Javan fauna. Noting that the pattern of distribution is similar to other groups with poorer dispersal powers, he continued:

> 'We learn from this fact, that the seas which separate the islands from each other are wide enough to prevent the passage of even flying animals, and that we must look to the same causes as having led to the present distribution of both groups. The only sufficient cause we can imagine is the former connexion of all the islands with the continent, and such a change is in perfect harmony with what we know of the earth's past history, and is rendered probable by the remarkable fact that a rise of only 300 feet would convert the wide seas that separate them into an immense winding valley or plain about 300 miles wide and 1200 long.' (Wallace 1880, p. 143)

He even used the idea of geological change to understand problematical distributions on a quite detailed scale. The 'Tin Isle' of Bangka, which lies just off the coast of Sumatra, had a number of endemic species not found on its larger neighbour but relatives of which are found on the Malay Peninsula. He speculated that Bangka is a geological extension of the Malay Peninsula, noting that granites are found on both but not on Sumatra, and that the east coast area of this latter island had been formed only in recent history. He continued:

> 'As the southern part of Sumatra extended eastward and formed the narrow straits of Banca, many birds and insects and some Mammalia would cross from one to the other, and thus produce a general similarity of productions, while a few of the older inhabitants remained, to reveal by their distinct forms their different origins'.

Wallace believed that evolution was a slow and gradual process and that a kind of 'morphological clock' existed. Thus he related genealogical relatedness of organisms to geological relatedness of areas, a remarkably 'modern' idea. It is precisely this connection that led him, on the basis of observable distributions, to reject the idea of long-distance dispersal, even for animals that could fly.

'Even the bats furnish an additional argument, if one were needed, to show that the islands could not have been peopled from each other and from the continent without some former connexion. For if such had been the mode of stocking them with animals, it is quite certain that creatures which can fly a long distance would be the first to spread from island to island, and thus produce an almost perfect uniformity of species over the whole region. But no such uniformity exists, and the bats of each island are almost, if not quite, as distinct as other mammals' (Wallace 1880, pp. 142–3).

These are hardly the sentiments of one who subscribes to the idea that present distributions can be understood in terms of a series of accidental events that occurred in the past. Indeed, rather than being regarded as the founder of 'dispersalist biogeography', a case could be made that he was a forerunner of modern biogeographic theories that use the idea of geological change as an explanatory principle (Nelson and Platnick 1981, p. 400).

**Wallace's Indonesian Biogeography**

A summary of Wallace's biogeographic conclusions is presented in Fig. 1. On the basis of his observations he divided the archipelago into five regions — the Indo-Malay Islands (Fig. 1 : 1), Celebes (Fig. 1 : 2), the Timor group (Fig. 1 : 3), the Moluccas (Fig. 1 : 4), and the Papuan group (Fig. 1 : 5). On the basis of the avian and mammalian faunas he also recognised a fundamental biotic discontinuity between Indo-Malay and Australian forms across the narrow Straits of Lombok. This discontinuity was later traced to the north along the Macassar Strait and became known as Wallace's Line (Fig. 1).

The general biogeographic characteristics of the Indo-Malay islands (Fig. 1 : 1) have already been noted. His explanations of distribution patterns among the islands of this group, based largely on sea-level changes, are more plausible here than anywhere else in the archipelago. There is, however, one interesting anomaly that he discusses at some length, concerning the complete absence of orangutan from Sarawak where apparently suitable habitat widely existed. He argued that because orangutan require swamp forest the inland mountains surrounding Sarawak acts as an effective ecological barrier to westward expansion.

Wallace regarded the fauna of Sulawesi (Fig. 1 : 2) as unusual, not only because it was rather impoverished, despite the island's central position in the archipelago, but also because of its uniqueness. The Sulawesian biota seemed to have more in common with the Philippines and Africa than its closest neighbours. The Sula Islands and Buton, which are geographically close to Sulawesi, were regarded by Wallace as essentially Moluccan.

The biota of the Timor region (Fig. 1 : 3) was characterised by a mixture of Indo-Malay and Australian elements. At the western end of this chain of islands there was a greater percentage of Indo-Malay forms, while Australian forms dominated in Timor. Several Moluccan bird species were also recorded. Wallace recorded no endemic bird genera from this region and thus viewed the avian fauna as derivative.

Wallace recognised two subregions in the Moluccas (Fig. 1 : 4 and 4′). The biota of the northern islands, centred around Halmahera, had affinities with the Papua region (Fig. 1 : 5), while the southern group, centred around Seram, had a biota related more to Australia.

Although this is a rather brief summary of the distribution patterns that Wallace recognised in Indonesia, it does, I believe, accurately reflect the major patterns he observed. His interpretations of these patterns, based as they were on an incomplete understanding of the dynamic nature of the earth's surface, are only really of historical interest. Wallace's attempt to understand distributional patterns in terms of geological change was doomed from the start because neither he nor anyone else at the time realised that land could move laterally as well as vertically. It wasn't until the publication of Wegener's 'Die Entstehung der Kontinentder Ozeane' in 1915 that anyone suggested otherwise (Wegener 1929). In any event the concept was largely ignored until the 1960s when Wegener's central thesis was confirmed and extensively developed.

**An Overview of the Tectonic Development of Indonesia**

Indonesia, like the Caribbean and Mediterranean, has developed in response to continental–continental collision. Such collisions are characterised by complex structural developments as plate boundaries adjust to changing forces. Although there are many difficulties, both practical and theoretical, in trying to unravel the tectonic development of an area such as Indonesia, the impetus to do so has been provided by the desire to develop the mineral potential of the region. As a result of this effort there exists an extensive literature dealing with the geology

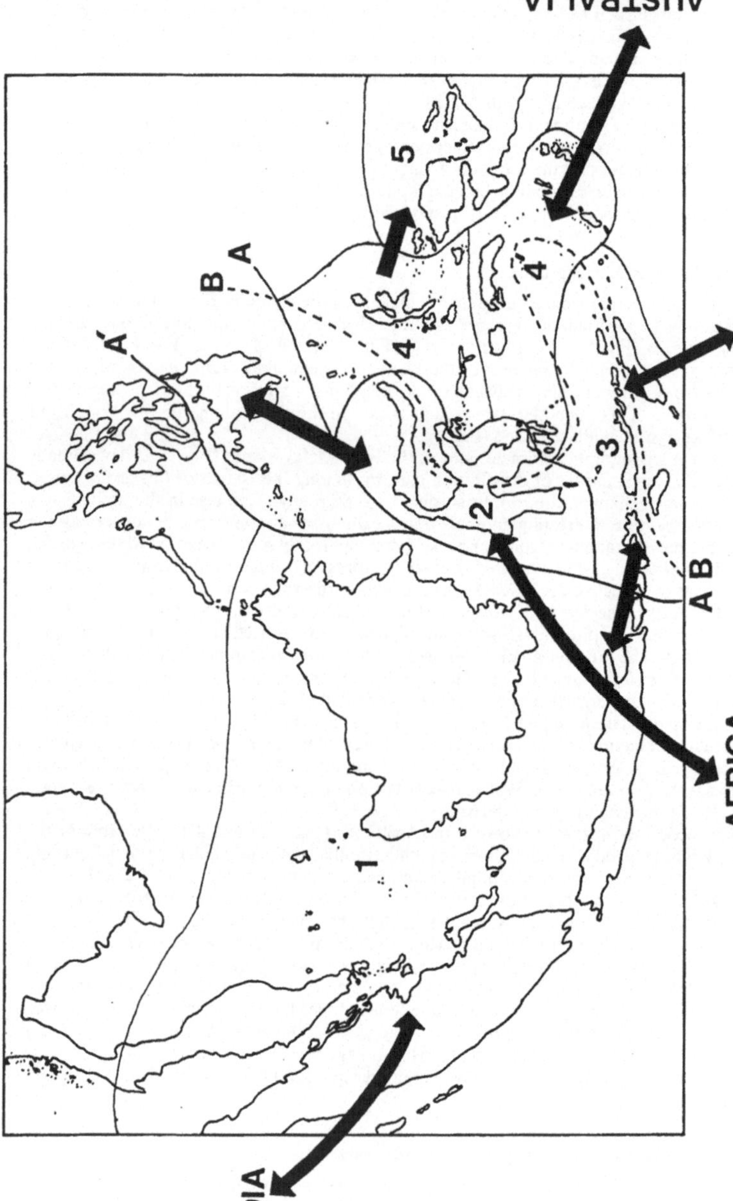


**Fig. 1.** A summary of the biogeographic regions recognised by Wallace in Indonesia. 1 = Indo-Malay Islands; 2 = Celebes (Sulawesi); 3 = Timor; 4 = northern Moluccas, 4′ = southern Moluccas; 5 = Papuan Region. A—A = Wallace's Line, A—A′ = modified Wallace's Line; B—B = geological partition of Asian and Australian land masses proposed by Carey in 1938 (Carey 1975).

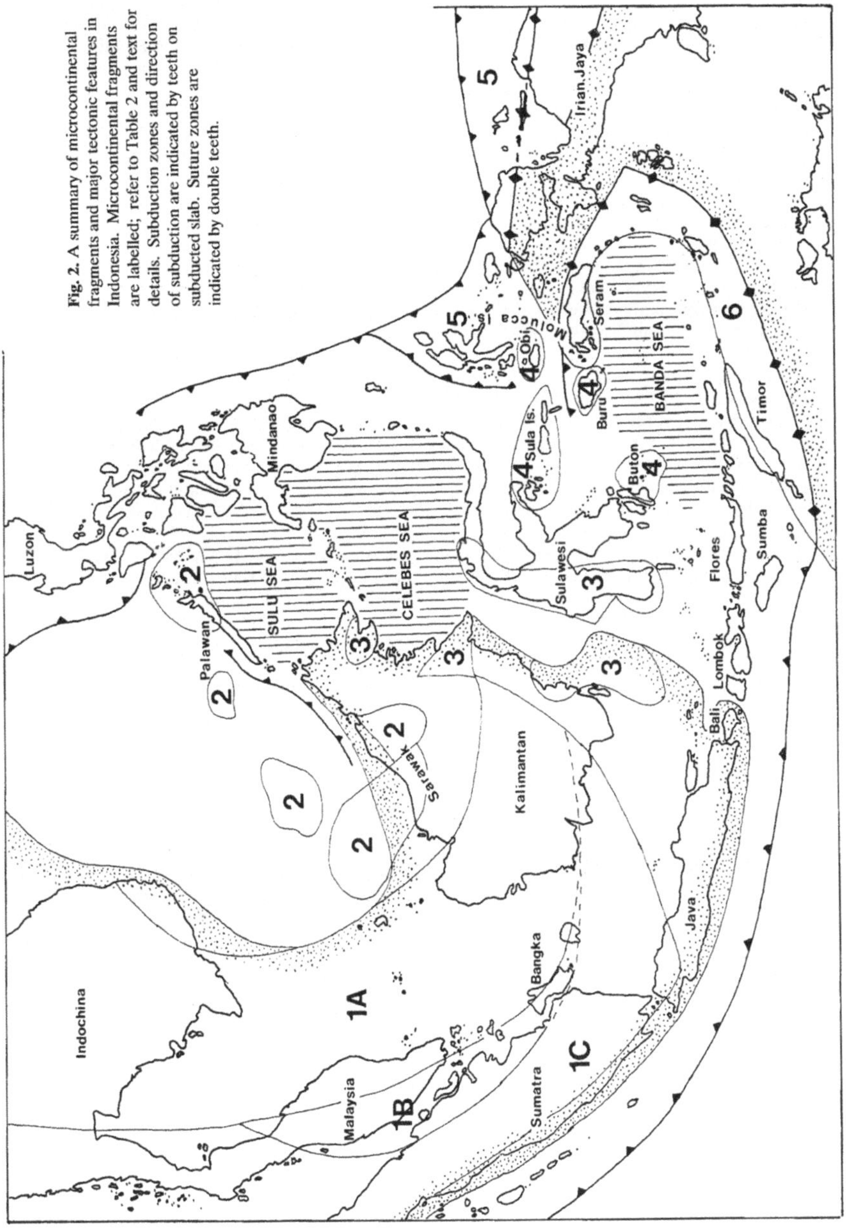


**Fig. 2.** A summary of microcontinental fragments and major tectonic features in Indonesia. Microcontinental fragments are labelled; refer to Table 2 and text for details. Subduction zones and direction of subduction are indicated by teeth on subducted slab. Suture zones are indicated by double teeth.

of the region and, more importantly for those of us interested in biogeography, with its tectonic interpretation.

Tectonic plates, despite being 100 km thick, do not act as rigid structures, particularly so at their margins. In structurally complex situations, such as exist in continental–continental collisions or during the breakup of continents, large plates become fragmented giving rise not so much to plate tectonics as 'flake tectonics'. Hutchinson (1989) has discussed the role that thermal heating has in this process of fragmentation. Convection currents within the upper mantle can cause strong thermal heating in overlying continental crust, which results in thermal relaxation. Rocks under these heat flow conditions act as a fluid and respond to extension or shear pressures by stretching. This can lead to rifting and the formation of marginal seas. During this breakup phase microcontinental fragments can be formed and transported elsewhere.

Indonesia is bounded by three major plates. The Pacific plate forms an eastern boundary, the Indo-Australian plate occurs to the west and south, and the Indo-China plate to the north. Rifting occurred along the edge of the Indo-China plate shortly after subduction of oceanic crust beneath it ceased at the end of the Cretaceous. As new oceanic crust formed and the rifts widened, continental fragments migrated southwards. To the south, on the Gondwanic margin, continental blocks had been rifted away and drifting to the north since the early Palaeozoic. Indonesia formed where these Asian and Gondwanic fragments collided. The eastern margin, formed by the Pacific plate, acted as a barrier containing these fragments within the Indonesian region. Australia and its associated rift systems separated from Antarctica during the early Tertiary. Australia's subsequent northward drift into the Indonesian region, a process still occurring today, resulted in further modifications and tectonic displacements within the region. A model for this three-plate system is provided by Charlton (1986). As Hutchinson (1989) remarked:

> 'on-going processes are now at work in Southeast Asia, and we see that as a major Australian Craton approaches China–Indochina, continental scraps rift off both cratons and drift away to meet in the middle ground characterized by island arcs. The complexity resulting from rifting and displacement of the terranes and their final compression together in a future orogenic belt will result in a jigsaw puzzle of insolvable complexity'.

**Table 2. A summary of the source areas and ages of derivation and collision of microcontinental fragments in Indonesia**

| Microcontinent | Source area | Derivation age | Collision age |
|---|---|---|---|
| 1A | Gondwana | Lower Palaeozoic | Early Triassic |
| 1B, 1C | Gondwana[a] | Middle Carboniferous | Late Triassic |
| | | Jurassic | Late Cretaceous[f] |
| 2 | Indochina | Late Cretaceous | Middle Miocene |
| 3 | Gondwana[b] | Jurassic/Cretaceous? | Late Cretaceous? |
| 4 | Gondwana[c] | Early Tertiary | Late Miocene |
| 5 | Gondwana[d] | Late Cretaceous | Late Miocene |
| 6 | Gondwana[e] | Jurassic | Pliocene |

[a] east coast of India (Burton 1970); [b] between India and Western Australia?; [c] western part of Irian Jaya (Audley-Charles *et al.* 1972; Hamilton 1979; Norvick 1979) or north-eastern New Guinea–Australia (Pigram and Panggabean 1984); [d] northern New Guinea–Australian margin; [e] north-western Australia; [f] Harbury *et al.* (1990).

Fig. 2 shows the approximate extent of the south-east Asian and Australian cratons (stippled) and the *melée* of rifted microcontinents, island arcs, and marginal seas in the middle. Microcontinental fragments are numbered. A summary of the source areas, and ages of derivation and collision of these microcontinental blocks is provided in Table 2. Inspection of Fig. 2 reveals that the 'regions' Wallace identified in Indonesia (Fig. 1) are geological composites, composed of either large continental blocks or smaller microcontinental fragments. These continental fragments were derived — at different times — from a number of distinct areas of Gondwana, or from Indo-China. Wallace's lack of any conception of lateral movement of land made informed interpretation of biogeographic patterns in Indonesia impossible. In the following section I would like to interpret what he observed in the light of the information summarised in Fig. 2, and thus to indulge in some speculation about what Wallace himself may have made of the data now available.

**Wallace Reinterpreted**

*Indo-Malay Islands*

The Indo-Malay Islands are part of continental south-east Asia. Three schools of thought exist as to how this region became assembled (Metcalfe 1988). All agree that the blocks from which south-east Asia have been assembled are Gondwanic. The evidence for Gondwanic origin includes extensive Permian glacial deposits thought to have formed in a rift system (Gatinsky and Hutchinson 1986), palaeontological evidence (Buffetaut 1982; Burrett and Stait 1985), and palaeomagnetic data (Achache *et al.* 1983). The source area was probably in that region of Gondwana between India and Australia ('Sinus Australis').

However, the different schools of thought differ markedly from one another on when these blocks assumed their present relative positions. Helmke (1985) considered that south-east Asia had amalgamated by the late Palaeozoic, while others consider that the blocks were derived during the early Triassic and assembled at their present latitudes at the end of the Triassic (e.g. Burton 1970; Acharyya 1979; Parker and Gealey 1983; Liew and McCulloch 1985; Metcalfe 1988; Hutchinson 1989). Thus, in the opinion of these latter authors, south-east Asia was assembled in the early Mesozoic. Yet a third view is favoured by Harbury *et al.* (1990), who suggest that the Sibumasu block (Fig. 2: terranes 1B and 1C) rifted from Gondwana much later than other south-east Asian blocks in the Jurassic, and docked with the rest of south-east Asia in the mid- to late Cretaceous. It is possible, as suggested by Burton (1970), that the Sibumasu block was adjacent to India prior to the Jurassic breakup of this part of Gondwana.

Wallace's biogeographic work indicated a close biotic relationship between the Indo-Malay islands and India. It is possible that this relationship dates from a period of mobilism initiated by mountain building during the collision of India with Eurasia. Under this hypothesis, animals and plants would have moved down newly created dispersal routes along these rising mountain chains.

A more intriguing possibility is that the Sibumasu block was, as Harbury *et al.* (1990) suggested, derived from Gondwana during the Jurassic. Biogeographic evidence would favour a source area close to the Indian sector of Gondwana. The east African–Madagascar–Indian–Malayan 'track' identified by Croizat (1964, fig. 24) is intelligible in the light of what is known of the breakup of Gondwana and the Cretaceous arrival of the Sibumasu block, with its associated biota, in south-east Asia.

*Borneo*

The island of Borneo is, as Croizat suggested, a geological composite (Croizat 1958, fig. 200, reproduced as Fig. 3). South-west Kalimantan is interpreted by geologists as either an extension of the Malay Peninsula (Hartono and Tjokrosapoetro 1986), or as a continental fragment derived from Indo-China either during the middle Cretaceous (Hutchinson 1989), or late Cretaceous (Gatinsky and Hutchinson 1986 ; Metcalfe 1988).

Hutchinson (1989) regarded south-east Kalimantan as a microcontinental fragment which, prior to the opening of the Macassar Strait during the mid-Tertiary (Oligocene according to McCabe and Cole 1989), formed a single unit with south-west Sulawesi. As indicated in Fig. 3, Croizat (1958) linked these latter two areas to Sumatra.

The third microcontinental block forms the state of Sarawak. For much of the Mesozoic the continental edge of south-east China and the east coast of Indochina was an Andean-type volcanic margin under which oceanic crust was being subducted (Hutchinson 1989 and references therein). By the close of the Mesozoic this subduction had ceased and portions of thinned continental crust had started to rift away from the margin. These rifted microcontinents (Fig. 2 : 2) were transported south during the early Miocene as a spreading centre developed to the north, forming what is now the South China Sea.

Older Mesozoic crust on the southern boundary of these rifted blocks was subducted under a trench system in the Palawan–northwest Borneo region. This trench system ceased to operate in the mid-Miocene when these microcontinents derived from the north collided with other continental fragments in the Borneo region (Hutchinson 1989). In Fig. 3 Croizat (1958) identified this event with a track connecting China with Sarawak. The occurrence of orangutan in Kalimantan and Sumatra but not in Sarawak, which Wallace (1880) interpreted in terms of ecological factors, may thus have its origin (as opposed to its perpetuation) in the different source areas from which the various terranes of Borneo were derived. Orangutan appear still to be restricted to areas derived from Gondwana.

*Sulawesi*

The island of Sulawesi is composite, with mid- to late Tertiary island arcs in the north and east sutured to a microcontinental fragment in the south. In addition, the Sula Islands and Buton are microcontinental fragments derived from the east (Hartono and Tjokrosapoetro 1986). The Sula Islands terrane is reported to have collided with eastern Sulawesi as it was translated westwards in the late Miocene (McCabe and Cole 1989). The source area for the Sula Islands has variously been identified as north of Misool (Audley-Charles *et al.* 1972), the Bird's Head of Irian Jaya (Norvick 1979), the Bird's Neck of Irian Jaya (Hamilton 1979), or from further east from central New Guinea (Pigram *et al.* 1985). Hartono and Tjokrosapoetro (1986) link Buton with the Buru–Seram region. Wallace's observation that the fauna of the Sula Islands and Buton were related to the 'Moluccas', rather than to Sulawesi, is therefore correct.

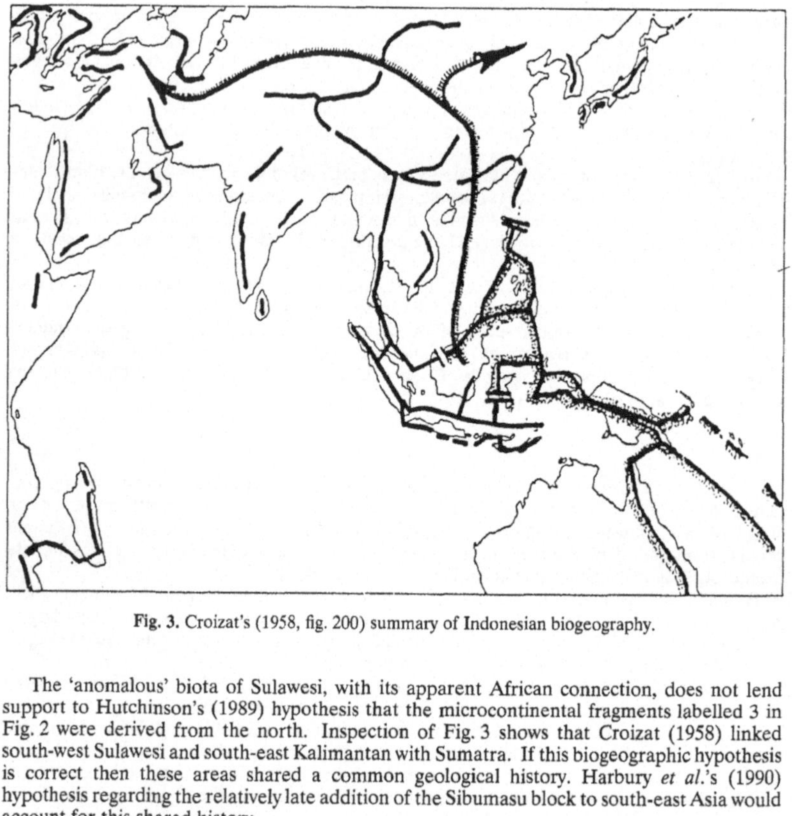

**Fig. 3.** Croizat's (1958, fig. 200) summary of Indonesian biogeography.

The 'anomalous' biota of Sulawesi, with its apparent African connection, does not lend support to Hutchinson's (1989) hypothesis that the microcontinental fragments labelled 3 in Fig. 2 were derived from the north. Inspection of Fig. 3 shows that Croizat (1958) linked south-west Sulawesi and south-east Kalimantan with Sumatra. If this biogeographic hypothesis is correct then these areas shared a common geological history. Harbury *et al.*'s (1990) hypothesis regarding the relatively late addition of the Sibumasu block to south-east Asia would account for this shared history.

One would have to hypothesise that an already differentiated biota occurred over that region of Gondwana composed of east Africa–Madagascar–India–Sibumasu prior to breakup. As Sibumasu rifted away from Gondwana the composition of its biota would have reflected this original pre-rift differentiation which, it seems reasonable to suggest, reflected the original geographic distances separating the component blocks. Because the Sibumasu block was originally nearer to India its biota was, in a general sense, more closely related to Indian forms. However, for more specific cases the genealogical similarities may have been with African forms, either reflecting the original pre-rift differentiation, or resulting from late phase mobilism along newly created rift systems formed as India moved away from Gondwana.

Wallace (1880) also suggested that the biotas of Sulawesi and the Philippines showed direct similarities. With the exception of the North Palawan microcontinent, the remainder of the Philippines archipelago is thought to be composed of Cretaceous to Recent oceanic sediments and volcanics (Karig 1983). This, together with the lack of sediments with a continental character, implies an island-related history since the Cretaceous (McCabe and Cole 1989). However, the diverse biota of the Philippines strongly suggests that the geological history of these islands may have been more complex than this. In particular, either the Philippines were adjacent to continental areas, or microcontinental fragments are present.

Croizat (1958) suggested that Sulawesi and the Philippines were linked by a generalised track via Halmehera (Fig. 3). Cladistic analyses of groups occurring in south-west Sulawesi–Halmahera–Mindanao may clarify the nature of the biotic relationship between Sulawesi and the southern Philippines. If a direct relationship is demonstrated to exist, as suggested by Wallace (1880), then one could speculate that parts of Mindanao may be continental in origin and derived from elsewhere. The Banda, Sulu, and Celebes Seas — the later of which is adjacent to Mindanao — are composed of old oceanic crust surrounded by younger, actively spreading oceanic crust (McCabe and Cole 1989). This anomalous situation indicates that these oceanic basins, which according to McCabe and Cole (1989) were originally a single unit, have become trapped in their present position. The implication is, of course, that they have also been derived from elsewhere. According to this hypothesis the three areas of old oceanic crust, the microcontinental fragments labelled 3 in Fig. 2, and parts of Mindanao formed a single tectonic unit.

### *Timor and the Banda Arc*

The Banda arc is a composite unit, the inner arc of islands being a volcanic extension of western Sumatra and Java, the outer arc being a fragment rifted from north-western Australia. On the island of Timor, representative rocks of both units are present (Hartono and Tjokrosapoetro 1986; McCabe and Cole 1989). Wallace observed that the biota of the Banda arc islands was a mixture of Australian and Malayan elements, and that the proportion of Australian elements increased on going eastwards while the proportion of Malayan elements decreased.

Once again the general biogeographic patterns of this area are quite intelligible in the light of what is known about the area's geological history. There is a strong possibility that the Australian element of the biota is temporarily composite (Michaux 1989), an original component linked to the separation of the Timor–Seram rift system during the Jurassic, and a younger element derived when this rift system collided with continental Australia during the Pliocene. As Timor was compressed between the inner Banda arc to the north and the Australian craton to the south, a period of mobilism was initiated as once disjunct regions became physically united. Audley-Charles and Hooijers' (1973) discussion of the migration of pigmy elephants between Flores and Timor during the Pleistocene is an example of this mobilism.

The islands of the inner Banda arc are a structural continuation of Java. Their biota shows a strong relationship with Java and this indicates that, although these islands are now physically isolated both from Java and each other, connections existed in the past. During the Pliocene–Pleistocene phase of mobilism initiated by the collision of the outer and inner Banda arcs, intermixing appears to have been restricted to the area of collision. This resulted in a 'purer' Malay biota in the west and a 'purer' Australian biota in the east. The faunal connection with the 'Moluccas' is a consequence of the geological relationship between Timor and Seram which occupy extreme ends of the same tectonic unit rifted from Australia in the Jurassic.

### *The Molucca Islands*

Considering the complex geological history of this 'area', which is composed of numerous terranes, it is not surprising that Wallace had difficulty in interpreting the biogeography. His recognition of two distinctive faunal provinces within this region is a consequence of the differences in geological relatedness of the 'northern' and 'southern' islands of the Moluccas. The northern group of islands, centred around Halmahera, is geologically related to the northern New Guinea/Australian margin, whilst the southern group, centred around Seram, is part of the rift system connected to north-western Australia.

In the light of what is now known, further refinements can be made. The Moluccas are composed of terranes from three distinct source areas. The island of Halmahera is part of Parker and Gealey's (1983) outer Pacific arc, and as such is related biotically to that part of Irian Jaya north of the Sorong fault (Dow and Sukamto 1984), or its structural equivalents, and thence eastwards to the Solomon Islands and Vanuatu. Many of the other islands of Wallace's northern province, such as Bacan (Hartono and Tjokrosapoetro 1986) and Obi, as well as Buru from Wallace's southern area, are related to what was once the foreland of the New Guinea/Australian craton. As already detailed, the island of Seram is part of a Jurassic rift system derived from north-western Australia.

**Conclusions**

There are a number of specific conclusions that can be drawn from this study. The major conclusion is that plant and animal distributions in Indonesia result from the transportation of ancestral taxa on continental fragments derived from both Gondwana and Indo-China during the Mesozoic, and Australia/New Guinea's mid-Tertiary collision with the region.

In many instances the distributional evidence and reconstructions of tectonic development are perfectly congruent. Thus Wallace's (1880) discussion of the biota of the Sula Islands or of Timor, and Croizat's (1958) identification of a Sarawak/Indo-China track and composite geological nature of Borneo, are in perfect agreement with independently derived tectonic data. In other instances the two sets of data are either contradictory or one or other is equivocal. These cases are of interest because they can help focus further research effort.

Examples from Indonesia where biogeographic and geological evidence contradict each other are few — the source area for the microcontinental fragments of south-east Kalimantan and south-west Sulawesi is one. Incongruencies appear more commonly to result from data that are inadequate to resolve complex questions. For example, detailed distributional data combined with cladistic analyses may help geologists resolve the source areas for those microcontinental fragments translated westwards from Australia/New Guinea during the mid-Tertiary (Fig. 2 : 4). A further example, again from eastern Indonesia, concerns the relationship between Mindanao, Sulawesi and the northern Moluccas. Resolution of the conflicting biogeographic evidence detailed previously may well provide important information to geologists in their attempts to understand, in detail, the tectonic development of eastern Indonesia.

The last specific conclusion concerns Wallace. As detailed in the introduction, it is an oversimplification to regard him as a dispersalist. The fundamental modern opposition in biogeography is not between dispersalism and vicariance *per se*, because for vicariant form making to occur an ancestral taxon must have become widely distributed during a period of mobilism. The distinction between mobilism and dispersal is clearly only one of degree. Rather, the fundamental opposition — as in all historical studies — is between those who wish to interpret history as a sequence of unique happenings and those who interpret history in terms of some underlying principle. Wallace attempted the latter.

Brundin (1966) regarded Wallace as a champion of the view that animal distributions were due to origins in northern centres with subsequent dispersal toward the southern antipodes. This analysis is correct, Wallace certainly championed this view in later writings. However, as I hope I have demonstrated, he originally attempted — unsuccessfully as it turned out — to interpret distributions in terms of geological history. Given the paucity of geological knowledge and a cultural context of imperialism and colonial expansion at the time, this change is not altogether surprising.

Finally, this study demonstrates that it is not possible to understand the distribution of plants and animals without a knowledge of how that distribution came about. Empirical studies concerned with describing distributions and theories concerned with reconstructing genealogical relationships between taxa are necessary, but not sufficient, to allow us to understand how the present has come to be. Area cladograms describe a sequence of splitting events and tracks link areas with common geological histories, but neither describes what has happened. This is so because of their 'one-to-many' relationship to geological development, i.e. one area cladogram or track is consistent with many geological histories. If biogeographers believe that geographic distributions are the result of tectonic events, then their studies should contain an in-depth analysis and discussion of pertinent geological evidence.

**Acknowledgments**

My thanks to Jack Grant-Mackie, Peter Ballance, David Lambert, Hans Duffels, Ronald Vonk and Garry Nelson for their help and encouragement, and for providing much appreciated feedback. The generous financial assistance provided by the Australian Systematic Botany Society and the Evolutionary Genetics Laboratory, which allowed me to attend the ASBS conference, is gratefully acknowledged. This is publication number 43 in conjunction with the EGL.

**References**

Achache, J., Courtillot, C., and Besse, J. (1983). Paleomagnetic constraints on the late Cretaceous and Cenozoic tectonics of southeastern Asia. *Earth and Planetary Science Letters* **63,** 123–36.

Acharyya, S. K. (1979). India and Southeast Asia in Gondwanaland fit. *Tectonophysics* **56,** 261–76.

Audley-Charles, M. G., Carter, D. J., and Milson, J. S. (1972). Tectonic development of eastern Indonesia in relation to Gondwana dispersal. *Nature* **239,** 35–9.

Audley-Charles, M. G., and Hooijer, D. A. (1973). Relation of Pleistocene migrations of pigmy stegodonts to island arc tectonics in eastern Indonesia. *Nature* **241,** 197–8.

Buffetaut, E. (1982). Mesozoic vertebrates from Thailand and their palaeobiological significance. *Terra Cognita* **2,** 27–34.

Burrett, C., and Stait, B. (1985). South East Asia as part of an Ordovician Gondwanaland — a palaeobiogeographic test of a tectonic hypothesis. *Earth and Planetary Science Letters* **75,** 184–90.

Burton, C. K. (1970). The palaeotectonic status of the Malay Peninsula. *Palaeogeography, Palaeoclimatology, Palaeoecology* **7,** 51–60.

Brundin, L. (1966). Transantarctic relationships and their significance as evidenced by chironomid midges. *Kunglia Svenska Vetenkapsakademiens Hanlingar, 4 Serien* **11,** 1–472.

Carey, S. W. (1975). Tectonic evolution of Southeast Asia. Fourth Indonesian Petroleum Association Congress, Jakarta, pp. 1–31.

Carr, E. H. (1970). 'What is History?' (Penguin: London.)

Charlton, T. R. (1986). A plate tectonic model of the eastern Indonesian collision zone. *Nature* **319,** 394–6.

Craw, R. C., and Weston, P. (1984). Panbiogeography: a progressive research program? *Systematic Zoology* **33,** 1–13.

Croizat, L. (1958) 'Panbiogeography.' Vols. 1, 2a, 2b. (Published by author: Caracas.)

Croizat, L. (1964). 'Space, Time, Form.' (Published by author: Caracas.)

Croizat, L. (1982). Vicariance/vicariism, panbiogeography, 'vicariance biogeography', etc.: a clarification. *Systematic Zoology* **31,** 291–304.

Dow, D. B., and Sukamto, R. (1984). Western Irian Jaya: the end product of oblique plate convergence in the Late Tertiary. *Tectonophysics* **106,** 109–39.

Gatinsky, Y. G., and Hutchinson, C. S. (1986). Cathaysia, Gondwanaland, and the paleotethys in the evolution of continental Southeast Asia. GEOSEA Proceedings Vol. 2, *Geological Society of Malaysia Bulletin* **20,** 179–99.

Hamilton, W. (1979). Tectonics of the Indonesian region. U.S. Geological Survey Professional Paper No. 1078, 345 pp.

Harbury, N. A., Jones, M. E., Audley-Charles, M. G., Metcalfe, I., and Mohamed, K. R. (1990). Structural evolution of Mesozoic Peninsula Malaysia. *Journal of the Geological Society of London* **147,** 11–26.

Hartono, H. M. S., and Tjokrosapoetro, S. (1986). Geological evolution of the Indonesian Archipelago. GEOSEA Proceedings Vol. 2, *Geological Society of Malaysia Bulletin* **20,** 97–136.

Helmke, D. (1985). The orogenic evolution (Permian–Triassic) of central Thailand. Implications on paleogeographic models for mainland SE-Asia. *Memoire du Société Géologique de France* **147,** 83–91.

Hutchinson, C. S. (1989). Displaced terranes of the southwest Pacific. In 'The Evolution of the Pacific Ocean Margins'. (Ed. Z. Ben-Avraham.) pp. 161–75. Oxford Monographs in Geology and Geophysics Volume 8. (Oxford University Press: New York.)

Karig, D. E. (1983). Accreted terranes in the northern part of the Philippine Archipelago. *Tectonics* **2,** 211–36.

Liew, T. C., and McCulloch, M. T. (1985). Genesis of granitoid batholiths of Peninsular Malaya and implications for models of crustal evolution: evidence from Nd–Sr isotopes and U–Pb zircon study. *Geochimica et Cosmochimica Acta* **49,** 587–600.

McCabe, R., and Cole, J. (1989). Speculations on the late Mesozoic and Cenozoic evolution of the Southeast Asian margin. In 'The Evolution of the Pacific Ocean Margins.' (Ed. Z. Ben-Avraham.) pp. 143–60. Oxford Monographs in Geology and Geophysics Volume 8. (Oxford University Press: New York.)

Metcalfe, I. (1988). Origin and assembly of south-east Asian continental terranes. In 'Gondwana and Tethys'. (Eds M. G. Audley-Charles and A. Hallam.) pp. 101–18. Geological Society Special Publication No. 37. (Oxford University Press: New York.)

Michaux, B. (1989). Generalized tracks and geology. *Systematic Zoology* **38,** 390–8.

Nelson, G., and Platnick, N. (1981). 'Systematics and Biogeography. Cladistics and Vicariance.' (Columbia University Press: New York.)

Norvick, M. S. (1979). The tectonic history of the Banda Arcs, eastern Indonesia, a review. *Journal of the Geological Society of London* **136,** 519–27.

Parker, E. S., and Gealey, W. K. (1983). Plate tectonic evolution of the western Pacific–Indian Ocean region. Proceedings of the EAPI/ASCOPE/CCOP/IOC workshop on the geology and hydrocarbon potential of the South China Sea and possibilities of joint development, Honolulu.
Pigram, C. J., and Panggabean, H. (1984). Rifting of the northeastern margin of the Australian continent and the origin of some microcontinents in eastern Indonesia. *Tectonophysics* **107**, 331–53.
Pigram, C. J., Surono, V., and Supandjono, J. B. (1985). Origin of the Sula platform, eastern Indonesia. *Geology* **13**, 246–8.
Wallace, A. R. (1880). 'The Malay Archipelago.' Seventh edition. (Macmillan: London.)
Wegener, A. (1929). 'The Origin of Continents and Oceans.' [Reprint of English translation, 1966.] (Dover Publications: New York.)

Manuscript received 3 September 1990, accepted 31 January 1991

# Attribution

**Michaux, B. 1991. Distributional patterns and tectonic development in Indonesia: Wallace reinterpreted. *Australian Systematic Botany*, 4:25–36.**

First published in 1991 in *Australian Systematic Botany*, 4:25–36, doi.org/10.1071/SB9910025. Reproduced with permission from CSIRO Publishing.

*chapter three*

# *Cleopatra's nose*

## *Serendipity at work*

This paper was my first publication about Alfred Russel Wallace and his biogeographical – I might even say biogeological – analysis of the Malay Archipelago. My interest in both Wallace and Indonesia started when I read a copy of *The Malay Archipelago* that I found in the historic Kaukapakapa Library. The book collection was started in 1865 by Morris Henley, a prominent member of the first European settlers of the district, and showed how important it was to these workingmen and women of this remote outpost of the British Empire to keep abreast of intellectual developments in the wider world. The book was a revelation to me; not only was it a great read but it also forced me to re-evaluate my views about Wallace himself. Was he really the great proponent of nineteenth-century dispersalist biogeography with its emphasis on centres of origin and waves of evolutionary advanced species spreading out from the northern hemisphere to displace less advanced forms to the south? Well, yes and no. You can certainly make a case that he did support and promote such a view, particularly later in life, but there were other sides to this complex character's intellectual breadth (Michaux 2008) that require a more nuanced evaluation to get to the truth of the matter.

Wallace specifically invoked geological changes to explain the similarities and differences in the floras and faunas of the Greater Sunda Islands, and thought that the peculiarities of the biota of Sulawesi were a product of past geography. Such views did not square with ideas of chance dispersal, not just in terms of explanation but in terms of world view. For a dispersalist, biogeographic explanations involve a series of contingent events and biotas of islands are the flotsam and jetsam of history. For a biogeologist, explanations of modern distributions involve historical tectonic change to both the present-day islands and the structures of which they are part. The paper started with a discussion of the role of contingent events in historical explanations, using the fallacy of Cleopatra's nose to make the point that contingent explanations cannot be generalised. The nature of explanation in historical studies is a theme that I think is worth expanding on, as I believe it has important implications for biogeography.

## *Explanation in historical studies*

> *All* causal explanations of a singular event can be said to be historical in so far as 'cause' is always described by singular initial conditions. And this agrees entirely with the popular idea that to explain a thing causally is to explain how and why it happened, that is to say, to tell its 'story'. But it is only in history that we are really interested in the causal explanation of a single event. In the theoretical sciences, such causal explanations are mainly means to a different end – the testing of universal laws. (Popper 1957:143–144)

The analysis of historical events has long been the subject of robust debate between those who support the interpretation of history by means of general principles and those who deny such principles exist. For example, Somers (1994) argued that relational accounts (narratives) in sociology could avoid general laws by using mechanisms to build causal explanations, an approach criticised by Goldstone (1998) who argued that explanations of this type would never be able to rise above the wholly contingent and unique Seussian explanation (refer to *The Cat in the Hat*): that things just happen – that this happened, then this, then that, and will not likely happen that way again, unless necessary or probable connections between events were posited, in which case unacknowledged general principles are being applied. The historian White (1984) suggested that although narrative is as universal as language itself and a mode of verbal representation so natural to human consciousness, its use in fields that aspire to be scientific is suspect. In his view, there is a spectrum of historical explanations ranging from telling a story, via a description of sequentially ordered historical facts, to an interpretation of these facts in terms of general principles. This continuum is based on a decreasing reliance on narrative and, according to White (1984), only explanations based on general principles can aspire to be scientific. Sahlins (1985) classified historical facts as events rather than mere happenings when they can be interpreted as instances of general principles. When Caesar crossed the Rubicon, it was an event because his action (at the head of his legions) was a direct challenge to the political establishment of Rome. Otherwise, crossing the Rubicon amounts to nothing more than wading across a rather insignificant waterway. For Sahlins,

> an event is a unique actualization of a general phenomenon, a contingent realization of the cultural pattern – which may be a good characterization of history *tout court*. (Sahlins 1985:vii)

Mark Antony's infatuation with Cleopatra was used as an example by Carr (1970) to illustrate the fallacy of drawing general conclusions from unique events. For Carr,

> the dual and reciprocal function of history ... [is] to promote our understanding of the past in the light of the present and of the present in the light of the past. Anything which, like Antony's infatuation with Cleopatra's nose, fails to contribute to this dual purpose is from the point of view of the historian dead and barren. (Carr 1970:141)

So was Cleopatra's pert nose the cause of Mark Antony's defeat at Actium, which was to set in motion a train of events whose effects were to reverberate down the centuries?

## *The battle of Actium*

The facts of the battle of Actium are clear enough: after a prolonged period of stalemate, the opposing fleets of Mark Antony and Octavian finally joined battle. Although the engagement was fierce, the outcome of the battle was far from settled and seemed destined to end in stalemate when Cleopatra's squadron of ships suddenly broke through Octavian's lines and, followed by Mark Antony, fled south to Egypt. The consequences of this seemingly inexplicable behaviour by their commander in abandoning the rest of his fleet and army were profound: for both Mark Antony and Cleopatra, the result was complete defeat and their suicides the following year; for Egypt, it was the loss of its independence and incorporation as a vassal state into the Roman Empire; for Octavian, it resulted in absolute power that allowed him to declare himself emperor and to break the hegemony of the old Roman aristocracy once and for all; for Rome, it resulted in a prolonged period of imperial expansion and conquest, which was to lay the seeds for the eventual flowering of European civilisation.

It was Pascal in his *Pensées* (Ober 2002) who first suggested that it was Mark Antony's infatuation with Cleopatra's beauty that clouded his judgement and led to his defeat at Actium. This narrative seems to be heavily influenced by Octavian's spin and propaganda, which took pains to paint Anthony as a lovesick dupe under the power of an eastern witch. But as Ober (2002) pointed out, this rather misogynistic reading doesn't stand up to inspection. For Ober (2002), the defeat of Mark Anthony at Actium can be traced directly to his disastrous campaign five years earlier against the Parthians, who were a direct threat to his power base in the Eastern Empire. The loss of both his prestige and apparent cloak of invincibility seriously weakened Mark Antony's standing among the legions and limited his options in dealing with the ambitions of his erstwhile ally

Octavian. According to Ober (2002), this forced Mark Antony into accepting Cleopatra as an active ally rather than giving her a more subservient role as a passive provider of money and supplies in what became an open conflict between the two men. This strategy of Mark Antony did not sit well with ordinary Romans or his army, and was further compounded by the naming of his children with Cleopatra as heirs to his Asian territories.

Octavian, ever the master politician, was quick to exploit this unease by conducting a propaganda campaign against Cleopatra, painting her as a dangerous and manipulative figure whose ultimate aim was to become the Queen of Rome, and insinuating that the coming war was more a crusade to maintain Roman values than a power grab. The effect of this propaganda and disinformation war was to isolate Mark Antony from Rome by forcing his allies to flee east, giving Octavian free rein in the western empire. Ober (2002) makes a case that Cleopatra and Mark Antony's flight to Egypt was a deliberate act, albeit a high-stakes gamble. The stalemate at Actium favoured Octavian because the situation was becoming increasingly desperate for Mark Antony whose army was suffering from both disease and defections exacerbated by Octavian's blockade of his supply routes from the south, and the effect of his propaganda war that Cleopatra's active presence reinforced among Antony's resentful troops. While Mark Antony was obviously prepared to sacrifice the bulk of his navy, he clearly intended to save his army that retreated southwards in good order. Unfortunately for Antony and Cleopatra, Octavian sued for peace and bought them off, thus leaving Egypt defenceless. So did Mark Anthony fail because his love for Cleopatra clouded his judgement, or did he fail because he was not an astute enough politician and lacked the necessary statecraft to counter his savvier opponent?

## *Narrative biogeography*

All would agree that plant and animal distributions have an historical component because what we see today has developed through time. Dispersalist biogeographers adopt a narrative form of explanation in which areas are colonised through chance dispersals, a series of contingent and unique occurrences unlikely to be repeated. In fact, if such dispersals are repeated, then they are neither unique nor chance events, and the hypothesis that these are either rare but normal occurrences in the life cycle of the organism in question, or that some general process is in operation, have to be considered. There are a number of consequences of adopting a narrative approach to biogeography.

1. Explanations of contingent events cannot be generalised as they apply only to the particular instance being explained. To do so is simply to fall into the fallacy that all generals lose battles because

they are infatuated with their lovers (Carr 1970). So, while individual cases can be explained by chance dispersals, one cannot generalise to conclude that all, or most, or even another species has done so. Each instance requires a unique explanation.

2. Narrative forms of explanation treat unique events as isolated instances devoid of context. The explanation that Mark Antony lost the battle of Actium because of Cleopatra's beauty loses whatever plausibility it may have had when the battle is viewed in its wider political and historical context. Dispersalist narratives ignore any biological context, such as common distribution patterns shown by other organisms, or the geological context provided by the tectonic history of the area(s) in question.
3. A narrative approach can never interpret or explain phenomena in terms of general principles, which limits its development into a truly scientific subject. Now dispersalist biogeographers might reply that this is all very well, but that's how the world is. Indeed, if this were so, then we would have to accept that biogeography would forever remain as narrative. But wouldn't we want some conclusive evidence that the New Zealand biota, for example, was derived by a series of chance, trans-oceanic dispersals before we consign biogeography to the fate of also-ran?
4. White (1984) distinguished narrative from story-telling based on whether the events being related are true or fictional. How do we know that explanations of historical events are true and that dispersalist biogeography is not just story-telling? This is a tricky question to answer when you are dealing with unique historical events. The best you can do is find corroborative evidence that confirms the explanation as the most probable. Contemporary New Zealand dispersalist biogeographers have used two lines of evidence to support their narratives. Firstly, the hypothesis of a great flood in the Oligocene (e.g. Trewick et al. 2007; Landis et al. 2008) that completely drowned the New Zealand landmass, wiping out all the original terrestrial biota, which meant that the modern biota *had* to be post-Oligocene in age and *had* to be derived by trans-oceanic dispersal. Secondly, calculated ages from molecular phylogenies indicate the recent (i.e. Neogene) arrival of most organisms in New Zealand.

The Great Flood hypothesis was never much more than a narrative with no possibility of verification and every probability of falsification, which was duly done in a special issue of the *New Zealand Journal of Geology and Geophysics* (volume 57, part 2, 2014). To me, the more interesting aspect of this idea was the motivation behind it. While it has long been known that the Oligocene was a period of maximum marine transgression in New Zealand, it is quite something else to suggest that the entire land surface was inundated. And it had to be total inundation to achieve the desired outcome – the creation of *de novo* land that could be recolonised afresh by trans-oceanic dispersal. Any small islands would have acted as refugia for terrestrial organisms during times of maximum sea level, which would then have expanded their ranges during the following marine regression that occurred at the start of the Miocene. In short, there would be no wiping of the slate clean.

The second line of evidence supporting the claim of post-Oligocene or trans-oceanic dispersal of much of New Zealand's biota is provided by the ages derived from molecular phylogenies (e.g. Waters and Craw 2006; Goldberg et al. 2008; Trewick and Gibb 2010). As Hipsley and Müller (2014) comment,

> This issue [clock calibration] is particularly relevant now, as time-calibrated phylogenies are used for more than dating evolutionary origins, but often serve as the backbone of investigations into biogeography, diversity dynamics and rates of phenotypic evolution.

As ages derived from molecular phylogenies are the only evidence to give credibility to dispersalist narratives, these dates should be subject to rigorous scrutiny, particularly in light of the critiques by Heads (2014) and Hipsley and Müller (2014). Finding ways of accurately dating phylogenies would represent a very significant step in advancing biogeographic and other time-dependent studies. Although I have advocated tectonic calibration as a preferred method, I suspect that unequivocal dates are likely to be achieved as a result of convergence of results obtained by different methods. There should also be a clear distinction made between dates of diversification and origin of clades. Post-Oligocene diversification dates are to be expected for many New Zealand taxa because mountain-building transformed New Zealand from a low-lying archipelago into a significant and mountainous landmass with concomitant habitat diversification and potential for allopatric speciation. But diversification of a clade is not the same as the origin of that clade, which is critical to any explanation of when the ancestral species first evolved in New Zealand. It would be interesting to see a statistical analysis of published molecular ages for New Zealand taxa that compares the distribution of these dates to some theoretical distribution based on a model incorporating parameters such as random arrival times, extinction rates and distance from source areas. I suspect the analysis would show a significant Neogene bias.

ELSEVIER Palaeogeography, Palaeoclimatology, Palaeoecology 112 (1994) 323–343

PALAEO

# Land movements and animal distributions in east Wallacea (eastern Indonesia, Papua New Guinea and Melanesia)

B. Michaux

*Private Bag, Kaukapakapa, New Zealand*

Received 15 September 1993; revised and accepted 2 May 1994

**Abstract**

Four generalized tracks are identified in eastern Indonesia–Papua New Guinea–Melanesia on the basis of animal distribution patterns. Two of these tracks are linked to a Melanesian rift system that developed to the north and east of Australian Gondwana during the late Cretaceous and early Tertiary. The other two tracks are related to Cretaceous rift systems formed off the west/northwest coast of Australian Gondwana. The histories of these rift systems are discussed, their present positions are tentatively located in terms of modern geography, and animal distributional patterns related to their formation and subsequent histories. Wallace's Line is reinterpreted in terms of a boundary between Asian and Australian derived terranes and microcontinents. Distributional patterns that reflect this discontinuity show incongruity with the geological pattern in Sulawesi and the Banda arcs because of the merging of Asian and Australian faunas as a direct result of Australia's collision with Asia. Indo-Australian distributions are interpreted in terms of the distribution of ancestral populations prior to the mid-Mesozoic breakup of the relevant sectors of Gondwana.

## 1. Introduction

I read in a recent review that the twentieth century would be remembered for only three intellectual ideas—relativity, quantum mechanics, and chaos theory (Jones, 1992). Whilst this thesis is clearly open to challenge, the effect of each of these ideas on our "way of seeing" the world has been significant. Theories that alter our perception in significant ways not only act as new unifying principles within particular fields, but also catalyse change in other disciplines.

An exciting and very twentieth century idea, which has had a profound effect on the way the world is viewed, is the theory of plate tectonics. Hamilton (1988) has reviewed the rapid development of plate tectonic concepts during the late 1960's, and the subsequent struggle that these mobilistic ideas had in penetrating established thinking in the USA. In its original form the theory envisaged that the Earth's lithosphere was split into a series of rigid plates that moved relative to each other. In more recent times the emphasis has changed with the recognition of the formation of microcontinental fragments during rifting (Nur and Ben-Avrahim, 1982) and other non-rigid responses of plates, particularly at their margins (Hamilton, 1988; Hutchison, 1989).

During the last two decades plate tectonic theory has not only unified diverse geological phenomena by relating them through process, but has also influenced some biogeographers to re-evaluate explanations of distribution patterns that do not explicitly recognise that continents have moved relative to each other and that through time their individuality has little reality. However, it is of


*SSDI* 0031-0182(94)00044-9

interest to note that it was a biogeographer, Leon Croizat, who first employed Wegener's mobilistic concepts in the 1950's prior to the publication of plate tectonic theory. Croizat, always the iconoclast, was scathing of "Wegenerism" because of his opposition to Pangea. However, his keen appreciation of Wegener's mobilistic concepts as explanatory principles has proved, by linking biology with geology, to be an important insight into global processes.

According to Veevers (1991), the global arrangement of continental crust appears to be cyclic, with a coalescent phase characterised by glacial climates alternating with a dispersive phase characterised by warm climates. The present arrangement of continents is approaching maximum coalescence. Australia is the latest and, with the exception of Antarctica, the last remaining Gondwanic block to become sutured to a northern landmass, a process that is underway at present as Australia collides with Asia.

Wallace (1880) recognised a fundamental biotic discontinuity between Bali and Lombok which separated Asian and Australian mammals and birds. However, Wallace's line proved difficult to fix on a map (Simpson, 1977) and for other taxa appeared not to exist at all. This rather enigmatic situation has traditionally been explained by recourse to narratives about dispersal events involving taxa island-hopping along sweepstake routes. Taxa that occur both west and east of Wallace's Line are thus interpreted as a result of successful dispersal events in the past. For example, Simpson (1961) interpreted the Australian murid fauna as derivatives of Asian murids that migrated to Australia via the Sunda and Banda Islands. Taxa that are restricted to either the west or east of this line have yet to pass this dispersal barrier.

Other biogeographic studies of the region have incorporated tectonic explanatory principles for distribution patterns (e.g. Whitmore, 1981; Holloway, 1987; Van Welzen, 1989; Humphries, 1990; Cracraft, 1990; Duffels and De Boer, 1990; Michaux, 1991). It is the purpose of this paper to try and relate the complex distribution patterns of this region to the details of the geological history of the collision between Australia and Asia.

## 2. Here be dragons

"If one were able to view Indonesia from a distance of 10,000 kilometres, it would look like mere blotches scattered between Asia and Australia. Look at detailed maps, and more islands appear—a total of 13,667, according to official statistics. Together these ecologically rich islands cover 1.9 million square kilometres, an area almost as large as Greenland. Pause for a few moments to ponder the origin of the islands and you can understand the reason for the wealth." (Hecht, 1992)

Indonesia is a region of exceptional tectonic complexity formed by Australia's northward drift into a *melée* of rifted Gondwanic and Cathaysian continental fragments and island arcs, and the subsequent lateral translation of diverse fragments within the collision zone (Hartono and Tjokrosapoetro, 1986; Pigram and Davies, 1987; Hutchison, 1989; McCabe and Cole, 1989; Rangin et al., 1990; Burrett et al., 1991).

This paper concentrates on the eastern part of this region which will henceforth be termed east Wallacea. It includes the Lesser Sunda Islands, Sulawesi, Maluku, New Guinea, and Melanesia. A locality map is shown in Fig. 1. As would be expected in a region with such a complex history, geological patterns may be equivocal and relationships between geological units unclear. However, it is clear that many of the geopolitical entities in east Wallacea are geological composites. In the following sections terranes are identified, and a brief summary of their geological histories is given.

### 2.1. The Banda arcs

The Banda arcs consists of two parallel, curvilinear island groups. The inner group of islands (which includes Lombok, Flores, Komodo, etc.) is a continuation of the Sunda Shelf islands of Sumatra and Java (IBA). The outer Banda arc islands (which include Timor and Seram) lie to the south and east (OBA). The Banda arcs are thus composed of islands that belong to both the Lesser Sundas and southern Maluku.

Hamilton (1988) characterised the IBA as a typical oceanic arc dating from the middle Tertiary.

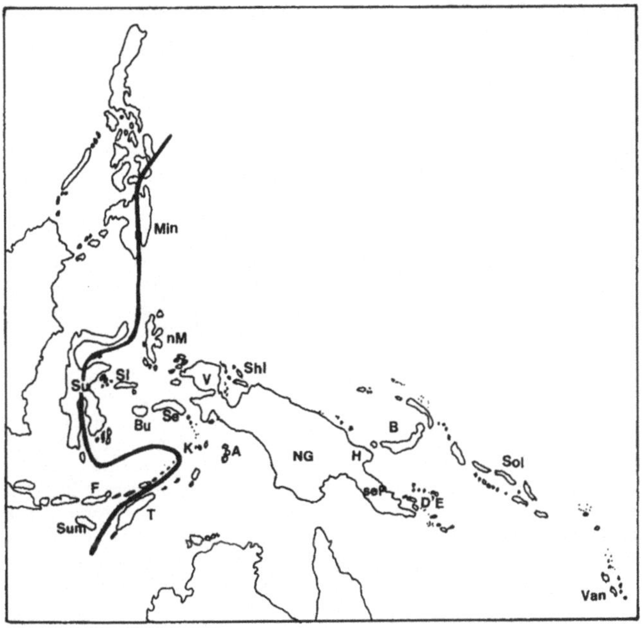


Fig. 1. Locality map. *A* = Aru Islands; *B* = Bismark Archipelago; *Bu* = Buru; *D'E* = D'Entrecasteaux Islands; *F* = Flores; *H* = Huon Peninsula; *K* = Kai Islands; *Min* = Mindainao; *NG* = New Guinea; *nM* = northern Maluku; *Se* = Seram; *seP* = southeast Papua; *ShI* = Shouten Islands; *SI* = Sula Islands; *Sol* = Solomon Islands; *Su* = Sulawesi; *Sum* = Sumba Islands; *T* = Timor; *Van* = Vanuatu. The line is a modified version of Carey's (1975) geological partition of Asian and Australian land masses. This line can also be interpreted as a boundary between Asian and Australian biotas. Tectonic events promoted mobilism in Sulawesi and in the Timor–Flores region which results in a blurring of the boundary in these regions. See text for full explanation.

He regards the OBA islands as emergent sections of a fore arc ridge (see Hamilton, 1988: fig. 3) which have subsequently collided with the leading edge of the north-drifting Australian continent forming a chaotic, imbricated mixture of accretionary wedge sediments, obducted oceanic crust material, and fragments of continental material. Hamilton (1988) regarded the IBA and OBA as a paired arc system that has migrated away from the Sunda Shelf with the concomitant formation of the south Banda Sea basin behind it. In straightforward terms Hamilton (1988) regards the position of the now aborted Banda trench to be east of the OBA.

Other workers interpret the position of the Banda trench to be between the OBA and IBA (e.g. Audley-Charles et al., 1975). These workers envisaged that the islands of Timor and Seram are geological composites comprising autochthonous Australian rocks overlain by Asian allochthonous overthrust masses. The juxtaposition of Asian and Australian geological units resulted from the mid-Tertiary collision of the Australian continental shelf with the islands of the Banda arcs (Hartono and Tjokrosapoetros, 1986; Hutchison, 1989). Thus, the autochthonous portions of the OBA are seen as Australian and separate from the IBA. The Banda Sea basin is interpreted as Mesozoic oceanic crust now trapped and surrounded by continental crust.

Duffels and De Boer (1990) included Buru as part of the OBA on the basis of a phylogenetic analysis of the cicada fauna. Michaux (1991), following the suggestion of Hartono and Tjokrosapoetro (1986) based on geological evidence, regarded Buru as a detached fragment of the Australian–New Guinea foreland. There are also conflicting views amongst geologists as to whether the island of Sumba in the south is part of the IBA (e.g. Burrett et al., 1991), or part of the OBA (e.g. Hartono and Tjokrosapoetro, 1986).

The IBA and OBA collided in the Flores–Timor sector during the late Miocene. Audley-Charles and Hooijer (1973) discussed the migration of pigmy stegodonts during the Pliocene/Pleistocene based on fossil evidence from Flores and Timor. Was this faunal exchange a general phenomenon with the effects still observable today, and if so is there any evidence for an exchange of Australian biota into Asian geological units?

### 2.2. Sulawesi

The island of Sulawesi occupies a pivotal position in Wallacea, not only because it is centrally placed geographically, but also because the origin of its highly endemic but depauperate fauna is problematic. Solving this problem may perhaps provide a key with which to understand distribution patterns in the region as a whole. A number of satellite islands surround Sulawesi, the largest of which are the Sula group to the east. Other island groups run from the north towards the southern Philippines, and from the south towards the islands of the IBA.

Hutchison (1989) identified the southwest arm and central Sulawesi as a microcontinental fragment to which an eastern arc became sutured during the mid-Tertiary, and a northern arc in the late Tertiary. Prior to the opening of the Macassar Strait the central and southwestern portion of the island was attached to southeastern Kalimantan. Both Duffels and De Boer (1990) and Michaux (1991) discussed the relevance of the composite geological nature of Sulawesi to patterns of relationship shown by the Sulawesian biota.

Michaux (1991) has linked the southern and central part of Sulawesi with a generalized track involving Sabah, east and southeast Kalimantan, and possibly Mindanao to eastern Java. Eastern Java is in turn linked to the islands of the IBA to the east, and western Sumatra and Burma to the north. Michaux (in press) suggested that these now geographically dispersed fragments may once have been contiguous, forming a southern extension of the Burma plate that collided with Southeast Asia in the late Mesozoic.

Duffels and De Boer (1990) and Vane-Wright (1991) suggested, on the basis of cicada and butterfly faunas respectively, that the eastern arc, which collided with Sulawesi in the late Miocene, was submerged and did not contribute any Australian elements to the Sulawesian fauna. Wallace's (1880) observation of marsupials in Sulawesi seems to indicate that a distinct Australian element is present. The Sula platform is generally regarded by geologists as a fragment of New Guinea. The suggested source area for this microcontinental terrane has been from various parts of the Vogelkop region of Irian Jaya, but Pigram and Supandjono (1985) favour a more easterly source. On the basis of pre-Cretaceous stratigraphic correlation, these authors place the source area for the Sula platform between 141° E and 145° E in central New Guinea. This geological hypothesis implies that the Sula platform is more closely related to the IMR than to the OBA. Such a hypothesis might be tested with independent phylogenetic data as they become available.

### 2.3. Northern Maluku

Wallace (1880) separated northern Maluku (Morotai, Halmahera, Ternate, Batjan, and Obi) as a subprovince of his Moluccan region. Although there were similarities between the northern biota and that of the southern islands of Buru, Ambon, and Seram, the former showed a closer relationship to New Guinea, whilst the latter seemed more strongly related to either Timor or Australia.

Duffels and De Boer (1990) confirmed that this biotic discontinuity existed for cicadas, although they regarded the position of Obi as ambiguous.

While it seems reasonably clear that Maluku as a whole is both geologically and faunally composite, the question is open as to whether northern Maluku is geologically composite. Duffels and De Boer (1990) have reviewed the distributional evidence linking northern Maluku to New Guinea, and similar patterns will later be demonstrated to exist for other groups.

### 2.4. Melanesian rift-systems

Croizat (1958: fig. 200) regarded New Guinea as geologically composite on the basis of distribution patterns. Michaux (1989) has discussed these ideas in detail, but the main point of Croizat's analysis was that New Guinea was involved with three tracks. The first linked southern New Guinea with north and east Australia, the second linked the central highlands with New Caledonia and New Zealand, and the third linked northern New Guinea and the Huon Peninsula with the Bismarck archipelago, Solomons, etc. Duffels and De Boer (1990) termed this latter unit the outer Melanesian arc, whilst they referred to the unit linking the central highlands of New Guinea with New Caledonia and New Zealand as the inner Melanesian arc. The use of the term "arc" to describe these units is problematical because they appear to have been rifted units rather than oceanic island arcs. Thus I shall refer to them as the outer and inner Melanesian rifts (OMR and IMR).

The geological history of these rift systems is described in detail by Parker and Gealey (1983). Both were derived from a single geological unit rifted from the north and east of Australia during the late Cretaceous or earliest Tertiary (Parker and Gealey, 1983) or latest Jurassic (Pigram and Davies, 1987). This unit subsequently split into two independent "units" (IMR and OMR) before being reunited to the northward drifting Australian craton.

Duffels and De Boer (1990) confirmed the overall picture suggested by Croizat (1958), but their work also indicated that the situation was more complex. Recent work indicates that numerous terranes exist in New Guinea (Pigram and Davies, 1987; identified 32), and so complex patterns would be expected. An aim of this work is to attempt to identify which areas in modern New Guinea were once parts of each of these rift systems.

## 3. Methods

In an ideal world, analysis of distributional data would be based on the area cladogram. Methods that lack phylogenetic constraint run the risk of linking areas on the basis of poly- or paraphyletic taxa. However, sufficient area cladograms are simply not available as yet for the region as a whole. Those that are available are discussed in the text. The approach adopted in this paper is to accept that sufficient data are unlikely to become available for some time, and instead to rely on congruence between independent data sets to establish preliminary hypotheses. These hypotheses can be tested as sound phylogenetic data become available. Michaux (in press) argued that this approach allows older taxonomic and distributional data to be utilized, and that future studies can be more precisely planned by concentrating effort in areas that may yield maximum information.

The method adopted in this study is Croizat's (1958) method of track drawing, which simply takes distributional data for a taxon at face value and looks for patterns that are common to as wide a variety of taxa as possible. It is assumed that the taxa used (generally taxonomically defined species rather than higher groupings) are monophyletic. Distributional data were obtained from the literature. These sources are listed in Table 1. Any congruence between distributional and geological patterns is then taken as meaningful, i.e. generalized tracks become intelligible in terms of hypothesized tectonic development.

## 4. Results

### 4.1. Banda arcs

The distribution results involving the Banda arc fauna are given in Tables 2 and 3. In Table 2, two

Table 1
Sources used in this study

| | | |
|---|---|---|
| **Murid rodents** | **Primates** | **Annalids** |
| Musser, 1977a | Musser & Dagosto, 1987 | Easton, 1976 |
| Musser, 1977b | | |
| Musser, 1979 | **Bats** | **Diptera** |
| Musser, 1981a | Hill, 1983 | Drew, 1989 |
| Musser, 1981b | Heaney, 1991 | |
| Musser, 1982a | | **Homoptera** |
| Musser, 1982b | **Mammals** | Duffels, 1990 |
| Musser & Califia, 1982 | Musser, 1981a | Duffels, 1991 |
| Taylor et al., 1982 | Musser, 1987 | Duffels &de Boer, 1990 |
| Musser & Newcombe, 1983 | | |
| Musser, 1985 | **Lepidoptera** | **Hemiptera** |
| Musser, Heaney, & Rabor,1985 | de Jong, 1983 | Moller Andersen, 1991 |
| Musser, 1986 | de Jong, 1990 | |
| Musser, 1990 | Vane-Wright, 1990 | **Birds of Paradise** |
| Musser, 1991 | Vane-Wright, 1991 | Gilliard, 1969 |
| Musser & Holden, 1991 | | |

distinct distributional patterns are apparent. The first links the southern islands of the OBA (Timor +) with Seram, Buru, the Sula islands, and northern Maluku. The second links the OBA to northern Australia/New Guinea and/or the Melanesian arc fragments.

From the data in Table 2 it is clear that Buru shows strong faunistic similarity to Seram and other northern OBA islands (Seram +). Other links between Seram and Buru occur to the south (Timor +), west (Sula islands), and north (northern Maluku). Duffels and De Boer (1990) have discussed cicada and lepidopteran distribution patterns that involve Maluku, OBA islands, and Sula islands. Like Wallace (1880) they recognised a significant biotic discontinuity between northern and southern Maluku, the latter islands (Buru and Seram +) being more closely related to the Sula islands than to northern Maluku. However, the widely distributed cicadas, *Hamza ciliaris* and *Baeturia macgillavryi* (Duffels, 1991), which are found throughout eastern Indonesia but avoid Sulawesi and New Guinea, suggest that Maluku + OBA + Sula islands may form an area of endemism.

Distributional evidence clearly links these eastern Indonesian islands to Australia rather than Asia. McCabe and Cole (1989) suggested, on the basis of both geological and geophysical evidence, that the Sula islands, Buru, Seram, Timor, and Sumba shared a common geological history. Rangin et al. (1990) also include the Lucipara Ridge (and associated islands) in the central Banda Sea as Australian rifted fragments. Rifting of continental crust from northwest Australia began in the late Jurassic, although the oldest oceanic crust is early Cretaceous in age (Burrett et al., 1991). According to Falvey and Mutter (1981) a period of up to 50 m.y. can separate the breakup and post-breakup stages of rifting. Hartono and Tjokrosapoetro (1986) envisaged that this rifted section formed islands close to the Australian shore. As a working hypothesis it is suggested that at least some parts of Timor, Seram, Buru, the Sula islands, and parts of northern Maluku, were originally part of this rift system. However, Pigram and Supandjono (1985), as discussed previously, relate the Sula platform to the IMR rather than the OBA.

This hypothesis is not necessarily in contradiction to the view held by Hamilton (1988) that the Banda arcs are a paired system, provided some of the continental fragments now incorporated into Timor and Seram prove to be derived from the Australian craton. One can imagine that as the Banda arcs migrated southwards these rift fragments were swept up and incorporated into the OBA. Subsequently this composite unit became

Table 2
Outer Banda arc (OBA) distributions. *X*=presence; *?*=possible presence; *Timor+* =southern islands from Timor to the Tanimbar Islands; *Seram+* =northern islands from the Kai Islands to Seram; *nMol*=northern Maluku; *Phil*=Philippines; *Aust/PNG*= Australian–New Guinea foreland; *IMR/OMR*=inner and outer Melanesian rifts; *ref*=references. *1*=Hill (1983); *2*=Laurie and Hill (1954); *3*=Vane-Wright (1990); *4*=Duffels (1991); *5*=De Boer (1989); *6*=Musser (1981a); *7*=Musser (1987); *8*=Duffels and De Boer (1990); *(a)*=not recorded from Sulawesi by Musser (1987); *(b)*=also recorded from Palau; *(c)*=also recorded from Obi; *(d)*=also recorded from Lesser Sunda islands; *(e)*=also recorded from Flores.

| | Sumba | Timor+ | Seram+ | Buru | Sula Is | Sulawesi | n Mol | Phil | Aust/PNG | IMR/OMR | Ref. |
|---|---|---|---|---|---|---|---|---|---|---|---|
| Rousettus a. amplexicaudatus | X | X | X | X | | | | X | | | 2 |
| Phalanger orientalis orientalis | | X | X | X | X | | | | | | 1 |
| Baeturia schulzi | | | X | X | X | | | | | | 8 |
| Diceropyga objecta | | | X | X | X | | | | | | 8 |
| Cosmopsaltria lata | | | X | X | | | | | | | 8 |
| Pteropus ocularis | | | X | X | | | | | | | 1 |
| Nyctimene minutus | | | X | X | | X | | | | | 1,7 |
| Pteropus chrysoproctus | | | X | X | | | | | | | 1 |
| (a)Dobsonia viridis complex | | | X | X | | ? | X | | | | 2,7 |
| Rhinolophus keyensis complex | | | X | | | | X | | | | 1 |
| Emballonura nigrescens nigrescens | | | X | X | | | | | | | 1 |
| Idea (Idea) idea | | | X | X | X | | | | | | 3 |
| (b)Hamza ciliaris | | X | X | X | X | | X | X | | | 4 |
| Baeturia macgillavryi | | X | X | X | | | X | | | | 5 |
| Hipposideros bicolor saevus | | | X | X | X | X | X | | | | 1 |
| Pteropus caniceps caniceps | | | | | X | X | X | | | | 1,7 |
| Echymipera rufescens | | | X | | | | | | X | | 1 |
| Miniopterus australis tibialis | | | X | | | X | X | | X | | 2 |
| Phalanger maculatus maculatus | | | X | X | | | | | X | | 1 |
| (c)Idea (Idea) idea | | | X | | | | X | | X | X | 3 |
| Melomys spp | | X | X | | | | X | | X | X | 6 |
| Pogonomelomys spp | | X | | | | | | | ? | X | 6 |
| Sus scrofa papuensis | | | X | | | | X | | X | X | 1 |
| Emballonura raffrayana complex | | | X | | | X | | | | X | 1,7 |
| Rhinolophus eurotis complex | | | X | X | | X | X | | ? | X | 2,7 |
| Miniopterus pusillus macroneme | | X | X | | | X | X | | ? | X | 2 |
| Miniopterus schreibersii oceanus | | X | X | | | | | | X | X | 2 |
| Pteropus melanopogon complex | | X | X | X | | | | | X | X | 1 |
| Pteropus temmincki complex | | X | X | X | | | | | | X | 1 |
| Dobsonia moluccensis complex | | ? | X | X | | | ? | | ? | X | 2 |
| Syconycteris crassa complex | | | X | X | | | | | X | X | 1 |
| (d)Nyctimene cephalotes complex | | X | X | X | | X | | | | X | 1 |
| (e)Murina florium complex | | | X | X | X | | | | X | X | 2,7 |

compressed, distorted, and dispersed as it became involved in the collision zone between Australia and Asia. Eastward movement along W–E transcurrent fault systems allowed Australia to continue its northward drift, and resulted in westward translation of continental fragments such as the Sula islands.

The second pattern evident from the data in Table 2 links the OBA with the Australian–New Guinea foreland and/or the Melanesian rifts. In terms of their geological histories, both the rifted sections of the OBA and Melanesian rift systems were originally attached to the Australian sector of Gondwana. Pigram and Davies (1987) favour a late Jurassic origin for the Melanesian rift system (IMR+OMR) but in subsequent events agree with Parker and Gealey (1983) who suggest they were derived as a single unit from the north and east of the Australian sector of Gondwana during the late Cretaceous. Smith and Hood (1981), on the basis of an analysis of the New Guinean bat fauna, suggested that OBA species have a sister-group relationship with those found in the Melanesian rifts. The geological history of these areas suggests other sister-group relationships involving Australian taxa may also be found, depending in part on the original pre-breakup distributions of ancestral taxa and the exact timing of rifting

Table 3
Inner Band arc (IBA) distributions. *X* = presence; *?* = possible presence; *west* = Malay Peninsula and/or Asia; *north* = Philippines; *Lesser Sund.* = Lesser Sunda Islands; *east* = Australia and/or New Guinea and/or Melanesian rifts; *Ref* = references; *1* = Laurie and Hill (1954); *2* = Musser (1987); *3* = Hill (1983); *4* = De Jong (1990); *5* = Musser (1981a); * = possibly introduced by man; *(f)* = fossil data

| | west | north | Sumatra | Java | Sulawesi | Lesser Sund. | Sumba | Timor | east | Ref |
|---|---|---|---|---|---|---|---|---|---|---|
| Acerodon mackloti complex | | | | | | X | X | X | X | 1 |
| Murina florium complex | | | | | | X | | X | X | 3 |
| Hipposideros diadema complex | | | | | | X | | X | X | 1 |
| (f)Spelaeomys + related genera | | | | | | X | | X | X | 5 |
| Dobsonia peronni complex | | | | | | X | X | X | | 3 |
| Cynopterus spp. | X | X | X | X | X | X | | X | | 3 |
| Taphozous melanopogon | X | | | X | X | X | | X | | 3 |
| Crocidura monticola | extra limital | | | X | | X | X | ? | | 1 |
| Crocidura tenuis | | | | | | | | X | | 1 |
| *Macaca irus | | | X | | | X | X | X | | 1 |
| Bibasis iluska | X | | | X | X | X | X | X | | 4 |
| Pteropus vampyrus complex | | X | X | X | | X | | X | | 1,2 |
| Rhinolophus celebensis complex | | | | X | X | | | X | | 2,3 |
| Tylonycteris robustula | X | X | X | X | X | | | X | | 1,2 |
| Scotophilus spp. | X | X | | X | X | X | X | X | | 1,2 |
| (f)Stegodon sompoensis | Similarities to | | | | X | X | | X | | 2 |
| (f)Stegodon trigonocephalus | Pleistocene fauna | | | X | X | X | | X | | 2 |
| (f)Stegodon sp. | of India or | | | | | | X | | | 2 |
| (f)Geochelone atlas | Pliocene of Java | | | X | X | | | X | | 2 |
| Taractrocera nigrolimbata | X | | X | X | X | X | X | | | 4 |
| Rhinolophus affinis | extra limital | | | X | | X | X | | | 1 |
| Pteropus alecto | | | | X | | X | X | | | 1,2 |
| Eonycteris speleae | X | X | X | X | | X | X | | | 1,2 |
| Kerivoula papillosa | X | X | X | X | | X | X | | | 1,2 |

events. For example, Duffels and de Boer (1990: fig. 21.4) report an OBA sister-group relationship with Australia-New Guinea foreland + Melanesian rift taxa in the cicada genus *Diceropyga*.

Holloway (1993) has discussed the relationships shown by Seram lepidoptera to New Guinea taxa. He includes the notodonti genera *Quaricalcarifera* and *Omichlis*, sections of the lymantriid genus *Euproctis*, and the geometrid genera *Paradromulia*, *Eucharidema*, and *Anisozyga* as examples of OBA-New Guinea distributions. It will be interesting to relate the phylogenetic pattern of member species of these genera to the geological hypotheses of Pigram and Davies (1987).

The data in Table 3 underline the pivotal importance of Timor in understanding distributions in east Wallacea. Two distinct patterns are also evident in the data presented in Table 3. The first pattern involves species with an eastern distribution (i.e. Australian) reaching the Lesser Sunda islands via Timor, but not extending any further west onto the Sunda Shelf. Holloway (1984) provides a particularly clear example of this type of pattern in the lepidopteran genus *Avitta*. *A. microsundana*, which occurs in both Timor and Flores, is part of an unresolved monophyletic group distributed over the Australian foreland and Melanesian rift regions. Van Welzen (1989), in an analysis of the plant genus *Guioa*, links *G. asquamosa*, found in Flores and Timor, with northern OBA and Australian foreland taxa. The second pattern links the Lesser Sunda islands to the Asian mainland via the Sunda Shelf and/or Sulawesi. Many of these Asian taxa also extend into Timor, but no further east into Australia or New Guinea.

Auffenberg (1980) cited geological evidence that the western part of the island of Komodo dates from the Jurassic (134 ± 19 m.y. B.P.), and that land became forested by perhaps the late Mesozoic. If this is the case then Rintja and western Flores share a similar geological history with Komodo. This contrasts with the generally held belief that the islands of the Lesser Sundas are entirely the result of relatively recent volcanic activity (Hamilton, 1988). Auffenberg's (1980) studies indicate that the herpetofauna of Komodo was a mixture of "old" endemics related to Sulawesian taxa, and a more modern element with relationships to taxa on the Sunda Shelf. The former pattern may also be evident in the endemic murid genera *Papagomys*, *Hoojeromys*, and *Floresomys* from Flores and *Komodomys* from Flores and Komodo, which Musser (1981a) tentatively links to the endemic Sulawesian genera *Lenomys* and *Eropeplus*.

The IBA is presently separated from Sulawesi by the south Banda Sea basin which is composed of oceanic crust. Whether this crust proves to be Cretaceous (McCabe and Cole, 1989; Burrett et al., 1991), or Tertiary in age (Hamilton, 1988), reconstruction of the relative positions of the IBA, OBA, and Sulawesi before the Banda Sea opened would result in the central parts of the IBA being closer to southwest Sulawesi, with the rifted sections of the OBA and continental Australia lying further to the south. The relationships noted by Auffenberg (1980) and Musser (1981a) linking endemics on Flores to Sulawesian taxa may date from this event.

The occurrence of Asian taxa in Timor and Australian taxa in the IBA suggests that a period of mobilism was initiated by the Pliocene collision between the Banda arcs and the Australian-derived rifted units detailed above. Reduced sea levels during the Pleistocene may have further promoted this process (Audley-Charles and Hooijer, 1973). The data in Table 3 indicate that the faunal exchange was asymmetric, with greater numbers of Asian taxa reaching Timor than Australian taxa reaching the islands of the IBA. Vermeij (1991) reports a similar asymmetric trans-Arctic exchange for molluscs.

Finally, the microcontinent of Sumba appears, on the basis of the data presented in Tables 2 and 3, to be related to the Asian IBA as suggested by McCabe and Cole (1989) and Burrett et al. (1991), rather than the Australian segments of the OBA as suggested by Hartono and Tjokrosapoetro (1986) and Hutchison (1989).

*4.2. Sulawesi*

The fauna of Sulawesi has long been recognised as enigmatic (Wallace, 1880). Despite its central location in the archipelago, the Sulawesian fauna is depauperate and exhibits an extraordinarily high degree of endemism. Sixty-eight percent of all Sulawesian genera of non-volant mammals are endemics (Musser, 1987). This pattern is common to other groups—for example Duffels and De Boer (1990) report six endemic cicada genera from Sulawesi. A further characteristic of Sulawesian mammals is that some have no known Sundaic relatives. Table 4 lists the native mammals of Sulawesi and includes general information on the distribution of related taxa.

Michaux (in press) has detailed a generalized track which includes the southern Philippines, Sabah, east and southeast Kalimantan, central and southwest Sulawesi, the IBA, east Java, west Sumatra and offshore islands, and Burma/Thailand. As a working hypothesis, this track has been related to the Burma plate, a rifted Gondwanic terrane that docked with SE Asia during the late Cretaceous. A number of mammalian distribution patterns, particularly for murid rodents, also mirror this and other geological patterns detailed in Michaux (in press).

Fig. 2 shows the relationships, deduced by Musser and Newcombe (1983), for the murid shelf-endemic genus *Sundamys*. While it should be stressed that the proposed relationships are not based on cladistic analysis, two interesting features are apparent. The first is the occurrence of *S. infraluteus* in Sabah and western Sumatra, which identifies this species as a component of the generalized track discussed above. The second feature is the linking of *P. muelleri* with *P. maxi*. A Malay Peninsula–Sumatra–west Java–Borneo distribution pattern is also congruent with a geological structure, the Shan-Thai terrane (Michaux, 1991).

A full analysis of the distribution of the Indo-Australian genus *Rattus* awaits further phylogenetic analysis. However, considerable progress has been made on the taxonomy of this genus during

332 B. Michaux/Palaeogeography, Palaeoclimatology, Palaeoecology 112 (1994) 323–343

Table 4
Native mammals of Sulawesi, modified from Musser (1987). The middle column briefly comments on the distribution of species listed or related taxa. $s$ = southern; $n$ = northern. Note (i) high degree of endemism, (ii) admixture of Asian and Australian groups, (iii) lack of close relatives on the Sunda Shelf for some taxa, (iv) depauperate with many groups absent.

| | | |
|---|---|---|
| **Marsupials** | | |
| Phalangers | | |
| Phalanger ursinus | most westerly occurrence of | Laurie & Hill, 1954 |
| P. pelengensis | Australian group. Phylogenetic | Musser. 1987 |
| P. celebensis | relationships unclear. | |
| **Placentals** | | |
| Shrews | | |
| Crocidura spp (8) | Sundaic and Indochina. Also | Laurie & Hill, 1954 |
| | found in the Philippines | Musser. 1987 |
| | | Musser & Heaney,1992 |
| Bats | | |
| 9 endemic spp. 59 | Indo-Australian | Hill, 1983 |
| species in total | | Musser, 1987 |
| Tarsiers | | |
| Tarsius spectrum | Tarsiurs found in s. Phil. | Musser, 1987 |
| T. pumilus | Borneo, and s. Sumatra. | Musser & Dagosto, 1987 |
| | Oligocene fossils from n. Africa | |
| Macaques | | |
| Macaca nigra | Sundaic and Mentawi Islands. | Musser, 1987 |
| M. tonkeana | Probably introduced into | |
| M. ochreata | Banda arc islands. | |
| M. maura | | |
| Squirrels | | |
| Rubrisciurus | Endemic genera. Asian but | Musser, 1987 |
| Hyosciurus | phylogenetic relationships | |
| Prosciurillus | unclear | |
| Rats and mice | | |
| 48 spp. in 14 genera | | Musser & Holden, 1991 |
| 10 genera endemic | | |
| Crunomys | Philippines | Musser, 1977a |
| Haeromys | Palawan, Borneo | Musser, 1987; 1990 |
| Maxomys | Sundaic, Indochina | Musser, 1981b |
| Rattus | Indo-Australian | Musser & Califia, 1982 |
| | | Musser & Heaney, 1985 |
| | | Musser, 1986 |
| | | Musser & Holden, 1987 |
| Civets | | |
| Macrogalidia | Sundaic. Other civets | Laurie & Hill, 1954 |
| musschenbroekii | probably introduced | Musser, 1987 |
| Pigs | | |
| Sus celebensis | Indo-Australian | Laurie & Hill, 1954 |
| Babyrusa babyrussa | No close relatives. Siwalik | Musser, 1987 |
| | beds of India (Pliocene) | |
| Anoas | | |
| Bubalus depressicornis | No Sundaic relatives | Laurie & Hill, 1954 |
| B. quarlesi | Siwalik beds of India | Musser, 1987 |

the past two decades, which prior to this had included a diverse and decidedly non-monophyletic assemblage of taxa. For example, Laurie and Hill (1954) recognised 35 murid species in six genera living on Sulawesi, of which 29 species were placed in the genus *Rattus*. Musser and Holden (1991) recognised 48 species in 14 genera, of which only nine species are in the genus *Rattus*.

In an interesting discussion on the relationship of *R. tawitawiensis* (Musser and Heaney, 1985),

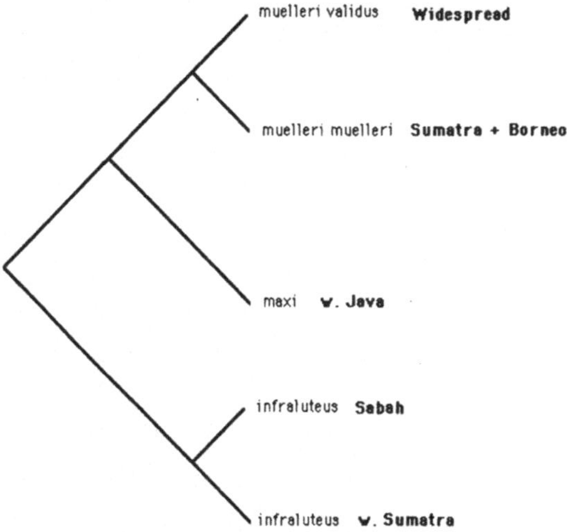


Fig. 2. Proposed relationships between species of the endemic murid genus *Sundamys* from Musser and Newcombe (1983). The figure is not a cladogram.

which is found on the island of Tawitawi in the Sulu group, these authors link this rat with species that are found in Mindoro (*R. mindorensis*), Maratua Island, off the east coast of Kalimantan (*R. tiomanicus mara*), and the *palarum*-group that are found on Enggano, the Mentawi Islands, Simeulue, and Nicobar Islands, all of which lie off the west coast of Sumatra. Van Welzen's (1989: fig. 77) description of the distribution of various morphological forms of *Guioa diplopetala* also indicate links between western Sumatra, eastern Java, south and east Kalimantan, Sulawesi, and Sabah. This type of distribution pattern mirrors the geological relationship between these areas proposed above.

Many characteristics of the Sulawesian fauna can be interpreted as a consequence of the southwest and central Sulawesi terrane (Hutchison, 1989) being a rather isolated fragment of Asian affinity. Although the evidence for Sulawesi's attachment to southeast Kalimantan until the opening of the Makassar Strait during the Miocene is good, this part of the island has clearly had a long period of isolation. However, one characteristic of the Sulawesian fauna which is not intelligible by reference to this aspect of Sulawesi's geological history is the presence of Australian elements in the fauna.

Wallace had great difficulty in placing the boundary of his Asian and Australian faunal regions to the north of Bali and Lombok. Initially Wallace's Line passed west of Sulawesi along the Makassar Strait, but later Weber was to draw his line of 'faunal balance' to the east. Similar uncertainties occurred in the southern Philippines region. The presence of marsupials on Sulawesi clearly gives the island's fauna a "mixed" flavour. Yet how common is this Australian element? Could it not be that these marsupials were self- or purposefully introduced? Neither Duffels and De Boer (1990) nor Vane-Wright (1990) could find Australian elements in the cicada or lepidopteran taxa they were discussing. Both suggested that eastern Sulawesi, an arc system which was sutured to the central and southwest Sulawesi during the Miocene and which was derived from the east, was not emergent until after its collision. Thus no terrestrial groups were imported into Sulawesi from Australasia.

De Jong (1990), based on his analysis of the Hesperiidae (Lepidoptera), commented:

"The cases of *Taractrocera* and *Cephrenes* show that the Australian element in the Sulawesi fauna may be stronger than thought at first sight. Studies are in progress to assess the contribution of the Australian element more precisely..."

The distribution of *Cephrenes augiades*, which occurs in eastern Australia, New Guinea, Solomons, Maluku, and Sulawesi (De Jong, 1990: fig. 5) is a particularly clear example.

Holloway's (1987, 1990) cladistic analysis of Sulawesian lepidoptera indicated that Sulawesian taxa not only show relationships to the west (Sundaland and Asia), but also that a number of genera include taxa that belong to Australian clades. For example, the geometrid genus *Bracca* has one species—*B. flavitaenia* from Sula—which forms a monophyletic group with OMR, OBA, and New Guinea taxa. Holloway discussed further examples from the genera *Idea*, *Narosa*, and *Probitha*.

There are also a small number of examples from the bat fauna of Sulawesi of species that show both strong Australian affinities and do not occur on the Sunda Shelf or SE Asia (Laurie and Hill, 1954; Hill, 1983; Musser, 1987). Both the bare-backed fruit bats (*Dobsonia* spp.) and the tube-nosed bats (*Nyctimene* spp.) may be Australasian elements in the Sulawesian fauna. In addition, a number of species of Indo-Australian genera, such as *Emballonura raffrayana*, *E. nigrescens*, *Miniopterus tristis*, *Murina florium*, and *Rhinolophus eurotis*, may prove, after detailed phylogenetic analysis, to be Australian derived.

Because marsupials do not appear to be completely isolated examples of Australian taxa occurring in the Sulawesian fauna, the hypothesis that they represent "accidental" introductions is weakened. The possibility that marsupials are part of a generalized track, perhaps resulting from the westward translation and collision of the Sula microcontinent and Sulawesi in the mid-Miocene, becomes a potential explanation awaiting further research.

### *4.3. Northern Maluku*

The similarity of the northern Maluku fauna to that of the islands of Sula, Buru, and Seram (northern OBA) has already been discussed. There are also similarities with New Guinea although, as noted by Duffels and De Boer (1990), this relationship often seems to bypass the Vogelkop peninsula and instead involves more easterly parts of New Guinea. The endemic mammals of Maluku are few, and only four involve northern Maluku and New Guinea without also including southern Maluku. An example of a mammalian distribution that involves northern Maluku but "misses" the Vogelkop is shown by bats of the *Nyctimene albiventer* complex. Two bat taxa, the *Pteropus conspiculatus* complex and *Emballonura nigrescens papuana*, and flying phalangers of the *Petaurus breviceps* complex, also include the Vogelkop. Perhaps the clearest linkage of northern Maluku with New Guinea, commented on originally by Wallace (1880), involves the birds of paradise. *Semioptera wallacei* and *Lycocorax pyrrhopterus*, found in northern Maluku, are the only members of this New Guinean-Australian group to be found west of the islands of Misoöl and Waigeo (Gilliard, 1969).

### *4.4. New Guinea and the Melanesian rifts*

As Duffels and De Boer (1990) have previously suggested on the basis of an analysis of the cicada fauna, New Guinea cannot be regarded as an area of endemism because the island is geologically composite. It is not known how many terranes will eventually be identified, but Pigram and Davies (1987) identify 32. Pigram and Davies' (1987) location of these terranes and discussion of their emplacement make fascinating reading from both biogeographic and geological viewpoints. Almost half the terranes are unequivocally continental in origin. Some appear to be deformed segments of the Australian foreland, whilst others are interpreted by these authors as being stray Gondwanic fragments (now incorporated in Misöol and the Vogelkop of Irian Jaya), or as parts of the IMR and OMR.

Pigram and Davies' (1987) fig. 2 located these

terranes in present-day New Guinea. The terranes that form the OMR occupy northern New Guinea east of Serera Bay. These terranes are isolated by mid Miocene sedimentary basins from IMR fragments which are now sutured to the Australian foreland forming the central mountain chain of New Guinea. The western part of the Huon Peninsula, which is part of the OMR, adjoins IMR terranes. The southeast peninsula of Papua New Guinea appears to be part of the IMR. A large composite terrane docked with the Australian foreland in this region later than those terranes further to the west, but prior to suturing of the Finesterre terrane which now forms the Huon Peninsula.

Table 5 shows the distribution of Birds of Paradise (Gilliard, 1969), Table 6 the distribution of fruit flies of the genus *Batrocera* (Drew, 1989) and marsupials (Laurie and Hill, 1954). Many taxa have distribution patterns linking north and eastern Australia with New Guinea, the Bismarck archipelago, Solomons, Vanuatu, and other Pacific islands. Within this rather extensive range, three distributional patterns may be recognised. In Tables 5 and 6 these patterns are related to the Australian–New Guinea foreland, and the outer and inner Melanesian rifts (OMR and IMR).

In Fig. 4, it is postulated that the Australian–New Guinea foreland consists of northwest, north, and eastern Australia with southern New Guinea, to which the Aru Islands and parts of the Vogelkop appear to be related. The IMR includes the Arfak mountains in the Vogelkop, the central mountainous region of New Guinea including the southeast peninsula, and New Caledonia. However, Pigram and Davies (1987) indicate that the Arfak Mountains are in part of Gondwanic origin. The IMR also extends into New Zealand (Michaux, 1989; Duffels and de Boer, 1990). The OMR consists of the Schouten Islands (and perhaps the northern coast of the Vogelkop), northern New Guinea, the Huon Peninsula, the Bismarck archipelago, and on to the Solomons, Vanuatu, Fiji, Samoa, Tonga, and New Zealand. The Huon Peninsula and the southeast peninsular region of New Guinea occupy an ambivalent position, in terms of mammalian and fruit-fly distributions, with links to both the IMR and OMR. The ambivalent position of the southeast peninsular is also reported by Van Welzen (1989), and Van Welzen et al. (1992). This is unsurprising in view of the terrane history described by Pigram and Davies (1987). However, other groups such as marine water-striders (Møller Andersen, 1990), and cicadas (Duffels and De Boer, 1990) link the Huon Peninsula to the OMR. For the present the former is linked with the OMR and the latter with the IMR, but further analyses are required. Research programs designed to produce area-cladograms relevant to the patterns described above are likely to yield important results.

Taylor et al. (1982) have provided a detailed analysis of native New Guinean *Rattus* spp. While this study does not pertain significantly to areas adjacent to mainland New Guinea (with only two species occurring in Australia, and one in the Bismarck archipelago, Solomons, and Vanuatu), it does provide a valuable insight into the composite nature of this island

She and her colleagues recognised three distinct groupings within New Guinean *Rattus* spp. on the basis of a UPGMA clustering of canonical variate scores derived from morphological measurements (Taylor et al., 1982: fig. 47). A summary of their conclusions is presented in Fig. 3. Fig. 3 is modified to exclude the non-endemic species *R. nitidus*, *R. rattus*, and *R. exulans*. This results in *R. giluwensis* becoming part of the *jobiensis-mordax-novaeguineae-praetor-leucopus-steini* group, which Taylor et al. (1982) favour.

The geographic range of each species is also included in Fig. 3. Although it is unfortunate that monophyly has not been established, each group occupies one of the three major geological divisions of New Guinea discussed above. The species of the largest of these three groups are, with the exception of *R. giluwensis* and *R. leucopus*, associated with the OMR. The phylogenetic position of *R. giluwensis* is at best equivocal, and the association of *R. leucopus* with other members of the OMR group may simply be a consequence of sympleisiomorphy. The OMR species show a sister group relationship to those of the IMR. These Melanesian rift taxa form a sister group to the Australian–New Guinean taxon *R. sordidus*. These

Table 5

Distribution of birds of pardise. North Maluku species not listed, see text. *X*=presence; *OMR*=outer Melanesian rift; *J+nV*=Japen and/or other Schouten islands and/or northern Vogelkop; *nNG*=northern lowlands/mountains of New Guinea; *HP*=Huon Peninsula; *IMR*=inner Melanesian rift; *Ar*=Arfak mountains; *CH*=central highland region that forms the spine of New Guinea; *seNG*=southeast peninsular of New Guinea including the D'Entrecasteaux archipelago; *NC*=New Caledonia; *Aust-NG foreland*=Australian–New Guinea foreland; *sNG*=southern lowlands of New Guinea includes Aru island; *wisl*=western islands; *Vog*=Vogelkop; *nwA*=northwest Australia; *nA*=northern Australia; *CY*=Cape York; *eA*=eastern Australia.

| | OMA | | | IMA | | | | Aust-NG foreland | | | | | | |
|---|---|---|---|---|---|---|---|---|---|---|---|---|---|---|
| | J+nV | nNG | HP | Ar | CH | seNG | NC | sNG | wisl | Vog | nwA | nA | CY | eA |
| Loria loriae | | | | | X | | | | | | | | | |
| Lobopradisea sericea subspp | | | | | X | | | | | | | | | |
| Cnemophilus macgregori subspp | | | | | X | | | | | | | | | |
| Paradigalla carunculata subspp | | | | X | X | | | | | | | | | |
| P. brevicauda | | | | | X | | | | | | | | | |
| Epimachus fastosus subspp | | | | X | X | | | | | | | | | |
| E. meyeri subspp | | | | | X | | | | | | | | | |
| Pteridophora alberti subspp | | | | | X | | | | | | | | | |
| Parotia wahensi | | | X | | | | | | | | | | | |
| P. sefilata | | | | X | | | | | | | | | | |
| P. carolae subspp | | | | | X | | | | | | | | | |
| P. lawesi | | | | | X | | | | | | | | | |
| Drepanornis albertisii subspp | | | X | X | X | | | | | | | | | |
| D. bruijnii | | X | | | | | | | | | | | | |
| Astrapia nigra | | | | X | | | | | | | | | | |
| A. splendidissima subspp | | | | | X | | | | | | | | | |
| A. mayeri | | | | | X | | | | | | | | | |
| A. stephanie subspp | | | | | X | | | | | | | | | |
| A. rothschildi | | | X | | | | | | | | | | | |
| Lophorina superba subspp | | X | X | X | X | | | | | | | | | |
| Manucodia ater ater | | X | X | | | | | X | | X | | | | |
| M. a. subalter | | | | | | | | X | X | | | | | |
| M. a. alter | | | | | | X | | | | | | | | |
| M. jobiensis subspp | X | X | | | | | | | | | | | | |
| M. chalybatus | | | X | X | X | | | | Misöol | | | | | |
| M. comrii subspp | | | | | | X | | | | | | | | |
| honygammus keraudrenii subspp | | X | | | X | X | | X | | X | | | X | |
| Diphyllodes magnificus subspp | X | X | X | X | | | | X | X | X | | | | |
| D. respublica | | | | | | | | | X | | | | | |
| Paradisaea apoda | | | | | | | | X | | | | | | |
| P. raggiana | | | | | | | | X | | | | | | |
| P. minor subspp | X | X | | | | | | | Misöol | X | | | | |
| P. decora | | | | | | X | | | | | | | | |
| P. rubra | | | | | | | | | X | | | | | |
| P. guilielmi | | | X | | | | | | | | | | | |
| P. rudolphi | | | | | X | | | | | | | | | |
| Cicinnurus regius regius | | | | | | | | Aru | | | | | | |
| C. r. rex | | | | | | | | X | X | X | | | | |
| C. r. gymnorhynchus | | | X | | | | | | | | | | | |
| C. r. similis | | X | | | | | | | | | | | | |
| C. r. cryptorhynchus | | X | | | | | | | | | | | | |
| C. r. coccineifrons | X | | | | | | | | | | | | | |
| Seleucidus m. melanoleuca | | | | | | | | X | | X | | | | |
| S. m. auripenis | | X | | | | | | | | | | | | |
| Ptiloris pradiseus | | | | | | | | | | | | | | X |
| P. victoriae | | | | | | | | | | | | | X | |
| P. magnificus subspp | | X | | | | | | X | | X | | | X | X |

Table 6
Distribution of Australian fruit fly species of the genus *Batrocera* and some marsupials in New Guinea and Australia. Abbreviations as in Table 5. *BA* = Bismark archipelago; *Sol* = Solomon Islands including Bougainville; *V* = Vanuatu; *P* = Pacific islands.

| | OMA | | | | | | | IMA | | | | Aus-NG foreland | | | | |
|---|---|---|---|---|---|---|---|---|---|---|---|---|---|---|---|---|
| | J+nV | nNG | HP | BA | Sol | V | P | Ar | CH | seNG | NC | sNG | nwA | nA | CY | eA |
| Batrocera (Batrocera) | | | | | | | | | | | | | | | | |
| atramentata | | | | X | | | | | | | | | | | | |
| longicornis | | | | X | X | | | | | | | | | | | |
| neonigrata | | | | X | X | | | | | | | | | | | |
| obliqua | | | | X | X | | | | | | | | | | | |
| obscura | | | | | | X | | | | | | | | | | |
| seguyi | | X | | X | | | | | | | | | | | | |
| kirki | | | | | | | X | | | | | | | | | |
| luteoli | | | | | | | X | | | | | | | | | |
| melanotus | | | | | | | X | | | | | | | | | |
| perfusa | | | | | | | X | | | | | | | | | |
| samoae | | | | | | X | | | | | | | | | | |
| Batrocera (Notodcus) spp | | | | | | X | X | | | | | | | | | |
| Echymiper kalubu subspp | X | X | X | X | Misöol, Waigeo | | | | | | | | | | | |
| Phalanger m. maculatus | X | X | Misöol, Salawati | | | | | | | | | | | | | |
| Thyogale bruijni bruijni | X | X | X | | | | | | | | | | | | | |
| Hipposideros calcaratus | X | X | X | X | | | | | | | | | | | | |
| Antechinus naso subspp | | | | | | | | X | X | X | | | | | | |
| Anisomys imitator | | | | | | | | X | X | | | | | | | |
| Phascosorex dorsalis subspp | | | | | | | | X | X | | | | | | | |
| Phalanger vestitus | | | X | | | | | X | X | | | | | | | |
| Pseudocheirus albertisi | X | X | | | | | | X | | | | | | | | |
| B. (B.) anthracina complex | | | | X | X | | | | X | | | | | | | |
| B. (B.) furfosa complex | | X | | | | | | | X | | | | | | | |
| B. (B.) biarctuata | | | | | X | | | | X | | | | | | | |
| B. (B.) curvuera | | X | | X | | | | | | X | | | | | | |
| B. (B.) furvescens | | | | | X | | | | X | | | | | | | |
| Batrocera (Zeugodacus) spp | | X | | | X | | | | X | | | | | | | |
| B. (B.) distincta complex | | | | X | X | X | | | X | X | | | | | X | |
| B. (B.) bryoniae complex | | ? | ? | X | X | X | | | X | X | | X | X | X | | |
| B. (B.) dorsalis complex | | | | X | X | | | | X | | | | | X | | |
| B. (B.) frauenfeldi complex | | | | X | X | X | | | | | X | X | | X | X | |
| B. (B.) tryoni complex | | | | | | | X | | | | X | | X | | X | X |
| B. (Polistomimetes) spp | | | | | X | | | | X | X | | | X | X | X | X |
| B. (Sinodacus) spp | | X | | X | X | | | | X | X | X | | | | X | |
| B. (B.) bidentata complex | | | | | | | | | | | | | | | X | X |
| B. (B.) fragraea complex | | | | | | | | | | | | | | | X | X |
| B. (B.) mayi complex | | | | | | | | | | | | | X | X | X | X |

distribution patterns are congruent with the geological history of the tectonic units outlined above.

An interesting pattern in intraspecific variation shown in a number of these murid species is that subspecies are distributed along an E–W axis. This pattern is also evident in both marsupial and birds-of-paradise taxa. As both the IMR and OMR were certainly island chains prior to their collision with the Australian-New Guinea craton, it is likely that such differentiation dates to this time.

The relationship of New Guinea *Rattus* spp. to those of the Sunda shelf and islands of east Wallacea is not known. Musser (1981a, b) has suggested that these two *Rattus* groups may prove to be generically distinct, with both independently deriving *Rattus*-like features from some pre-*Rattus*

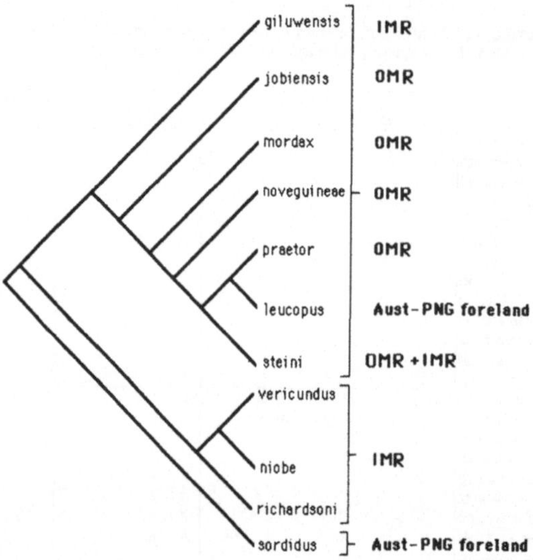


Fig. 3. Relationships of endemic New Guinean species of the murid genus *Rattus*. Modified from Taylor et al. (1983) and based on a UPGMA clustering of canonical variate scores from morphological features.

ancestor. From the perspective developed in this paper these two *Rattus* groups would have shared a common ancestor at a time prior to the separation of the Burma plate from Gondwana. Their present juxtaposition east and west of Wallace's Line is the result of subsequent tectonic events.

## 5. Discussion

Distributional evidence links Timor with Seram, Buru, the Sula islands, and north Maluku. North Maluku has a mixed fauna, part of which, for example cicadas and Birds of Paradise, show closer phylogenetic similarity to taxa of northern New Guinea and the islands of the OMR (see below). In Fig. 4, parts of Timor, Seram, Buru, the Sula islands and northern Maluku are all hypothesized to be related to an early Cretaceous rift system derived from the northwest of Australian Gondwana.

Evidently the spreading system responsible for this rifting aborted leaving a series of satellite islands off the coast of northwest Australia. This rift system acted as a leading edge in the collision between Australia and Asia. In the north, the Sula Islands became detached and were translated westwards to collide with Sulawesi in the late Miocene or early Pliocene. This may have promoted a period of mobilism which would be an explanation for the presence of any Australian element in the fauna of Sulawesi.

In the south, the Australian segments of the OBA and the Asian IBA became sutured in the Flores–Timor region during the Pliocene/Pleistocene where evidence for a similar faunal exchange is stronger. The IBA, on the basis of the distributional data presented here, also includes the island of Sumba (Fig. 4). Asian taxa which migrated into Timor from the IBA include extinct stegodonts and reptiles, and a variety of mammalian groups (Table 3). Australian groups which

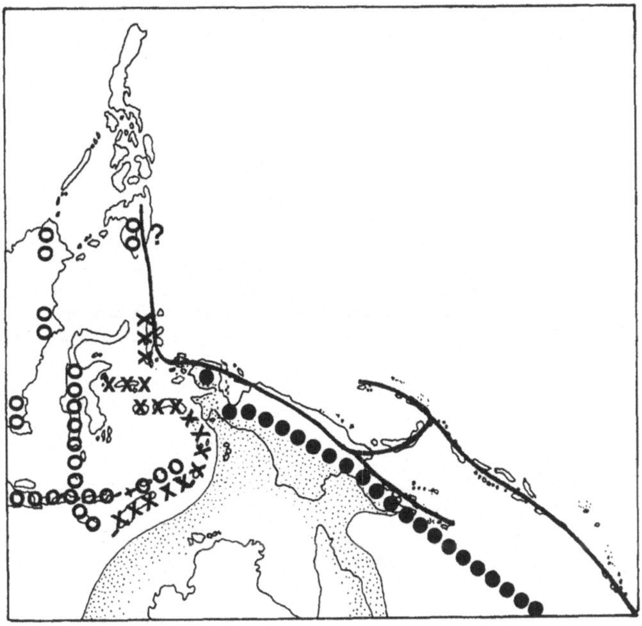

Fig. 4. Four generalized tracks in the New Guinea–eastern Indonesian region. Solid line = track 1 associated with the outer Melanesian rift (OMR); Filled circles = track 2 associated with the inner Melanesian rift (IMR); Crosses = track 3 associated with the outer Banda arc (OBA); Open circles = track 4 associated with the inner Banda arc + Java etc. Stippled = Australia–New Guinea foreland. Note the dispersed nature of track 4, a consequence of the opening of the Celebes Sea basin and Makasser Straight during the Tertiary. Sulawesi, Timor, north Maluku, the southeast peninsular of New Guinea, the Vogelkop, and possibly Mindanao, are all complex areas that are geologically composite and where different tracks meet.

migrated into the IBA appear fewer and, in the absence of clear phylogenetic data, are identified by having relatives to the east and being restricted to the Lesser Sunda islands. The track involving the IBA (Fig. 4) is discussed in detail in Michaux (in press). This track may represent the southernmost extension of a Gondwanic rift system linked to the Burma plate.

The Melanesian rifts were formed to the north and east of Australian Gondwana during the late Cretaceous (Michaux, 1989; Duffels and De Boer, 1990). This rift system became divided into the IMR and OMR by the Eocene. The IMR collided with the Australian/New Guinea craton in the late Oligocene forming the central highlands of New Guinea. In Fig. 4 it is suggested that northern Vogelkop may have been part of both these rift systems. Other parts of the Vogelkop are clearly related to the Australian craton. The OMR became sutured to New Guinea in late Miocene times forming the Huon Peninsula, northern New Guinea, the Schouten Islands, possibly that part of the Vogelkop north of the Sorong fault, parts of northern Maluku, and possibly parts of Mindanao in the southern Philippines. On the basis of mammalian and fruit fly distribution patterns, the north coast of southeast Papua and the D'Entrecasteaux Islands are included as part of the OMR.

The rifted segments of the OBA and the Melanesian rifts share common faunal elements because of their original geographic proximity as adjacent(?) coastal areas of Gondwana. In Fig. 5, which is based purely on geological evidence, it is suggested that OBA taxa may show a sister-group relationship to taxa of the Australian/New Guinea foreland plus the Melanesian rifts. However, it is quite possible that other types of relationship may exist in different groups, depending on the original, pre-breakup Gondwanic distribution of ancestral taxa. Notice that some marsupials, such as the phalangers and bandicoots (or their ancestors), appear to have been present prior to the detachment of the Banda rift system.

Carey's line (Carey, 1975), shown in Fig. 1, is the best representation of a geological partition of Asian and Australian land masses in east Wallacea. In Fig. 1 Carey's line has been modified to pass to the east of Sumba as discussed above, and tentatively in the north to pass through Mindanao. If Hamilton (1988) is correct in his assessment of the geology of Timor (and perhaps other OBA islands), then Carey's line could be drawn through Timor to indicate that a chaotic mixture of Australian and Asian elements exists there.

Wallace's Line, which represents a discontinuity

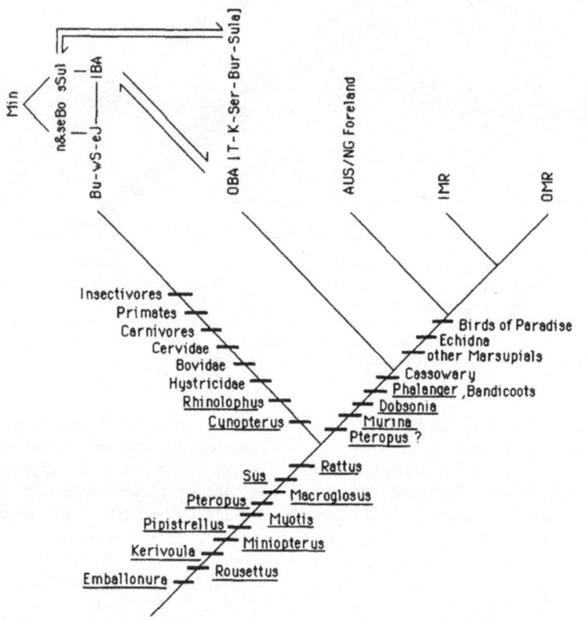


Fig. 5. Geological "cladogram" showing eastern Gondwanic splitting sequence and faunal elements associated with each branch. Arrows indicate faunal exchanges between IBA and OBA, and Sulawesi and Sula/Banggai terrane. *Bu* = Burma; *wS* = western Sumatra; *eJ* = eastern Java; *n and se Bo* = north and southeast Borneo; *sSul* = southern Sulawesi; *Min* = Mindinao; *T* = Timor; *Ser* = Seram; *Bur* = Buru; *Aus/NG foreland* = Australia–New Guinea foreland. See text for details.

between some Asian and Australian faunas, approximates this geological partition. The faunal discontinuity is blurred around Sulawesi and Timor/Flores to account for the faunal exchanges discussed above. Indo-Australian groups ignore Wallace's line. It would appear that representatives of these groups (or their ancestors) were already widespread in eastern Gondwana prior to its fragmentation during the mid-Mesozoic. Subsequent earth movements have juxtaposed them once again in modern times.

Whether concepts such as geological partitions or faunal discontinuities are appropriate is another matter. In many ways they are derived from a stabilist viewpoint. In collision zones, such as Indonesia, where numerous mobile microcontinents and migrating island arc complexes interact both with each other and more stable cratonic areas, the idea of fixity is an illusion. This contribution is an attempt to demonstrate that faunal distributions can be made intelligible by recourse to the dynamics of process, rather than to the outdated stabilist concept of fixed faunal provinces.

## Acknowledgements

Thanks are due to Guy Musser for his generosity in providing many reprints of his and others' work on murid rodents, Peter Patterson for bringing the mammalian and birds-of-paradise literature to my notice, and Jerry Holloway for reprints of his work on lepidoptera. Warren Hamilton, Chris Humphries, Hans Duffels, and Arnold de Boer provided constructive criticism of earlier drafts of this paper.

## References

Audley-Charles, M.G. and Hooijer, D.A., 1973. Relation of, Pleistocene migrations of pigmy stegodonts to island arc tectonics in eastern Indonesia. Nature, 241: 197–198.

Audley-Charles, M.G., Carter, D.J. and Barber, A.J., 1975. Stratigraphic basis for tectonic interpretation of the outer Banda arc, eastern Indonesia. Indon. Pet. Assoc. 3rd Annu. Gov. Proc., pp. 25–44.

Auffenberg, W., 1980. The herpetofauna of Komodo, with notes on adjacent areas. Bull. Fla. State Mus. Biol. Sci., 25: 39–156.

Burrett, C., Duhig, N., Berry, R. and Varne, R., 1991. Asian and south-western Pacific continental terranes derived from Gondwana, and their biogeographic significance. Aust. Syst. Bot., 4: 13–24.

Carey, S.W., 1975. Tectonic evolution of Southeast Asia. In: Proc. 4th Indon. Pet. Assoc. Congr., Jakarta, pp. 1–31.

Cracraft, J., 1990. From Malaysia to New Guinea: evolutionary biogeography within a complex continent–island arc contact zone. Proc. 19th Int. Ornithol. Congr., 2: 2581–2593.

Croizat, L., 1958. Panbiogeography. Vols. 1, 2a, 2b. Published by author, Caracas.

De Jong, R., 1983. Revision of the Oriental genus *Matapa* Moore (Lepidoptera, Hesperiidae) with discussion of its phylogeny and geographic history. Zool. Meded., 57: 243–270.

De Jong, R., 1990. Some aspects of the biogeography of the Hesperiidae (Lepidoptera, Rhopalocera) of Sulawesi. In: W.J. Knight and J.D. Holloway (Editors), Insects and the Rain Forests of South East Asia (Wallacea). R. Entomol. Soc. London, pp. 35–43.

Drew, R.A.I., 1989. The tropical fruit flies (Diptera: Tephritidae: Daninae) of the Australasian and Oceanian regions. Mem. Queensl. Mus., 26: 1–521.

Duffels, J.P., 1990. Biogeography of Sulawesi cicadas (Homoptera: Cicadoidea). In: W.J. Knight and J.D. Holloway (Editors), Insects and the Rain Forests of South East Asia (Wallacea). R. Entomol. Soc., London, pp. 63–72.

Duffels, J.P., 1991. The eye-catching cicada *Hamza ciliaris* (Linnaeus, 1758) comb. n. in Indonesia and the Pacific: taxonomic status, synonymy, and distribution (Homoptera, Cicadoidea). Bijdr. Dierkd., 62: 119–130.

Duffels, J.P. and De Boer, A.J., 1990. Areas of endemism and composite areas in east Malesia. In: P. Baas et al. (Editors), The Plant Diversity of Malesia. Kluwer, Dordrecht, pp. 249–272.

Easton, E.G., 1976. Taxonomy and distribution of the *Metapheratima elongata* species-complex of Indo-Australasian earthworms (Megascolecidae: Oligochaeta). Bull. Br. Mus. (Nat. Hist.), 30: 31–53.

Flavey, D.A. and Mutter, J.C., 1981. Regional plate tectonics and the evolution of Australia's passive continental margins. BMR J. Aust. Geol. Geophys., 6: 1–29.

Gilliard, E.T., 1969. Birds of Paradise and Bower Birds. Weidenfeld and Nicolson, London.

Hamilton, W.S., 1988. Plate tectonics and island arcs. Geol. Soc. Am. Bull., 100: 1503–27.

Hartono, H.M.S. and Tjokrosapoetro, S., 1986. Geological evolution of the Indonesian Archipelago. GEOSEA Proc., 2. Geol. Soc. Malaysia Bull., 20: 97–136.

Hecht, J., 1992. Here be dragons. New Sci., 1851: 46.

Hill, J.E., 1983. Bats (Mammalia: Chiroptera) from Indo-Australia. Bull. Br. Mus. Nat. Hist., 45: 103–208.

Holloway, J.D., 1984. Lepidoptera and the Melanesian arcs. In: F.J. Radovsky et al. (Editors), Biogeography of the Tropical Pacific. Bishop Mus. Spec. Publ., 72: 129–169.

Holloway, J.D., 1987. Lepidoptera patterns involving Sulawesi: what do they indicate of past geography? In: T.C. Whitmore (Editor), Biogeographical Evolution of the Malay Archipelago. Clarendon Press, Oxford, pp. 103–118.

Holloway, J.D., 1990. Patterns of moth speciation in the Indo-Australian Archipelago. In: E.C. Dudley (Editor), The Unity of Evolutionary Biology. Dioscorides Press, Portland, OR, pp. 340–370.

Holloway, J.D., 1993. Aspects of the biogeography and ecology of the Seram moth fauna. In: I.D. Edwards, A.A. Macdonald and J. Proctor (Editors), Natural History of Seram, Maluku, Indonesia. Intercept, Andover, pp. 91–114.

Humphries, C.J., 1990. The importance of Wallacea to biogeographical thinking. In: W.J. Knight and J.D. Holloway (Editors), Insects and the Rain Forests of South East Asia (Wallacea). R. Entomol. Soc. London, pp. 7–18.

Hutchison, C.S., 1989. Displaced terranes of the southwest Pacific. In: Z. Ben-Avrahim (Editor), The Evolution of the Pacific Ocean Margins (Oxford Monogr. Geol. Geophys., 8). Oxford Univ. Press, New York, pp. 161–75.

Jones, B., 1992. Chaos. New Sci. 135 (1831): 48.

Laurie, E.M.O. and Hill, J.E., 1954. Land mammals of New Guinea, Celebes, and adjacent islands. British Museum, London, 175 pp.

McCabe, R. and Cole, J., 1989. Speculations on the late Mesozoic and Cenozoic evolution of the Southeast Asianmargin. In: Z. Ben-Avrahim (Editor), The Evolution of the Pacific Ocean Margins (Oxford Monogr. Geol. Geophys., 8). Oxford Univ. Press, New York, pp. 143–60.

Michaux, B., 1989. Generalized tracks and geology. Syst. Zool., 38: 390–398.

Michaux, B., 1991. Distributional patterns and tectonic development in Indonesia: Wallace reinterpreted. Aust. Syst. Bot., 4: 25–36.

Michaux, B., in press. Distributional patterns in west Wallacea and their relationship to regional tectonic structures. Sarawak Mus. J.

Møller Andersen, N., 1991. Cladistic biogeography of marine water-striders (Insecta, Hemiptera) in the Indo-Pacific. Aust. Syst. Bot., 4: 151–163.

Musser, G.G., 1977a. *Epimys benguetensis*, a composite, and one zoogeographic view of rat and mouse faunas in the Philippines and Celebes. Am. Mus. Novitiates, 2624: 1–15.

Musser, G.G., 1977b. Results of the Archbold expeditions. No. 100. Notes on the Philippine rat, *Limnomys*, and the identity of *Limnomys picinus*, a composite. Am. Mus. Novitiates, 2636: 1–14.

Musser, G.G., 1979. Results of the Archbold expeditions. No. 102. The species of *Chiropodomys*, arboreal mice of Indochina and the Malay Archipelago. Bull. Am. Mus. Nat. Hist., 162: 381–445.

Musser, G.G., 1981a. The giant rat of Flores and its relatives east of Borneo and Bali. Bull. Am.. Mus. Nat. Hist., 169: 71–175.

Musser, G.G., 1981b. Results of the Archbold expeditions. No. 105. Notes on the systematics of Indo-Malay murid rodents, and descriptions of new genera and species from Ceylon, Sulawesi, and the Philippines. Bull. Am. Mus. Nat. Hist., 168: 229–334.

Musser, G.G., 1982a. Results of the Archbold expeditions. No. 108. The definition of *Apomys*, a native rat of the Philippine Islands. Am. Mus. Novitiates, 2746: 1–43.

Musser, G.G., 1982b. Results of the Archbold expeditions. No. 107. A new genus of arboreal rat from Luzon Island in the Philippines. Am. Mus. Novitiates, 2630: 1–23.

Musser, G.G., 1986. Sundaic *Rattus*: Definitions of *Rattus baluensis* and *Rattus korinichi*. Am. Mus. Novitiates, 2862: 1–24.

Musser, G.G., 1987. The mammals of Sulawesi. In: T.C. Whitmore (Editor), Biogeographical Evolution of the Malay Archipelago. Clarendon Press, Oxford, pp. 73–94.

Musser, G.G., 1990. Sulawesi rodents: species traits and chromosomes of *Haeromys minahassae* and *Echiothrix leucura* (Muridae: Murinae). Am. Mus. Novitiates, 2989: 1–18.

Musser, G.G., 1991. Sulawesi rodents: descriptions of new species of *Bunomys* and *Maxomys* (Muridae, Murinae). Am. Mus. Novitiates, 3001: 1–41.

Musser, G.G. and Califia, D., 1982. Results of the Archbold expeditions. No. 106. Identities of rats from Palau Maratua and other islands off east Borneo. Am. Mus. Novitiates, 2726: 1–30.

Musser, G.G. and Newcomb, C., 1983. Malaysian murids and the giant rat of Sumatra. Bull. Am. Mus. Nat. Hist., 174: 329–598.

Musser, G.G. and Heaney, L.R., 1985. Philippine *Rattus*: a new species from the Sulu Archipelago. Am. Mus. Novitiates, 2818: 1–32.

Musser, G.G., Heaney, L.R. and Rabor, D.S., 1985. Philippine rats: a new species of *Crateromys* from Dinagat Island. Am. Mus. Novitiates, 2821: 1–25.

Musser, G.G. and Dagosto, M., 1987. The identity of *Tarsius pumilus*, a pygmy species endemic to the montane mossy forests of central Sulawesi. Am. Mus. Novitiates, 2867: 1–53.

Musser, G.G. and Holden, M.E., 1991. Sulawesi rodents (Muridae: Murinae): morphological and geographic boundaries of species in the *Rattus hoffmanni* group and a new species from Palau Peleng. Bull. Am. Mus Nat. Hist., 206: 322–413.

Nur, A. and Ben-Avraham, Z., 1982. Oceanic plateaus, the fragmentation of continents, and mountain building. J. Geophys. Res., 87: 3644–3661.

Parker, E.S. and Gealey, W.K., 1983. Plate tectonic evolution of the western Pacific–Indian Ocean region. Proc. EAPI/ASCOPE/CCOP/IOC Workshop on the Geology and Hydrocarbon Potential of the South China Sea and Possibilities of Joint Development, Honolulu.

Pigram, C.J. and Supandjono, J.B., 1985. Origin of the Sula Platform, eastern Indonesia. Geology, 13: 246–48.

Pigram, C.J. and Davies, H.L., 1987. Terranes and accretion history of the New Guinea orogen. BMR J. Aust. Geol. Geophys., 10: 193–211.

Rangin, C., Jolivet, L. and Pubellier, M., 1990. A simple model for the tectonic evolution of southeast Asia and Indonesia region for the past 43 m.y. Bull. Soc. Géol. Fr., 6: 889–905.

Simpson, G.G., 1961. Historical zoogeography of Australian mammals. Evolution, 15: 431–46.

Simpson, G.G., 1977. Too many lines: the limits of the Oriental and Australian zoogeographic regions. Proc. Am. Philos. Soc., 121: 107–120.

Smith, J.D. and Hood, C.S., 1981. Preliminary notes on the bats from the Bismarck Archipelago (Mammalia: Chiroptera). Sci. New Guinea, 8: 81–121.

Taylor, J.M., Calaby, J.H. and Van Deusen, H.M., 1982. A revision of the genus *Rattus* (Rodentia, Muridae) in the New Guinea region. Bull. Am. Mus. Nat. Hist., 173: 178–336.

Vane-Wright, R.I., 1990. The Philippines—key to the biogeography of Wallacea? In: W.J. Knight and J.D. Holloway (Editors), Insects and the Rain Forests of South East Asia (Wallacea). R. Entomol. Soc. London, pp. 183–197.

Vane-Wright, R.I., 1991. Transcending the Wallace line: do the western edges of the Australian region and the Australian plate coincide? Aust. Syst. Bot., 4: 183–97.

Van Welzen, P.C., 1989. *Guioa* Cav. (Sapindaceae): Taxonomy, phylogeny, and historical biogeography. Leiden Bot. Ser., 12.

Van Welzen, P.C., Piskaut, P. and Windadri, F.I., 1992. *Lepidopetalum* Blume (Sapindaceae): Taxonomy, phylogeny, and historical biogeography. Blumea, 36: 439–465.

Veevers, J.J., 1991. Phanerozoic Australia in the changing configuration of proto-Pangea through Gondwanaland and Pangea to the present dispersed continents. Aust. Syst. Bot., 4: 1–11.

Vermeij, G.J., 1991. Anatomy of an invasion: the trans-Arctic interchange. Paleobiology, 17: 281–307.

Wallace, A.R., 1880. The Malay Archipelago. MacMillan, London.

Whitmore, T.C. (Editor), 1981. Wallace's Line and Plate Tectonics. Clarendon Press, Oxford.

# Attribution

**Michaux, B. 1994. Land movements and animal distributions in east Wallacea (eastern Indonesia, Papua New Guinea and Melanesia). *Palaeogeography, Palaeoclimate and Palaeogeography,* 112:323–343.**

First published in 1994 in *Palaeogeography, Palaeoclimate and Palaeogeography,* 112:323–343. Reproduced with permission from Elsevier Publishing.

# *chapter four*

# *New Guinea revisited*

## *Mammals, birds, cicadas and fruit flies*

One of the challenges of being an independent researcher and a biogeographer in the 1980s and 1990s was access to publications for research purposes. This paper was written long before the Internet made such data easily accessible. Guy Musser (American Museum of Natural History) had very generously sent me copies of his publications and these formed the backbone of this paper. Other workers were also generous in sending reprints of key papers, particularly Hans Duffels and Arnold de Boer of the Amsterdam Cicada Group and Hubert Turner (Leiden) who also had research interests in Indo-Pacific biogeography. Other sources of data included monographs available through specialist book dealers, which is where the fruit fly distributions came from, and of course bird books. Tom Gilliard's *Birds of Paradise and Bower Birds* (Gilliard 1969) was an important reference for New Guinea because of the detailed distributional maps it contained. Birds have always formed an important component of my work because they are well studied and the information is readily available as published regional guides or taxonomic treatments. In some respects, little had changed since Wallace's time as he also relied heavily on bird and mammal data.

The data provided in the paper are extensive but descriptive only. There are no phylogenetic data analysed because there were so few phylogenies available then. New Guinean taxonomy was still focussed on collecting and describing because much of the fauna and flora, particularly of the more isolated parts of the island, was poorly known and many species remained uncollected or undescribed. The detailed comparative morphological work required for any cladistics analysis was available for only a limited number of taxa. The revolution in molecular systematics, which now provides a rapid and largely automated method of generating data for phylogenetic analyses, was still a long way off. Allozyme data were commonly used in the 1980s and 1990s in an early form of molecular systematics, but the collection of such data was significantly hindered by the necessity of keeping fresh specimens stored in liquid nitrogen until they could be returned to the laboratory. In reality, there was only a ten-day window for collecting in the field before the liquid nitrogen had evaporated and specimens were ruined. Nowadays, DNA can even be extracted from museum specimens and usable sequences still obtained (Besnard et al. 2016).

## New Guinea tectonics

While Pigram and Davies (1987) remains a landmark publication, there has been a considerable research effort in the last 30 years that has clarified and refined ideas about the geological development of New Guinea (Abbott et al. 1994; Davies et al. 1997, Davies 2012; Polhemus and Polhemus 1998; Hill and Hall 2003; Sutriyono 2003; Baldwin et al. 2004, 2012; Wallace et al. 2004; Satyana et al. 2008; Woodhead et al. 2010). A summary of New Guinean geology is shown in Figure 4.1. New Guinea can be thought of as being divided into three parts (Davies 2012) – a western province that includes the Bird's Head region, a large central province, and Peninsular

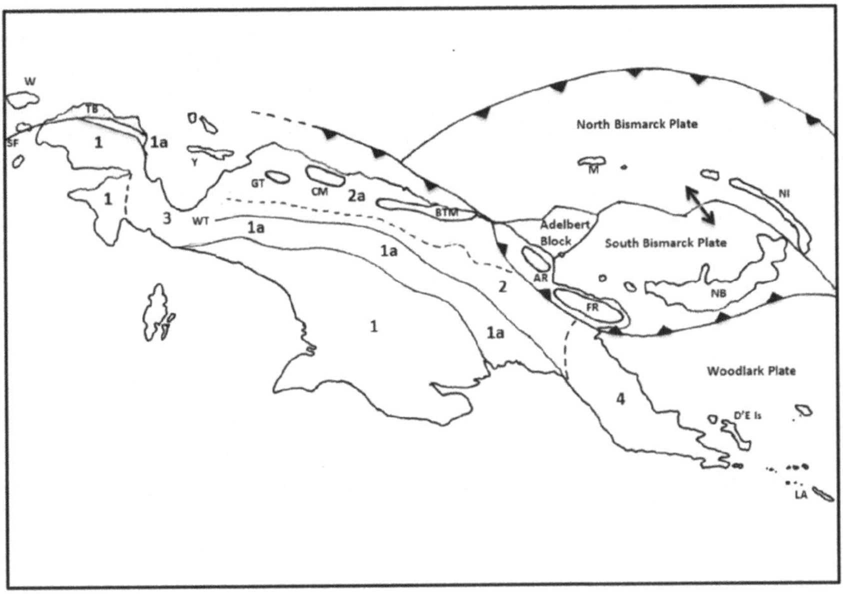


*Figure 4.1* Simplified geological map of New Guinea. 1 = Australian cratonic rocks, 1a = Papuan fold and thrust belt; 2 = Early Palaeogene arc and continental terranes, 2a = Neogene arc terranes; 3 = Transition Zone; 4 = Peninsula New Guinea. AR = Adelbert Range, BTM = Bewani-Torricelli Mountains, CM = Cyclops Mountains, D'E Is = D'Entrecasteaux Islands, FR = Finisterre Range, GT = Gauttier terrane, LA = Louisade Archipelago, M = Manus, NB = New Britain, NI = New Ireland, SF = Sorong Fault, TB = Tosem Block, W = Waigeo, WT = Weyland terrane, Y = Yapen Island. Arrows show direction of basin opening; teeth show direction of downgoing slab.

New Guinea plus the D'Entrecasteaux and Louisiade archipelagos (Figure 4.1: 4, D'E, LA). There is a transition zone between the Bird's Head and central New Guinea (Figure 4.1: 3).

## *Bird's Head Province*

The Bird's Head consists of two distinct geologies – an oceanic province north of the Sorong Fault (Figure 4.1: SF) and a continental province to the south. Baldwin et al. (2012) called the volcanic province the Tosem block (Figure 4.1: TB), which is composed of Palaeogene arc volcanics overlain by younger clastic and carbonate sediments. The presence of ophiolites on Waigeo (Figure 4.1: W) is thought to be associated with the island arc's collision with the New Guinea mainland. Baldwin et al. (2012) considered the Tosem block to be part of an extended arc system that included the Weyland terrane (Figure 4.1: WT) and other island arcs now sutured to the northern margin of central New Guinea (Figure 4.1: GT, CM, BTM, AR and FR). The Tosem block was probably accreted to the New Guinea margin during the Late Oligocene (c 25 Ma), and possibly at some distance east of its present position.

The continental block south of the Sorong fault (Figure 4.1: 1) consists of Silurian and Devonian slates and quartzites intruded by Devonian granites and overlain by later Palaeozoic and Mesozoic sediments (Pieters et al. 1979). These continental rocks are composed of a number of cratonic terranes, the largest of which is the Kemum terrane (Pigram and Davies 1987). Pigram and Davies (1988) suggested that the Kemum terrane was detached from the Australian craton during the Cretaceous and attained its present position by the Miocene. However, there is little consensus about this, and other workers interpret it as autochthonous or para-autochthonous (Hill and Hall 2003; Baldwin et al. 2012) with little movement relative to Australia apart from anticlockwise rotation during the past six million years. The Arfak and Taurau mountains (Figure 4.1: 1a) are folded basement sediments, metamorphics and igneous intrusives. There are arc volcanics exposed on the coastal section of the Arfak Mountains.

## *Central New Guinea Province*

The Central Province consists of a mountainous spine (the Central Highlands or Central Cordillera) rising to almost 5000 m with flanking plains to the north and south. There are isolated ranges along the north coast that rise to over 2000 m. The southern plain (Figure 4.1: 1) is the only unequivocally autochthonous region in New Guinea. It comprises two large sedimentary basins overlying an Australian cratonic basement.

The western basin contains up to 16 km of sediments of Late Precambrian to Cenozoic age, the eastern basin some 4 km of sediments of Permian to Cenozoic age. Hill and Hall (2003) considered the boundary between the basins to mark the Tasman Line that separates the Australian craton from terranes accreted to Gondwana during the Palaeozoic. To the north of the southern plain lies the Papuan fold and thrust belt (PFTB, Figure 4.1: 1a), which represents the crumpled edge of the Australian craton formed when Australia collided with numerous island arcs and continental fragments to the north. The interaction was oblique (approximately 60° according to Watkinson and Hall (2017)) with the western section of the leading edge entering the collision zone before the eastern section, resulting in earlier collision ages and higher mountains in the west.

Two broad phases of this collision process can be identified. The earlier phase consisted of ophiolite emplacement in the west as an island arc (Irian Volcanic Arc of Davies et al. 1997; West Papuan Volcanic Arc of Davies 2012) was sutured to the leading edge of the craton during the latest Cretaceous/Palaeocene (Davies et al. 1997). Further east, a series of small continental fragments (Bena Bena, Jimi, Kubor, Pale, Border and Landslip terranes), island arcs (Marum and Schrader terranes) and associated ophiolites (April Ophiolite, Sepik terrane) were sutured to the margin during the Eocene to Oligocene. These terranes form the northern foothills of the Central Highlands (Figure 4.1: 2). It is thought that the continental terranes were rifted from the Australian craton during the Late Cretaceous/Palaeocene opening of the Coral Sea (Weissel and Watts 1979). In the reconstruction of Davies et al. (1997), they are shown to be allochthonous, although there is uncertainty about how far some have been displaced (Van Wyck and Williams 2002). The Border terrane, for example, may be autochthonous (Hill and Hall 2003) and others may be para-autochthonous.

The second phase of accretion of island arcs to the northern margin of New Guinea occurred from the Oligocene to the present day (New Britain). These arcs and associated ophiolites (Figure 4.1: Y, GT, CM, BTM, AR, FR) now form the northern margin of New Guinea, and are part of an island arc system that includes the Bismarck Archipelago and the Solomon Islands. Their collision with the New Guinea margin caused uplift of the Central Ranges from about 12 Ma. The plain to the north of the Central Highlands (Figure 4.1: 2a) is composed of some 10 km of Middle Miocene to Quaternary sediments eroded from the Central Highlands to the south and the coastal mountains to the north. The sediments have been subjected to compressional deformation since the Pliocene, resulting in folding and thrusting towards the north (Davies 2012). The area north of the PFTB, that is the area of accreted terranes, is known as the Papuan Mobile Belt (Figure 4.1: 2 and 2a).

### *Transition zone*

A geologically complex transition zone is found between the Bird's Head and the Central Province (Figure 4.1: 3). The Lengguru fold and thrust belt (LFTB) in the west consists of metamorphosed Palaeozoic sediments, Early Jurassic granites and Plio-Pleistocene subduction zone metamorphics. There are a number of interpretations concerning the origin of the LFTB (Satyana et al. 2008), including it being an extension of the Papuan fold and thrust belt (Davies 2012) or a later structure formed when the Weyland terrane was sutured in the Miocene. The Weyland terrane consists of high-grade metamorphic rocks and ophiolites of Palaeocene age and Miocene arc volcanics. An extension of the Papuan Basin lies to the south of the Weyland terrane.

### *Peninsula New Guinea*

Peninsula New Guinea (Figure 4.1: 4) and offshore islands of the Louisiade and D'Entrecasteux archipelagos (Figure 4.1: LA and D'En) are composed of a single composite terrane. A continental rift – the Owen Stanley Metamorphic Complex (OSMC) – is sutured to an oceanic terrane of Cretaceous ophiolite overlain by Middle Eocene arc volcanics (East Papuan Ultramafic Belt). The two became sutured in the Palaeocene and collided with the rest of New Guinea in the Oligocene circa 30 Ma (Davies et al. 1997). The OSMC is thought to have originated with other continental terranes of the Papuan Mobile Belt during the Late Cretaceous/Palaeocene opening of the Coral Sea.

The geological model of Michaux (1994: Figure 4) requires modification. The idea of an Inner Melanesian Rift is no longer tenable as a natural (monophyletic) area. It is better to view the areas connected by this track as autochthonous or para-autochthonous parts of the Gondwana margin, which are primarily related to Australia rather than to each other. Any similarities that the continental terranes of the Papuan Mobile Belt and Peninsula New Guinea have to each other, or to other rifted fragments further south such as New Caledonia, are the result of their derivation from a common source area at similar times.

## *New Guinea plants and animals*

### *Reanalysis of original distributional data*

There is a way of analysing distribution data known as parsimony analysis of endemism (PAE). In PAE, areas are treated as taxa and species as characters. A matrix of species versus areas is constructed by scoring 1 when a species is present in an area and 0 when absent. This matrix can then be analysed by any parsimony algorithm in the same way that one would a taxon/character matrix. While I accept the criticism that PAE is

A

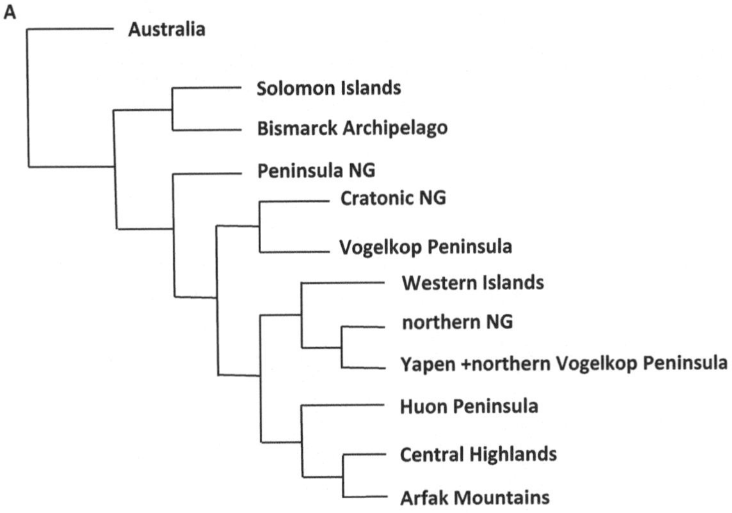


B

| | |
|---|---|
| **Australia** | 000000010000010000000000000000000000111111 |
| **Solomon Islands** | 000000000000001110000000000010101111111111 |
| **Bismark Archipelago** | 000000000000001111100100000100100111101 |
| **Peninsula NG** | 000000010000000000000001000000100110011 |
| **Cratonic NG** | 000011011001110000000000000000000000101000 |
| **Vogelkop Peninsula** | 00001001101111000000000000000000000000000 |
| **Western Islands** | 0000011010110000001100000000000000000000 |
| **Northern NG** | 0001100111100100011111000010101010?0001 |
| **Yapen +nVogelkop** | 000000001110000000111100001000000000000 |
| **Huon Peninsula** | 001110101000000000101100010000000000?0000 |
| **Central Highlands** | 111100110000000000000001111011101111111011 |
| **Arfak Mountains** | 111100001000000000000001111100000000000000 |

***Figure 4.2*** A. PAE analysis of birds-of-paradise and *Bactrocera* distributions from Michaux 1994. Single most parsimonious tree length = 78 steps. B. Data matrix 1 = presence 0 = absence ? = unknown or missing data.

not based on phylogenetic data, needs must. In the absence of phylogenetic information, do we simply ignore distribution data or can it be utilised in a form of preliminary analysis? Figure 4.2 shows the data matrix constructed from the distributions listed in Table 5 (birds-of-paradise) and Table 6 (*Bactrocera* fruit flies) of Michaux (1994). The data were analysed with the phylogenetic programme TNT (Goloboff et al. 2008) using an implicit enumeration algorithm (*ienum*) that is guaranteed to find the most parsimonious solution(s). A single most parsimonious tree (length 78 steps) is shown in Figure 4.2. Interesting features of this tree include: the sister group relationship of the Bismarck Archipelago with the Solomon Islands rather than any New Guinean area; the basal position of the

Peninsula New Guinea; the grouping of the two cratonic areas (cratonic NG and Vogelkop Peninsula) as sister areas; a clade containing the three accreted arc fragments along the north coast (Western Islands, Yapen and northern Vogelkop, Northern Mountains); and the linkage of the Central Highlands and Arfak Mountains.

The sister-area relationship between the Vogelkop (Bird's Head) and southern New Guinea suggests that the two areas can be regarded as a single area of endemism, and does not support the view that the Bird's Head has been separated from the rest of the Australian craton since the Cretaceous. The placement of the island arc fragments of northern New Guinea in a clade clearly makes sense in light of these areas' geological connection, although the position of the Huon Peninsula is anomalous. A possible connection between the Central Highlands and the Arfak Mountains raises some interesting questions. The Arfak Mountains consist of Silurian and Devonian metamorphosed sediments intruded by Permo-Triassic granites that are clearly part of the Kemum terrane of the Bird's Head. These continental rocks are separated from island arc volcanics by the Ransiski fault zone. This zone is up to several hundred metres wide and shows signs of extensive shearing (Pieters et al. 1979). The Arfak Mountains clearly represent a deformed continental margin with an accreted arc, a pattern repeated along the Central Ranges and also further south in the Bird's Head across the Lengguru Fold Belt, although Decker et al. (2017) have drawn attention to the difference in zircon age profiles between the Bird's Head and Lengguru Fold Belt indicating they are not related. The age of the Arfak arc is not altogether clear. If it belongs to the early phase of accretion (i.e. Palaeocene/Eocene), then it would indicate that the Arfak Mountains are a clockwise rotated fragment of the Central Highlands.

## *Molecular phylogenies*

A literature search found 56 published phylogenies of New Guinean species (the majority from *Molecular Phylogenetics and Evolution* between 2005 and 2017). Of these papers, 20 provided a total of 22 usable areagrams. The reasons for rejecting 36 of the phylogenies were: species' distributions were too geographically widespread; studies included too many areas outside of New Guinea, which increased the number of 'taxa', reducing the resolving power of any analysis; reduction to two-area statements when species' distributions were assigned to terranes; or insufficient distributional information. Areagrams are derived from phylogenies (or clades within phylogenies) by substituting areas for species and are shown in Table 4.1.

The areagrams shown in Table 4.1 were used as input data into LisBeth ver. 1.3 (Bagils et al. 2012; http://infosyslab.fr/downloadlisbeth/LisBeth.exe, accessed 1 February 2018). LisBeth has been developed to analyse three-item statements for biogeographic studies. LisBeth first

***Table 4.1*** Phylogenies used in the analysis of New Guinean terranes

| Trees | References |
|---|---|
| (BH(nM+nL, sNG)) | Georges et al. (2014) |
| (AUS(nM+CM+PNG(CM+PNG,BH+nM))) | Meredith et al. (2010) |
| (AUS(sNG(nM,BH+CM+nL))) | Zwiers et al. (2008) |
| ((B,V)(AUS,sNG)) | Nyári et al. (2009) |
| (nM(V(V,B))) | Austin et al. (2010) |
| (nM(B,V)) | Schweizer et al. (2015) |
| ((CM,sNG)(CM,nM)) | Schweizer et al. (2015) |
| (AUS(BH(CM,sNG))) | Driskell et al. (2011) |
| ((BH,sNG)(nL,PNG)) | Bruxaux et al. (2018) |
| (sNG(nM(nM(sNG,CM)))) | Bruxaux et al. (2018) |
| ((CM,sNG)(nM,B)) | Cibois et al. (2017) |
| (((AUS,sNG+nL)(AUS(AUS,sNG+AUS)))(sNG+AUS(sNG+BH+nM+PNG,AUS))) | Kearns et al. (2013) |
| (PNG(PNG(PNG(CM(PNG,nM))))) | Oliver et al. (2013) |
| ((nL,AUS)nL(PNG(PNG(PNG,V)))) | Wood et al. (2012) |
| (AUS(BH(nL,sNG+AUS))) | Unmack et al. (2013) |
| ((PNG,V)(CM((PNG,CM)(PNG((AUS,V)(B,V)))))) | Johnson et al. (2017) |
| ((V,nM)(PNG,BH)) | Jønnson et al. (2018) |
| (((sNG,AUS)(V(V(PNG,AUS))))((B,PNG)(PNG+CM(AUS(V(B(PNG+CM,AUS))))))) | Filardi and Smith (2005) |
| (((AUS(sNG,AUS))(PNG(nM(nM,B)))(sNG,nM)(sNG,CM))) | Macqueen et al. (2010) |
| ((nM,CM)(V(nL(MIC(V,PNG))))) | Austin (2000) |
| ((CM,sNG)((nM(AUS(nL(PNG(CM,nM)))))((CM,sNG)(PNG(CM,nM))))) | Dumbacher et al. (2003) |
| (sNG(nM(nM(CM,nM)(nM(nM,BH))))) | Dumbacher and Fleischer (2001) |

Cratonic areas

BH = Bird's Head + Misool

sNG = southern New Guinea

Aus = Australia

Mobile belt

CM (Central Mountains) = folded cratonic margin, Phase 1 terranes, Lengguru FTB

PNG = Peninsular New Guinea

nL = northern lowlands

Arc terranes

nM (northern mountains) = nBH, Yapen, Waigeo, Halmahera, Phil, Weyland terrane

B = Bismarck Archipelago

V (Vityaz Arc) = Solomon Islands, Vanuatu, Fiji

*Note:* Analysis run by V. Ung, whose help is gratefully acknowledged.

produces paralogy-free subtrees (Nelson and Ladiges 1996) before applying the transparent method of Ebach et al. (2005) to resolve multiple areas as terminal taxa, finding all possible three-item statements implicit in the original data. Three-item statements are rooted trees that represent area relationships directly rather than in a matrix form. Optimal trees are constructed from all the three-item statements, employing an exhaustive branch and bound algorithm using compatibility analysis. An intersection tree is formed from all the three-item statements common to the optimal trees, and is equivalent to a consensus tree. The intersection tree produced by the analysis is shown in Figure 4.3 and has a completeness index of 0.39. The completeness index is the proportion of three-item statements derived from the paralogy-free subtrees that were used to construct the intersection tree. The low value of the completeness index shows that only 39% of the three-item statements were informative, that is about two fifths of the 'characters' support the areagram.

Because the general areagram is poorly supported, only a limited interpretation is possible. The placement of southern New Guinea and the Vityaz Arc as unresolved outareas indicates that many of the three-item statements not used in the construction of the intersection tree involve these two areas; in other words, the data simply weren't sufficient to place these areas. Australia is basal to the clade containing all other New Guinean areas. Of these, the Central Mountains are basal to two sister clades, one containing the northern coastal mountains and Bismarck Archipelago as sister areas, the other showing the northern lowlands as sister area to peninsula New Guinea, which in turn is sister area to the Bird's Head. Comparison with Figure 4.2 shows that the two areagrams have few

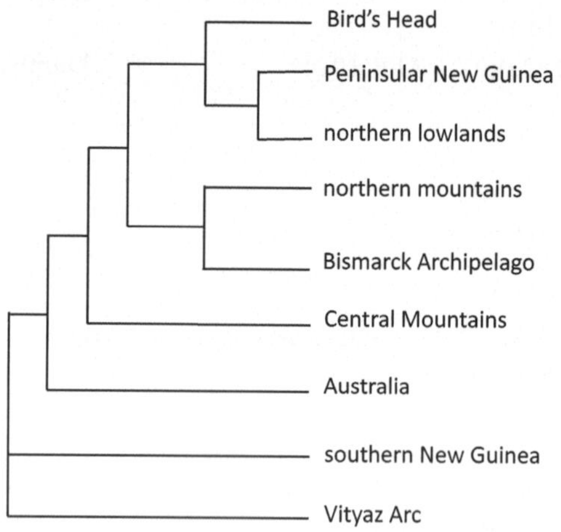


*Figure 4.3* Intersection tree derived from the molecular phylogenies detailed in Table 4.1 and analysed using LisBeth. Completeness index = 0.39.

features in common, although direct comparability is confounded by the use of different areas in each analysis. Until there are sufficient phylogenies available to resolve the positions of southern New Guinea and the Vityaz Arc within the general areagram, the relationship between terranes will remain unresolved.

## *Single taxon studies*

The preceding analysis, based on papers published since 2000, illustrates the problem in using data designed for one purpose (systematics) for another purpose (biogeology). The majority of published biogeographical studies are adjuncts to molecular systematic treatments of a single taxon, normally either the constituent species of a genus or genera within a family. Treatments of higher taxa are rarely of any use for biogeological studies because the terminal taxa are too widely distributed and exacerbate the computational and resolution problems associated with multiple areas on a single terminal branch (MASTs). Even those dealing with fewer taxa that are more geographically restricted are usually only partially useful because, for example, included taxa don't occur in many of the areas of interest. But the major problem with single taxon studies is that individual phylogenies are not *biogeologically* interpretable. I outline the reasons in the following section.

### *The informal fallacy or anecdotal syndrome*

Anecdotal evidence, while possibly interesting and indicative of future research lines, cannot be regarded as scientifically legitimate because single instances don't conform to the criterion of verifiability and cannot validate a hypothesis. Areagrams based on a single phylogeny are analogous to anecdotal evidence when used for *biogeological* studies because they too are single instances, which may be unreliable or non-representative in some way. Put bluntly, are any hypothesised area relationships correct, mostly correct, or incorrect? While various metrics give some indication of how well an areagram is supported by the data, this should not be confused with an areagram's accuracy in reconstructing area relationships because the data may be at fault or insufficient to allow such a resolution or are in some way peculiar to that taxon.

### *Data inflation and the problem of too many terminal taxa*

Molecular studies are often guilty of data inflation, because such phylogenies are based on a vanishingly small proportion of the total molecular data available from the organism. Modern studies now use a number of different nuclear as well as mitochondrial gene sequences, often amounting to thousands of base pairs in total. This represents a considerable data set, especially when compared to morphological data matrices. But this appearance is an illusion. Are a handful of gene sequences representative of a genome composed of many thousand functional genes and tens or hundreds of million base pairs? Would other samples of the same size taken from the genome yield similar or different results? The problem of reliability is exacerbated because as the number of taxa in the study increases, so does the unreliability of the resulting areagram. It is not uncommon to see many tens or even hundreds of taxa included in molecular studies, and while trees are produced because of the apparent size of the data set, the real information content is too small to reliably resolve relationships to the accuracy needed in biogeological studies. I suspect that major clades are usually identified correctly, but would question the reliability of any proposed relationships between taxa within clades and of the relationship between clades when too many taxa are included.

## *Over-interpretation*

The study of Toussaint et al. (2014), which is based on the freshwater diving beetles *Exocelina* (Coleoptera: Dytiscidae), illustrates the dangers of over-interpreting an areagram based on a single phylogeny. These authors suggested that the history of *Exocelina* in New Guinea is recent (within the past 5 million years) and is characterised by frequent colonisations out of the Central Highlands and multiple altitudinal changes of habitat preference accompanying speciation. They present supporting evidence for their interpretation in the form of molecular dating (based on modern substitution rates) and the supposed lack of land in central New Guinea (based on widespread limestone deposits covering much of New Guinea) until the Central Highlands became elevated at circa 5 Ma.

The problems with dating methods in general, and rates extrapolated from extant and often distantly related groups, have already been discussed. The claim of a lack of suitable terrestrial habitats is reminiscent of the Great Flood hypothesis in which land is effectively sterilised before recolonisation at times consistent with molecular dates. Neither the geological evidence on the ground nor modern analogues such as the Great Barrier Reef support the hypothesis of complete inundation of land based on the occurrence of extensive Cenozoic limestone over much of New Guinea. Pieters et al. (1979) provided an account of the geology of Kepala Burung (Bird's Head). According to these authors, the limestones of the Bird's Head were deposited in shallow water and include clastic horizons indicative of a marine regression during Oligocene and Miocene times. The Miocene limestone south of Senopi, Kepala Burung,

> was accumulated on an unstable shelf as forereef, backreef, and littoral deposits. The occurrence of limestone conglomerate with pebbles derived from the Kemum Formation suggest nearby hinterland with high relief. (Pieters et al. 1979:26)

A modern analogue for a platform reef is the Great Barrier Reef of Australia. The reef is covered in shallow water and dotted with numerous islands that were once part of the Australian mainland ('continental islands') as well as atolls. The continental islands support a diverse flora and fauna that include 2195 plant species, 118 species of Lepidoptera (30% of all Australian butterflies) and numerous other invertebrates representing 39 families in ten orders, 7 amphibians (all frogs), 40 reptiles (9 snakes and 31 lizards), and a variety of volant and non-volant mammals (Stokes et al. 2005). The occurrence of widespread reef limestones actually indicates the presence of terrestrial habitats that clearly act as refugia for a diverse flora and fauna. As uplift occurs, this diverse flora and fauna will simply recolonise newly available habitats.

The areagram from which Toussaint et al. (2014) draw their conclusions is based on three mitochondrial and five nuclear gene fragments (4,299 bp), and contains 94 terminal areas. Their conclusion that there have been repeated invasions of Central Highland taxa into other parts of New Guinea in a 'complex and dynamic process' may simply be a consequence of poorly resolved relationships or pre-vicariance diversification (Heads 2017). Collapsing their areagram to its simplest form yields the relationship (northern arc terranes (cratonic areas, northern arc terranes)). In any case, it's difficult to argue, as they do, that the fauna was derived from the Central Highlands when the most basal taxon and the earliest resolved clade are found on northern arc terranes. Arguing that the diversification in the basal clade is shallow does not alter the fact that the oldest species (now either extinct or uncollected) was present on one of the Eocene arc fragments sutured to New Guinea sometime after 25 Ma (Davies 2012). In the following chapter, I will outline a procedure that I think is needed to plan, analyse and interpret a biogeological study.

ELSEVIER Palaeogeography, Palaeoclimatology, Palaeoecology 122 (1996) 167–183

PALAEO

# The origin of southwest Sulawesi and other Indonesian terranes: a biological view

B. Michaux

*Private Bag, Kaukapakapa, New Zealand*

Received 24 August 1994; revised and accepted 13 October 1995

**Abstract**

A generalised track linking geographically dispersed areas in Indonesia with Burma is identified from moth and bird distributional data. These areas include Sulawesi, the Philippines, Sabah (northern Borneo), Java, Sumatra, Mentawai, the Nicobars, and Andamman Islands. A cladistic analysis of Sulawesian *Dacus* (Diptera: Tehritidae) shows that Sabah and Sulawesi are sister areas and together form a sister area to Sulawesi + Java + Sumatra + Thailand. Cladograms for other taxa are reviewed and these also lend support to the hypothesis that these now diverse Indonesian fragments share a closer biotic relationship to each other than they do to other areas. The geological history of the region is discussed and evidence that these Indonesian fragments are terranes which became sutured to the continental margin of Sundaland in the late Cretaceous or early Tertiary is discussed. The congruence between distributional patterns and tectonic history requires further investigation, but a working hypothesis suggests that these Indonesian terranes are linked to Mount Victoria Land, and that all were derived from a Gondwanic source area close to India.

## 1. Introduction

Over the past two decades or so the geological complexity of Indonesia has been brought into clearer and clearer focus. It has been established that numerous continental fragments, sutured island arcs, and obducted or trapped oceanic crust exist in this region (Audley-Charles, 1983; Pigram and Supandjono, 1985; Belov et al., 1986; Hartono and Tjokrosapoetro, 1986; Hall, 1987; Pigram and Davies, 1987; Audley-Charles et al., 1988; Hamilton, 1988; Hutchison, 1989, 1993; McCabe and Cole, 1989; Sungupta et al., 1990; Burrett et al., 1991; Metcalfe, 1991; Rangin et al., 1991). It is also thought that the majority of continental fragments are Gondwanic and that the processes of fragmentation, migration, and collision have been operating over a considerable time scale up to and including the present (Metcalfe, 1991). The identification of terranes and microcontinental blocks has been based on fossil, palaeomagnetic, stratigraphic, and structural data. By these means geologists can describe, with varying detail and degree of confidence, the history of each terrane and in doing so identify relationships between different terranes.

An area of research that may be less familiar to geologists, but which may have an important bearing on the question of relationships between areas, concerns the analysis of geographical distribution patterns of extant organisms. Prior to a resurgence in the field of biogeography during the 1980's, animal and plant distributions were routinely interpreted in terms of individual dispersal histories.


*SSDI* 0031-0182(95)00100-X

Such explanations are no longer accepted uncritically and research effort has instead focussed on biotic patterns (the area cladogram of vicariance biogeography or the generalised track of panbiogeography) and their possible relationship to geological processes.

Relating distributions to geological processes stems from the work of the late Leon Croizat. Croizat's main conclusions from his extensive studies (Croizat, 1958) were that animal and plant distributions are repetitious, that similar patterns of distribution appear to be indepedent of individual plant and animal vagility or general ecological tolerances, and that these general patterns (generalised tracks) appear to be related to large scale geological features. In my view, Croizat's most profound insight was to establish a direct link between biogeography and geology. For Croizat, generalised tracks resulted from large scale tectonic processes and could, potentially at least, be used in the reconstruction of those processes.

My interest in Indonesia dates from reading A.R. Wallace's (1880) *Malay Archipelago* in which he gives an account of his travels in the region between 1854 and 1862. I was particularly interested in Wallace's attempt to understand distributional patterns in terms of general geological principles. Wallace viewed the earth's surface as dynamic, but only as far as vertical change was concerned. The possibility of extensive horizontal movement of land masses simply didn't occur to the Victorians. Wallace could hardly be blamed for this lack, but it proved, in my opinion, to be a fundamental weakness in his attempts to understand distributional patterns in Southeast Asia, Indonesia, New Guinea, and Melanesia. What, I wondered, would Wallace have made of modern geological evidence from Indonesia? The initial results of such a reinterpretation of Wallace's data are presented in Michaux (1991).

Wallace (1880) regarded the biota of Sulawesi as anomolous. He did so because despite Sulawesi's central location in the archipelago the biota is depauperate, highly endemic, and often shows closest phylogenetic affinities to areas outside of Indonesia. The native mammals of Sulawesi, listed in Table 1, illustrate these points. Musser (1987) commented on the absence of many Sundaic mammalian families from Sulawesi (Musser, 1987: table 7.6) and calculated that bats and murid rodents account for 70% of all endemics. Musser (1987) noted that oceanic islands have similarly depauperate and unbalanced faunas. The degree of endemism is high for mammalian genera as it is for cicadas (Duffels, 1990) and butterflies (Vane-Wright, 1991). The phylogenetic relationships shown by mammalian endemics is often obscure. Table 1 also shows why the position of Wallace's Line, which supposedly sepatates Asian from Australian biotas, is confused in Sulawesi where the mammalian fauna is a mixture of Asian placentals and Australian marsupials. Michaux (1994) discussed other Sulawesian groups that appear to be a mixture of Asian and Australasian taxa.

Sulawesi is clearly a pivitol region in any interpretation of Indonesian biogeography. This contribution details a distributional pattern apparent among both bird and moth data which link Sulawesi and/or Sabah (north Borneo) with areas that were marginal to a late-Cretaceous to early Tertary margin of the Sunda Shelf (see Fig. 1: 1–8). An important feature of this pattern is that despite many of the species being widespread, they are not found on the Malay Peninsula or central Thailand. A second pattern evident in the bird distributional data links Sulawesi with Australasia and is more fully described in Michaux (1994). An area cladogram based on a cladistic analysis of Sulawesian *Dacus* (Diptera: Tephritidae) is presented and compared to published cladograms involving Sulawesian taxa. Finally, the geological histories of various marginal areas to the Sunda Shelf will be outlined and the relevance of these histories to distributional patterns discussed.

## 2. Patterns in bird and moth distributional data

Bird distributional data for Southeast Asia, Indonesia, New Guinea, and Australia came from Delacour (1947), Du Pont (1971), Walters (1980), Tikader (1984), Beehler et al. (1986), Holmes and Nash (1990), Macdonald and Slater (1992), MacKinnon and Phillipps (1993), and Strange and Jeyarajasingam (1993). One hundred and thirty-

Table 1
Native mammals of Sulawesi

| | | |
|---|---|---|
| **Marsupials** | | |
| PHALANGERS | | |
| *Phalanger ursinus* | Most westerly occurrence of | Laurie and Hill (1954) |
| *P. pelengensis* | Australian group. Phylogenetic | Musser (1987) |
| *P. celebensis* | relationships unclear | |
| **Placentals** | | |
| SHREWS | | |
| *Crocidura* spp. (8) | Sundaic and Indochinese. | Laurie and Hill (1954) |
| | Also found in the Philippines | Musser (1987) |
| | | Musser and Heaney (1992) |
| BATS | | |
| 9 endemic spp., | Indo-Australian | Hill (1983) |
| 59 species in total | | Musser (1987) |
| TARSIERS | | |
| *Tarsius spectrum* | Tarsiers found in S. Phil. | Musser (1987) |
| *T. pumilus* | Borneo, and S. Sumatra. | Musser and Dagosto (1987) |
| | Oligocene fossils from Africa | |
| MACAQUES | | |
| *Macaca nigra* | Sundaic and Mentawai Islands. | Musser (1987) |
| *M. tonkeana* | Probably introduced into | |
| *M. ochreata* | Banda arc islands. | |
| *M. maura* | | |
| SQUIRRELS | | |
| *Rubrisciurus* | Endemic genera. Asian but | Musser (1987) |
| *Hyosciurus* | phylogenetic relationships | |
| *Prosciurillus* | unclear | |
| RATS and MICE | | |
| 48 spp. in 14 genera | | Musser and Holden (1991) |
| 10 genera endemic | | Musser (1991) |
| *Crunomys* | Philippines | Musser (1977) |
| *Haeromys* | Palawan, Borneo | Musser (1987, 1990) |
| *Maxomys* | Sundaic, Indochina | Musser (1981) |
| *Rattus* | Indo-Australian | Musser and Califia (1982) |
| | | Musser and Heaney (1985) |
| | | Musser (1986) |
| CIVETS | | |
| *Macrogalidia musschenbroekii* | Sundaic. Other civets | Laurie and Hill (1954) |
| | probably introduced | Musser (1987) |
| PIGS | | |
| *Sus celebensis* | Indo-Australian | Laurie and Hill (1954) |
| *Babyrusa babyrussa* | No close relatives. Siwalik | Musser (1987) |
| | beds of India (Pliocene) | |
| ANOAS | | |
| *Bubalus quarlesi* | No Sundaic relatives. | Laurie and Hill (1954) |
| *B. depressicornis* | Siwalik beds of India | Musser (1987) |

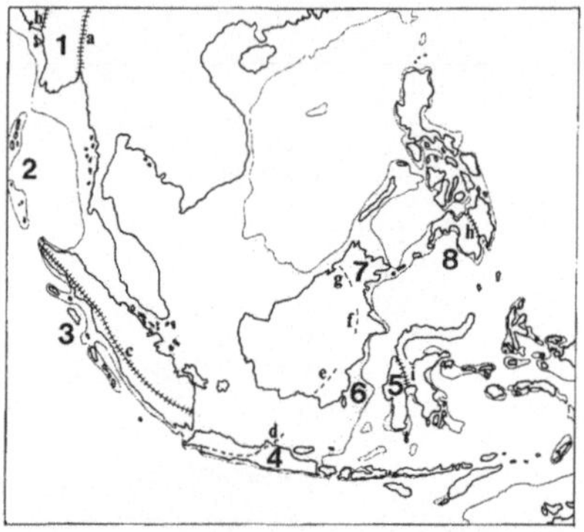


Fig. 1. Map of study area. Sunda Shelf is stippled. *1* = Mount Victoria Land sensu Metcalfe (1991), *2* = Andaman and Nicobar islands, *3* = Sumatra west of the Sumatran fault + western islands, *4* = eastern Java, *5* = southwest Sulawesi, *6* = southeast and eastern Kalimantan, *7* = Sabah, *8* = Mindanao, *a* = Shan suture, *b* = Indo-Burma ranges, *c* = Sumatran fault, *d* = Ciletum and Lukulo melanges, *e* = Meratus suture, *f* = Adio suture, *g* = Kinabalu suture, *h* = Philippine fault, *i* = Palu fault. All faults and boundaries are schematic.

five bird species occurring in the Southeast Asia/Indonesian region have distributional patterns that avoid the Malay Peninsula and Thailand but include such areas as the Andaman and Nicobar islands, the western islands of Sumatra, Java, south east and east Kalimantan, Sabah, Sulawesi, and Mindanao, which are all areas marginal to the Sunda shelf (see Fig. 1). These species' distributions are listed in Table 2 together with endemic bird species found in the Andamans/Nicobars (14 species), western islands (4 species), south east Kalimantan (1 species), and Sabah (22 species). In addition, the number of Sulawesian endemics are also given in brackets after the family names listed in Table 2. The moth distributional data for Borneo came from Holloway (1976). A total of 548 species were included in the analysis, of which 213 (40% of the non-endemic species) are absent from the Malay Peninsula. These species' distributions are listed in Table 3.

### 2.1. *Sulawesian birds*

Sulawesian birds show similar characteristics to Sulawesian mammals. According to Walters (1980) there are 83 endemic bird species and 9 endemic bird genera found on Sulawesi. This represents an endemicity of approximately 50%. As with mammals, many genera found in surrounding areas appear to be absent from the Sulawesian fauna. Fifty-five non-endemic Sulawesian bird species that are also absent from the Malay Peninsula or Thailand, representing 60% of all non-endemic species, are listed in Table 2. Approximately one third of these species (18/55) are distributed across Wallace's line. One quarter are predominantly Indian species at the eastern extremity of their ranges (14/55), while a similar number are Australasian species at the western extremity of their range (15/55). The remainder are Sundaic endemics (8/55).

Two distinct distributional patterns are evident in these data. The first pattern links Sulawesi to Australasia, a generalised track discussed in detail by Michaux (1994). The second pattern links Sulawesi and parts of Borneo and the Philippines with India via areas marginal to a late-Cretaceous or early-Tertiary Sunda Shelf. For example, 51% of Sulawesian species are shared with the Philippines, 38% with Borneo (15% with Sabah or south east Kalimantan), 56% with Java, 35% with Sumatra, and 29% with India. The high proportion of Indo-Australian distributions indicates that the two generalised tracks are linked at some level.

### 2.2. *Moths of Mount Kinabalu*

The second pattern described for Sulawesian birds is also demonstrated by Holloway's (1976) moth data for Borneo, particularly Mount Kinabalu. Table 3 lists the distributional details of 213 moths that are found on Mount Kinabalu or Borneo. The Indian connection is strong with 81% of all Kinabalu moth species being Indian (including Punjab, Nepal, Sikkim, Bhutan, Assam, Sri Lanka). Approximately one quarter of all species listed in Table 3 are also found in the Phillippines, Sulawesi, and Burma/Andamans. Higher percentages are found in Sumatra (31%) and Java (42%).

*B. Michaux/Palaeogeography, Palaeoclimatology, Palaeoecology 122 (1996) 167–183* 171

Table 2
Distribution data for Indonesian birds that do not occur in the Malay Peninsula. *India* = Himalayas + India + Sri Lanka, *SEA* = Southeast Asia but not Thailand or the Malay Peninsula, *Mentawai+* = western islands off Sumatra. *X* = present, *n* = north, *s* = south, *se* = south east. Endemic species are italicised, stippled entries denote Sulawesian species. Numbers in brackets after family names refer to the number of Sulawesian endemics

| Family | Species | Africa | Eurasia | India | sChina | Andamans/Nicobars | SEA | Sumatra | Mentawai + | Java | n Borneo | se Borneo | Borneo | Sulawesi | Philippines | Lesser Sundas | Maluku | New Guinea | Australia | Pacific |
|---|---|---|---|---|---|---|---|---|---|---|---|---|---|---|---|---|---|---|---|---|
| Sulidae | Sula sula | | | | | | | X | | X | X | | | | | | | | | |
| | S. dactylatra | | | X | | | | | | X | X | | | | | | | | | |
| Phalacrocoracidae | Phalacrocorax carbo | | | | | | | X | | | X | | | | | | | | | |
| | P. melanoleucos | | | | | | | | | | | | | X | | | ? | ? | X | X |
| | P. sulcirostris | | | | | | | X | | | | X | | | | ? | ? | X | X | |
| | P. niger | | | | | | | | | X | | X | | | | | | | | |
| | Anhinga melanogaster | | | X | | | X | X | | X | | | X | X | X | | | | | |
| Ardeidae | Ardeola bacchus | | | X | X | X | X | | | | | | | | | | | | | |
| | A. striatus | | | X | X | X | | | | | | | | | | | | | | |
| | A. grayii | | X | X | | X | X | | | | | | | | | | | | | |
| | Ardea purpurea | | X | X | X | X | | X | | X | | | X | X | X | | | | | |
| | A. speciosa | | | | | | X | X | | X | | | X | X | X | | | | | |
| | Egretta garzetta | | | X | | X | X | X | | X | | | X | | | | | | | |
| | E. sacra | | | ? | | X | X | X | | X | | | X | | | | | X | X | X |
| | E. intermedia | X | | X | X | X | X | X | X | X | | | X | | | | | | | |
| | Bubulcus ibis | | | X | X | X | X | X | | X | | | | | | | | | | |
| | Nycticorax caledonicus | | | | | | | | | X | X | | | | X | | | X | X | X |
| | Ixobrychus flavicollis | | | X | X | | X | X | X | | X | | | | X | ? | ? | X | X | |
| Threskiornithidae | Pseudibis davisoni | | | | X | | X | | | | | X | | | | | | | | |
| | Plegadis perigrinus | | | | | | | | | X | | | | X | X | | | | | |
| | Platalea regia | | | | | | | | | | | | | X | | | | X | X | X |
| Anatidae | Dendrocygna arcuata | | | | | | | X | | X | | X | | X | | | | X | X | |
| | D. guttata | | | | | | | | | | | | | X | X | | | | | |
| | Anas poecilorhyncha | | | X | X | X | X | | | | | | | | | | | | | |
| | A. gibberifrons | | | | | X | | | | X | | | | X | | | | | | |
| | Netapus pulchellus | | | | | | | | | | | | | X | X | | X | X | X | |
| Accipitridae (5) | *Spilornis kinabaluensis* | | | | | | | | | | X | | | | | | | | | |
| | *S. elgini* | | | | | X | | | | | | | | | | | | | | |
| | *S. klossi* | | | | | X | | | | | | | | | | | | | | |
| | Accipiter badius | X | | X | X | X | X | | | | | | | | | | | | | |
| | A. virgatus | | | X | X | X | X | X | | X | X | | | | | | | | | |
| | A. leuphotes | | | ? | | X | | | | | | | | | | | | | | |
| | Spizaetus cirrhatus | | | X | | X | | | | | | | | | | X | | | | |
| | Avicida jerdoni | | | X | X | | X | X | | | | | X | X | | | | | | |
| | Elanus hypoleucus | | | | | | | | | | | | | X | X | | | | | |
| | Butaster liventer | | | | X | | X | | | X | | | | X | | | | | | |
| | Hieraatus kienerii | | | X | | | X | | | | | | | X | X | | | | | |
| | Haliaeetus leucogaster | | | X | X | X | X | X | | X | | | X | | | | | X | X | X |
| | Circus assimilis | | | | | | | | | | | | | X | | | | | X | |
| Turnicidae | Turnix tanki | | | X | ? | X | | | | | | | | | | | | | | |
| Falconidae | Falco moluccensis | | | | | | | | | X | | X | | X | | X | X | | | |
| | *Microhierax latifrons* | | | | | | | | | | X | | | | | | | | | |
| Megapodidae | Megapodius cumingii | | | | | | | | | | X | | | X | X | | | | | |
| | M. freycinet | | | | | X | | | | | | | | | | X | ? | X | X | |
| | M. reinwardt | | | | | | | | X | | | | | | | X | ? | X | X | |
| Phasianidae | *Arborophila hyperythra* | | | | | | | | | | X | | | | | | | | | |
| Rallidae (2) | Porzana pusilla | X | X | | | X | | X | | | | X | | | | | | X | X | |
| | G. tenebrosa | | | | | | | | | | | X | | ? | | ? | ? | X | X | |
| | Porphyrio porphyrio | | | | | X | | X | | X | | X | | | | | | X | X | X |
| | *Rallina canningi* | | | | | X | | | | | | | | | | | | | | |
| | R. eurizonoides | | | X | X | | X | | | | | | | X | X | | | | | |
| | Fulica atra | | X | X | X | | X | | | X | | | | X | | | | X | X | |
| Jacanidae | Irediparra gallinacea | | | | | | | | | | | X | | | X | | | X | X | |
| | Hydrophasianus chirurgus | | | X | X | | X | | | | | X | | | X | | | | | |
| Sternidae | Sterna bengalensis | | X | X | | X | X | X | | X | | | X | | X | | | X | X | |
| | S. bergii | X | X | X | | X | | | ? | X | | | | | | | | X | X | X |
| | S. sumantrana | | | X | | X | | X | X | X | | | X | | | | | X | X | X |
| Recurvirostridae | Himanotropus leucocephalus | | | | | | | | | X | | | | X | | X | X | X | X | X |
| Burhinidae | Burhinus giganteus | | | | | | X | X | | X | | | X | X | | X | X | X | X | |
| Columbidae (8) | Treron pompadore | | | X | X | X | X | | | | | | | | | | | | | |
| | T. griseicauda | | | | | | | X | | X | | | | X | | | | | | |
| | Ptilinopus melanospila | | | | | | | | | X | X | | | X | X | | X | | | |
| | P. superbus | | | | | | | | | | | | | X | X | | | X | X | |
| | Ducula pickeringi | | | | | | | | | | X | | | X | X | | | | | |
| | D. bicolor | | | | | X | | X | X | X | | | X | X | X | X | X | X | | |
| | Columba argentina | | | | | | | X | | | X | | | | | | | | | |
| | *C. palumboides* | | | | | X | | | | | | | | | | | | | | |
| | Streptopelia bitorquata | | | | | | | | | X | X | | | | X | X | | | | |

Table 2 (*continued*)

| Family | Species | Africa | Eurasia | India | sChina | Andamans/Nicobars | SEA | Sumatra | Mentawai + | Java | n Borneo | se Borneo | Borneo | Sulawesi | Philippines | Lesser Sundas | Maluku | New Guinea | Australia | Pacific |
|---|---|---|---|---|---|---|---|---|---|---|---|---|---|---|---|---|---|---|---|---|
| | S. tranquebarica | | | X | X | X | | | | | | | | | | | | | | |
| | *Macropygia rufipennis* | | | | | X | | | | | | | | | | | | | | |
| | M. emiliana | | | | | | | | X | X | | | X | | | | | | | |
| | M. phasianella | | | | | | | | X | | | | | | X | | | | X | |
| | M. amboinensis | | | | | | | | | | | | | X | | | X | X | X | |
| | Chalcophaps stephani | | | | | | | | | | | | | X | | | | X | | X |
| | Caloenas nicobarica | | | | | X | | | X | X | X | | | | | | | X | | X |
| Psittacidae (7) | Cacatua sulphurea | | | | | | | | | | | | | X | | X | | | | |
| | Psittacula alexandri | | | X | X | X | X | | X | X | | X | | | | | | | | |
| | P. eupatria | | | X | | X | X | | | | | | | | | | | | | |
| | *P. caniceps* | | | | | X | | | | | | | | | | | | | | |
| | Loriculus vernalis | | | X | X | X | X | | | | | | | | | | | | | |
| Cuculidae (3) | Cuculus sparveroides | | | X | X | | | X | | | X | | | | X | | | | | |
| | C. russatus | | | | | | | | | | X | | | X | X | X | | X | X | |
| | C. micropterus | | | X | | X | X | X | X | X | | | X | | X | | | | | |
| | Clamator coromandus | | | X | X | | X | X | X | X | | | X | X | X | | | | | |
| | Cacomantis sonneratii | | | X | X | | | X | X | X | | | X | | X | | | | | |
| | C. merulinus | | | X | X | | | X | | X | | | X | X | X | | | | | |
| | Surniculus lugubris | | | X | X | | | X | X | X | | | X | X | X | | X | | | |
| | *Centropus andamanensis* | | | | | X | | | | | | | | | | | | | | |
| | C. bengalensis | | | X | X | | X | X | | X | | | X | X | X | X | X | | | |
| | Scythrops novaehollandiae | | | | | | | | | | | | | X | | | | X | X | |
| Strigidae (4) | Otus brookii | | | | | | | X | | X | X | | | | | | | | | |
| | *O. umbra* | | | | | | | | X | | | | | | | | | | | |
| | *O. balli* | | | | | X | | | | | | | | | | | | | | |
| | *O. mentawi* | | | | | | | | X | | | | | | | | | | | |
| | *O. enganensis* | | | | | | | | X | | | | | | | | | | | |
| | O. mantananensis | | | | | | | | | | X | | | | X | | | | | |
| | Ninox affinis | | | | | X | | | | | | | | | | | | | | |
| Caprimulgidae (1) | Caprimulgus affinis | | | X | X | | X | X | | X | | | X | X | X | | | | | |
| Podargidae | *Batrachostomus harterti* | | | | | | | | | | X | | | | | | | | | |
| Apodidae (3) | Collocalia salangana | | | | | | | X | | X | X | | | | | | | | | |
| | C. brevirostris | | | X | X | ? | X | | | X | | | | | | | | | | |
| | C. linchi | | | | | | | X | | X | X | | | | | X | | | | |
| Trogonidae | *H. whiteheadi* | | | | | | | | | | X | | | | | | | | | |
| Alcedinidae (5) | Halcyon pileata | | | X | X | X | X | | | | | | | | | | | | | |
| Meropidae (1) | Merops philippinus | | | X | X | X | | ? | | ? | | | ? | X | X | | | X | | X |
| Capitonidae | *Megalaima monticola* | | | | | | | | | | X | | | | | | | | | |
| | *M. eximia* | | | | | | | | | | X | | | | | | | | | |
| | *M. pulcherrima* | | | | | | | | | | X | | | | | | | | | |
| | M. australis | | | X | X | | X | X | X | X | | | X | | | | | | | |
| Bucerotidae (2) | *Rhyticeros narcondami* | | | | | X | | | | | | | | | | | | | | |
| Picidae (2) | Chrysocolaptes lucidus | | | X | X | | | X | | X | X | | | | X | | | | | |
| | Dendrocopus macei | | | X | | X | X | X | | X | | | | | | | | | | |
| Pittidae | Pitta erythrogaster | | | | | | | | | | | | | X | | | | X | X | |
| | P. sordida | | | X | X | | | X | | X | | | X | X | X | | | | | |
| Eurylaimidae | *Calyptomena whiteheadi* | | | | | | | | | | X | | | | | | | | | |
| Alaudidae | Mirafra javanica | X | | X | X | | X | | | X | | X | | | X | ? | ? | X | X | |
| Campephagidae (5) | Coracina larvata | | | | | | | X | | X | X | | | | | | | | | |
| | C. tenuirostris | | | | | | | | | | | | | X | | X | X | X | X | X |
| | C. morio | | | | | | | | | | | | | X | X | | X | X | | |
| | Lalage sueurii | | | | | | | | | X | | | | X | | X | | X | X | |
| | Pericrocotus cinnamomeus | | | X | | X | X | | | X | | | X | | | | | | | |
| Pycnonotidae | P. flavescens | | | X | X | | X | | | | X | | | | | | | | | |
| | P. aurigaster | | | | X | | X | | | X | | X | | ? | | | | | | |
| | P. nieuwenhuisi | | | | | | | X | | | | X | | | | | | | | |
| | *Hypsipetes nicobarensis* | | | | | X | | | | | | | | | | | | | | |
| Dicruridae (1) | Dicrurus hottentottus | | | X | X | | | X | X | X | | | X | | | | | | | |
| | *D. andamanensis* | | | | | X | | | | | | | | | | | | | | |
| Oriolidae | Oriolus xanthornus | | | X | X | X | X | X | | | X | | | | | | | | | |
| | *O. hosei* | | | | | | | | | | X | | | | | | | | | |
| Artamidae (1) | Artamus leucorhynchus | | | | | X | | X | X | X | | | X | | X | | | X | X | X |
| Corvidae (2) | Cissa thalassina | | | | X | | X | | | X | X | | | | | | | | | |
| | *Dendrocitta bayleyi* | | | | | X | | | | | | | | | | | | | | |
| Timaliidae (2) | Malacocincla perspicillata | | | | | | | | | | | X | | | | | | | | |
| | *Napothera crassa* | | | | | | | | | | X | | | | | | | | | |
| | Garrulax palliatus | | | | | | | X | | | X | | | | | | | | | |
| Turdidae (3) | *Copsychus stricklandi* | | | | | | | | | | X | | | | | | | | | |

Table 2 (*continued*)

| | | Africa | Eurasia | India | sChina | Andamans/Nicobars | SEA | Sumatra | Mentawai + | Java | n Borneo | se Borneo | Borneo | Sulawesi | Philippines | Lesser Sundas | Maluku | New Guinea | Australia | Pacific |
|---|---|---|---|---|---|---|---|---|---|---|---|---|---|---|---|---|---|---|---|---|
| | Saxicola caprata | | X | ? | X | | X | | | X | X | | | X | X | X | | X | | |
| | *Zoothera everetti* | | | | | | | | | | X | | | | | | | | | |
| | Z. andromedea | | | | | | | X | X | X | | | | | X | X | | | | |
| | Z. citrina | | | X | X | X | X | | | X | X | | | | | | | | | |
| | Z. dauma | | X | X | X | | X | X | | X | X | | | | X | X | | | | |
| | Turdus poliocephalus | | | | | | | X | | X | X | | | X | | X | | X | | X |
| | *Chlamydochaera jefferyi* | | | | | | | | | | X | | | | | | | | | |
| Sylviidae (1) | *Bradypterus accentor* | | | | | | | | | | X | | | | | | | | | |
| | Acrocephalus stentoreus | X | X | X | X | X | X | | | X | | | X | | | X | X | X | X | |
| | Phylloscopus magnirostris | | | X | X | X | X | | | | | | | | | | | | | |
| | Cettia vulcania | | | | | | | X | | X | X | | | | | X | | | | |
| | C. pallidipes | | | X | X | X | X | | | | | | | | | | | | | |
| | Cisticola exilis | | | X | X | | X | | | X | | | | X | | X | X | X | X | |
| | Gerygone sulphurea | | | | | | | X | X | X | | | X | X | X | | | | | |
| | *Urosphena whiteheadi* | | | | | | | | | | X | | | | | | | | | |
| Muscicapidae (7) | Rhinomyias ruficauda | | | | | | | | | | X | | | | X | | | | | |
| | R. gularis | | | | | | | | | | X | | | | X | | | | | |
| | *Muscicapa indigo* | | | | | | | | | | X | | | | | | | | | |
| | Eumyias indigo | | | | | | | X | | X | X | | | | | | | | | |
| Pachycephalidae (3) | Pachycephala homeyeri | | | | | | | | | | X | | | | X | | | | | |
| | *P. hypoxantha* | | | | | | | | | | X | | | | | | | | | |
| Laniidae | Lanius schach | | X | X | X | | X | X | | X | | X | | | X | | | X | | |
| Sturnidae (6) | Sarcops calvus | | | | | | | | | | X | | | | X | | | | | |
| | Aplonis minor | | | | | | | | | X | | | | X | X | X | | | | |
| | *Sturnus erythropygius* | | | | | X | | | | | | | | | | | | | | |
| | Acridotheras javanicus | | | ? | X | | X | | | X | | | | X | | | | | | |
| Nectariniidae | *Arachnothera everetti* | | | | | | | | | | X | | | | | | | | | |
| | *A. juliae* | | | | | | | | | | X | | | | | | | | | |
| | Nectarinia aspasia | | | | | | | | | | | | | X | | | X | X | | X |
| Meliphagidae (2) | Myzomela sanguinolenta | | | | | | | | | | | | | X | | | | | X | X |
| Dicaeidae (3) | Dicaeum trochileum | | | | | | | X | | X | | X | | | | X | | | | |
| | D. monticolum | | | | | | | | | | X | | | | | | | | | |
| Zosteropidae | Zosterops flavus | | | | | | | | | X | | X | | | | | | | | |
| | Z. atricapilla | | | | | | | | | | X | | | | X | | | | | |
| | Z. montanus | | | | | | | X | | X | | | | X | X | X | X | | | |
| | *Z. salvadorii* | | | | | | | | X | | | | | | | | | | | |
| | Z. chloris | | | | | | | | | X | | | | X | | X | X | X | | X |
| | Z. everetti | | | | | | | | | | X | | | | X | | | | | |
| | *Oculocincta squamifrons* | | | | | | | | | | X | | | | | | | | | |
| Ploceidae | Erythrura trichroa | | | | | | | | | | | | | X | | | | X | X | X |
| | Lonchura molucca | | | | | | | | | | | | | X | | X | X | | | |
| Fringillidae | Serinus estherae | | | | | | | X | | X | | | | X | X | | | | | |

As in the bird data approximately one third of distributions cross Wallace's line (73/213).

### 3. Sulawesian *Dacus*

The data for Sulawesian *Dacus* (Diptera: Tephritidae) came from Hardy (1982, 1983). A preliminary cladistic analysis of the *Batrocera*-group of subgenera was undertaken based on the data provided in Drew (1989). The list of subgenera and character states used are shown in Table 4. Character states A, B, C, and H are unordered. These data were analysed by PAUP (Macintosh v.3.2.2) using the heuristic option. Starting trees were generated by a stepwise simple addition sequence and branch-swapping by nearest-neighbour interchange (NNI).

As expected, when fewer characters are used to analyse a greater number of taxa, many most parsimonious trees were generated. The majority rule consensus tree of 72 most parsimonious trees (length 24, $CI = 0.458$, $RI = 0.745$) is shown in Fig. 2. The important feature of this tree is that the subgenera *Paradacus* and *Sinodacus* form a monophyletic group which appears stable as it occurs in two thirds of all most parsimonious trees.

Table 3

Distribution data for Bornean moths that do not occur in the Malay Peninsula. *Australasia* = Australia + New Guinea + Melanesia, *X** = north Borneo or west Sumatra. Extralimital distributions given at right of Table, stippled entries denote Sulawesian species. See Table 2 legend for other details

| | | India/Sri Lanka | s or w China | SEA | Burma/Andamans | Borneo | Sumatra | Java/Bali | Sulawesi | Philippines | Lesser Sundas | Maluku | Australasia | Pacific | |
|---|---|---|---|---|---|---|---|---|---|---|---|---|---|---|---|
| Arctiidae | Agylla mjobergi | | | | | X | | | X | | | | | | |
| | A. divisa | X | | X | | X* | | | | | | | | | Taiwan |
| | Tigroides puncticollis | | | | | X | | X | | | X | | | | |
| | Chionaema perornata | X | | | | X | X | X | | | | | | | |
| | Creatonotos transiens | X | X | | | X* | | X | | X | | | | | |
| | Rhodogastria omissa | X | | | | X* | | | | | | | | | |
| Hypsidae | Asota plana | X | X | | | X* | | | | | | | X | | |
| | A. producta | X | X | X | | X | | X | | | | | | | |
| | A. elegans | X | X | X | | X* | | | | | | | X | | Japan |
| | A. javanaca | | | | | X* | X | X | X | X | | | | | |
| | Nyctemera coleta | X | X | | | X* | | | | | | | X | | |
| | N. latistriga | X | X | | | X* | | | | X | X | | | | |
| | N. adversata | X | X | | | X* | X | | | | | | | | Japan |
| Noctuidae | Diarsia nigrosigna | X | X | | X | X* | | | | X | | | | | |
| | D. strictica | X | X | | | X* | | | | | | | | | |
| | D. ochracea | X | | | | X* | | X | X | | | | | | |
| | Leucania roseilineatrigata | | | | X | X | | | X | X | | | X | | |
| | Pseudaletia albicosta | X | X | | | X* | | X | | | X | | | | Taiwan |
| | Hypopteridae reversa | X | | | | X* | | X | | X | | | X | | |
| | Dypterygia indica | X | | | | X* | | | | | | | | | |
| | Platyprosopa nigros | | | | | X* | X | X | X | | | | X | | |
| | Data eriopoides | | | | | X* | X* | X | | | | | | | |
| | Callopistria rivularis | X | | | X | X* | | | | | | | X | | Japan |
| | Diptherocome muscosa | X | | | | X* | X | X | | | | | | | |
| | Thalatha conjecturalis | X | | | | X | X | | X | | | | | | |
| | T. sinens | X | | | X | X* | X* | | | | | | | | |
| | Acronicta pruinosa | X | | | | X* | | | | | | | | | |
| | A. rubiginosa | X | | | | X* | X | | | | | | | | |
| | Platysenta oroana | | X | | | X | | X | X | X | | | | | |
| | P. magna | X | | | | X* | | X | | | | | | | |
| | Sphetta apicalis | X | | | | X* | | X | | | | | | | |
| | Apsarasa radians | X | | X | | X* | X | | X | | | | | | |
| | Eutelia fulvipicta | X | | | X | X | | | X | X | | X | | | |
| | Phlegetonia transversa | | | X | | X | | | | | | | X | | |
| | P. delatrix | X | X | X | | X | | | | | | | X | X | |
| | Anigraea cinctipalpis | X | | | | X* | | | | X | | | | | |
| | *Stictoptera cucullioides* | X | | | X | X* | | | | | | | X | | |
| | S. olivascens | X | | | | X* | | | | | | | | | |
| | S. describens | X | | | | X* | | | | | | | | X | |
| | Lophoptera costata | X | X | | X | X* | | X | X | | | | X | | Taiwan |
| | L. acuda | | | | X | X* | X | | | | | | | | |
| | Sadarsa longipennis | X | | | | X | X | X | | | | | | | Taiwan |
| | Gyrtona nama | X | | | | X* | X* | | X | | | X | X | | |
| | Characoma basisuffusa | | | | | X* | | X | | | | | X | | |
| | Gadirtha exacta | | | | | X* | | | X | X | | | X | | |
| | Risoba avola | | | | | X* | | | | X | | X | X | | |
| | Macrobarasa xantholopha | | | | | X* | | | | X | | X | X | | |
| | Hylophilodes rubromarginata | X | | | X | | X | X | | X | | X | X | | |
| | Paracrama dulcissima | X | | | X | X | | | | | | | X | | |
| | Aiteta deminutiva | | | | | X* | | | X | | | | X | | |
| | A. brooksi | | | | X | X* | X | | | | | | | | |
| | Carea endophaea | X | | | | X | | X | | X | | | | | |
| | Orthocruspis acypera | | | | | X* | X | X | | X | | | X | | |
| | Ulotrichopus macula | X | | | | X* | | X | | | | X | | | |
| | Erebus gemmans | X | | | X | X* | | | | | | | | | |
| | E. macrops | X | X | | X | X* | | | | | | | | | |
| | Thyas honesta | X | | | X | X* | | | | X | | X | | | Christmas Is. |

Table 3 (*continued*)

| | | India/Sri Lanka | s or w China | SEA | Burma/Andamans | Borneo | Sumatra | Java/Bali | Sulawesi | Philippines | Lesser Sundas | Maluku | Australasia | Pacific | |
|---|---|---|---|---|---|---|---|---|---|---|---|---|---|---|---|
| | Anua coronata | X | | ? | X | X* | | | | | | | X | X | |
| | Parallelia joviana | | ? | X | | X* | | | | | | X | X | | |
| | P. maturescens | X | | X | | X* | X | X | | | | | | | |
| | P. fulvotaenia | X | | | | X* | | | | | X | X | | | Taiwan |
| | P. arctotaenia | X | X | | X | X* | | X | | X | X | | X | X | Taiwan, Japan, Korea |
| | Chalciope mygdon | X | X | ? | | X | | X | | | | | | | |
| | Daddala lucilla | X | | | X | X* | X | X | | | | | X | | |
| | Plusia nigriluna | X | | | X | X* | | X | | | | | | | Taiwan |
| | P. lectula | X | | | | X | | X | | | | | X | | |
| | Ramadasa pavo | X | | | X | X* | | X | X | X | | | | | |
| | Baorisa heiroglyphica | X | | X | | X* | X* | X | | X | | | | | |
| | Dordura aliena | X | | | X | X | | X | X | | | | | | |
| | Orthreis cajeta | X | | X | | X | X | X | X | | | | | | |
| | O. abathyglypta | X | | | X | X* | X | | | | | | | | |
| | Lineopalpa horsfieldi | X | | | | X* | X* | X | X | X | | X | | | |
| | L. orsa | | | | | X* | | | X | X | | X | X | | |
| | Anomis mesogona | X | | | | X* | X | X | | X | | | | | |
| | Calpe minuticornis | X | | | | X* | | | | | | | X | | Taiwan |
| | Singara diversalis | X | X | | | X* | | | | | | | X | | |
| | Sarobe ceylonica | X | | | X | X* | | | | | | | | | |
| | S. albopunctata | | | | | X* | X | X | | X | | | X | | |
| | Hamodes propitia | X | | | X | X | | | | X | | | | | |
| | Hyposemansis singha | X | | | X | X* | X | | | | | | | | |
| | Cryptastria fuscomarginata | X | | | | X* | | | X | | | | X | | |
| | Hypena modesta | X | | | | X* | | | | | | X | X | | |
| | Bomolocha obductalis | X | | | | X* | | | | | | | | | |
| Lymantriidae | Dendrophleps semihyalina | X | | | | X* | | | | | | | | | |
| | Euproctis seitzi | X | X | | | X* | X | | | | | | | | |
| | Lymantria inordinata | | | | | X | X | X | X | | | | | | |
| | Cassidia kinabulensis | | | | | X* | | | X | | | | | | |
| | Dasychira dudgeoni | X | X | | | X* | X | X | | | | | | | |
| Sphingidae | Megacorma obliqua | X | | | X | X* | X | X | | | | | X | | |
| | Acherontia lachesis | X | X | | | X* | | | | | | X | | | |
| | A. styx | X | X | ? | | X* | | | | | X | X | | | Japan |
| | Oxyambulyx substrigilis | X | | | | X* | | X | | X | | | | | |
| | O. subocellata | X | X | | | X | | X | | | | | | | |
| | Marumba spectablis | X | X | | | X* | X | | X | | | | | | |
| | Acosmeryx socrates | X | | | | X* | | X | | X | | | | | |
| | Hippotion echeclies | X | | | | X* | | | X | X | X | | | | |
| | Rhagastis aurifera | X | | | | X* | X | | | | | | | | |
| Thyatridae | Horithyatria decorata | X | | | | X* | X* | | | | | | | | Taiwan |
| | Tethea oberthuri | X | X | | X | X* | | | | | | | | | |
| Notodontidae | Besida xylinata | | | X | | X | X | X | | | | | | | |
| | Gangarides vardena | X | | | | X | X | | | | | | X | | |
| | Norraca xanthophila | X | | | | X* | X | X | | X | | | | | |
| | Surgiga suriga | | | | | X* | | | X | X | | | | | |
| | Somera viridifusca | X | | | | X* | | X | | X | | | | | |
| | Chadisra bipars | X | | | | X | | X | | X | | X | X | | |
| | Ginshachia gemmifera | X | | | X | X* | X | X | | | | | | | |
| | Clostera pallida | X | X | X | X | X* | | X | | | | | | | Japan |
| | C. restitura | X | | | | X* | | | | | | | | | |
| Geometridae | Eumelia rubrifusa | | | | | X* | | | X | X | | | | | |
| | Noreia unilineata | X | | | X | X | X | X | | | | | | | |
| | Canolophia nigripuncta | X | | X | X | X | | | | | | | | | |
| | Pingasa venusta | X | | | | X | X | X | X | | | X | X | | |
| | Argyrocosma phrixopa | | | | | X* | | X | X | | | X | | | |
| | Dooabia puncticosta | | | | | X | | X | X | | | | | | |
| | Tanaorhinus reciprocata | X | X | | X | X* | | | | | | | | | |

176 B. Michaux/Palaeogeography, Palaeoclimatology, Palaeoecology 122 (1996) 167–183

Table 3 (*continued*)

| | India/Sri Lanka | s or w China | SEA | Burma/Andamans | Borneo | Sumatra | Java/Bali | Sulawesi | Philippines | Lesser Sundas | Maluku | Australasia | Pacific | |
|---|---|---|---|---|---|---|---|---|---|---|---|---|---|---|
| Mixochlora vittata | X | X | | | X* | X* | X | | | | | | | Japan |
| Spaniocentra apetella | | | | | X | | | X | X | | | | | |
| Comibaena albimarginata | X | | | | X* | | | | | | | | | |
| Gelasma thetydaria | X | | | | X* | | | | | | | | | |
| Oenospila strix | X | | | | X* | X | X | | | | | | | |
| Paramaxates posterecta | X | | | X | X* | X* | X | | | | | | | |
| Comostola meritaria | X | | | | X | | X | | | | | | | |
| Prasinocyma floresaria | X | | | | X | | X | | X | X | | | | |
| Organopoda carnearia | X | | X | | X | X | X | | | | | | | |
| Anisodes penumbrata | | | | | | | | X | X | | X | | | |
| A. dimerites | | | | | X* | | X | | | | | X | | |
| A. flavissima | | | | | X* | | | | | | X | X | | |
| A. denticulata | X | | | | X | X | | X | | | X | X | | |
| Scopula insolata | X | X | | | X* | | X | | | | | | | |
| S. mecysma | X | | X | | X* | | X | | | | | X | | |
| Idaea comparanda | X | | | | X | | X | | X | | | | | |
| Bythana uniformis | | | | | X | X | X | X | | | | | | |
| Horisme hypererythra | X | X | | X | X* | | X | X | | | | | | |
| Eupithecia melanolopha | X | | | | X | | X | | | | | | | |
| Chloroclystis filicata | X | | | | | | | X | | | X | | | |
| C. palpata | X | X | | | X* | | X | | | | | | | |
| C. xanthocomes | X | | | X | X | | X | | | | | | | |
| Gymnoscelis phoenicopis | | | | | X* | | | X | | | X | | | |
| G. polyodonta | X | | | | X | | | | | | | X | | |
| Mariaba convoluta | X | | | X | X* | | | X | | X | | | | |
| Hybridoneura metachlora | X | | | | X | | | X | | X | | | | |
| Symmimetis cristata | X | | | X | X* | | X | X | | | | X | | |
| Antimimistis attenuata | X | | | | X* | | | X | | | | X | | |
| Trichopterigia pulcherrima | X | | | X | X* | | | | X | | | | | |
| Hypocometa clauda | X | | | X | X* | | | | | | | | | |
| Sauris marginepunctata | | | | | X | | | | X | | X | | | |
| Hydrelia flexilinea | | | | | X | | | X | X | | | | | |
| Petelia circularia | | | | X | X | X | X | X | | | | | | |
| P. capitata | X | | | | X* | | | | | | | | | |
| Iridoplecta ferrifera | X | | | | X | | | | | | | | | |
| Hyposidra talaca | X | | | X | X* | | | X | | | | X | | |
| Dalima patularia | X | X | | | X* | | | X | | | | | | |
| Hypochrosis herois | | | | | X* | | | X | | | | | | |
| H. hyadaria | X | | | | X* | | | | | | | | | |
| H. festivaria | X | | | X | X | | | | | | | | | |
| H. muscicdor | | | | X | X | | | | | | | | | |
| Sabaria incitata | X | | | | X* | X* | | | | | | | | |
| S. multidentata | | | | | X | X | | | X | | | | | |
| Garaeus apicata | X | | | X | X* | X* | | | | | | | | Taiwan |
| Luxiaria tephrosaria | X | | | | X | | | X | | | | | | |
| L. acutaria | X | | | | X* | X | | | | | | | | |
| L. mitorrhapes | X | | | X | X* | | | | | | | | | |
| L. phyllosaria | X | | | | X* | X | | X | X | | | | | |
| L. submonstrata | | | | | X | | X | X | | | | X | | |
| L. amasa | X | | | | X | | X | X | | | | | | Japan |
| Semiothisa nora | X | | X | | X* | | | X | | | | | | |
| S. ozararia | X | | | | X | X | X | | | | | | | |
| S. temeraria | X | | | | X* | | X | | | | | | | Japan,Taiwan |
| S. oliva | X | | | | X* | | X | | | | | | | |
| Ourapteryx clara | X | | | | X* | | X | | | | | | | |
| Thinopteryx nebulosa | X | | | | X* | | X | | | | | | | |
| Menophra subterminalis | X | X | | | X* | | | | | | | | | |
| Ectropis intermedia | | | X | | X* | | X | | X | | | | | |

Table 3 (*continued*)

| | | India/Sri Lanka | s or w China | SEA | Burma/Andamans | Borneo | Sumatra | Java/Bali | Sulawesi | Philippines | Lesser Sundas | Maluku | Australasia | Pacific | |
|---|---|---|---|---|---|---|---|---|---|---|---|---|---|---|---|
| | Ruttellerona lithina | | | | | X* | | | X | | | | X | | |
| | Diplurodes parvularia | X | | | | X* | | | | | | | | | Japan/Taiwan |
| | D. vestita | X | | | X | X* | | | | | | | | | |
| | Racotis boarmiaria | X | X | | | X* | | | | | | | | | Japan/Taiwan |
| | Cleora pupillata | | | X | X | X | X | X | X | X | | | | | |
| | C. inoffensa | X | | | | X | X | X | X | X | | | X | | |
| | Boarmia cineracea | X | | | | X* | X | | | | | | | | |
| | B. glochinophora | X | | | | X* | | | | | | | | | |
| | B. bisinuata | X | X | | X | X* | | | | | | | | | Taiwan |
| | Myrioblephara flexilinea | | | | | X* | | X | X | | | X | X | | |
| | M. simplaria | X | | | | X* | | X | | X | | X | | | |
| | Abaciscus tristis | X | X | | | X* | | | | | | | | | |
| | Borbacha pardaria | X | | | | X | X | X | | | | | X | | |
| | Lomographa indistincta | X | | | X | X* | | | | | | | | | |
| | Arichanna maculata | X | | | | X* | X* | | | | | | | | |
| | Plutodes flavescens | X | | | | X* | | X | | | | | | | |
| Saturniidae | Leopa katinka | X | X | | X | X* | X | X | X | X | | | | | |
| | Antheraea frithi | X | | X | X | X* | X | X | | | | | | | |
| | Cricula andrei | X | | X | X | X* | X | X | | | | | | | |
| | C. trifenestrata | X | | | | X* | | X | | X | | | | | |
| Brahmaeidae | Brahmidia hearseyi | X | X | | X | X* | | X | X | X | | | | | |
| Bombycidae | Theophila religiosae | X | | | | X* | | | | | | | | | |
| | Ocinara signifera | X | | | X | X* | X | | | X | | | X | | |
| Uraniidae | Acropteris ciniferaria | X | | | X | X* | X | | | | | | | | |
| | Pseudomicronia coelata | X | X | | | X | | X | | | | | | | Japan |
| | Metorthocheilus emarginatus | X | | | | X* | | X | | | | | | | Japan |
| | Epiplema moza | X | X | | | X* | | | | | | | | | Japan |
| | E. flavistriga | X | X | | | X* | | | | | | | | | Japan |
| | Dirades conifera | X | | | | X | | | | | | | | | |
| | Gathynia fumicosta | X | X | | | X* | | | X | | | | | | |
| | G. pernigrata | X | | | | X* | | X | | | | | | | |
| Cossidae | Zeuzera caudata | | | | | X* | X | | | | | X | X | | |
| | Xyleutes persona | X | X | | | X* | | | | X | | | X | | |
| Lasiocampidae | Opsirhina vinata | X | X | | | X* | X* | | | | | | | | |
| | Odontocrapsis hasora | X | | | | X* | X | | | | | | | | |
| Limacocidae | Birthamoides junctura | X | | | X | X | | | | X | | | | | |
| Zygaenidae | Agalope glacialis | X | X | | X | X* | | | | | | | | | |
| Drepanidae | Nordstroemia duplicata | X | X | | | X* | | | | | | | | | |
| | Teldenia specca | X | X | | | X | X | | | | | X | X | | Palawan |
| | Canucha specularis | X | X | | | X | X | X | X | | | | | | |

The taxa and character states for Sulawesian *Dacus* are shown in Table 5 and were derived from Hardy (1982, 1983) and Drew (1989). The subgenera *Asiadacus* and *Pacifodacus* that Hardy (1982, 1983) refers to have been synonomised with *Sinodacus* by Drew (1989) The initial analysis

178 B. Michaux/Palaeogeography, Palaeoclimatology, Palaeoecology 122 (1996) 167–183

Table 4
Characters and character states for the *Batrocera*-group of subgenera of the genus *Dacus* (Diptera: Tephritidae). *0* = presumed primitive condition

| | Characters | | | | | | | |
|---|---|---|---|---|---|---|---|---|
| | A | B | C | D | E | F | G | H |
| Afrodacus | 2 | 1 | 1 | 1 | 0 | 0 | 1 | 1 |
| Batrocera | 2 | 1 | 1 | 1 | 0 | 1 | 1 | 1 |
| Gymnodacus | 2 | 1 | 1 | 0 | 0 | 1 | 1 | 1 |
| Notodacus | 2 | 1 | 1 | 1 | 1 | 1 | 1 | 1 |
| Polistomimetes | 2 | 1 | 1 | 1 | 0 | 0 | 0 | 1 |
| Trypetidacus | 2 | 1 | 1 | 0 | 0 | 0 | 0 | 1 |
| Hemisurstylus | 2 | 1 | 2 | 0 | 0 | 1 | 0 | 2 |
| Hemizeudacus | 2 | 1 | 2 | 1 | 0 | 0 | 1 | 2 |
| Melanodacus | 2 | 1 | 2 | 0 | 0 | 1 | 1 | 2 |
| Queenslandacus | 2 | 2 | 1 | 0 | 0 | 0 | 0 | 1 |
| Austrodacus | 2 | 2 | 2 | 0 | 0 | 0 | 0 | 2 |
| Diplodacus | 2 | 2 | 2 | 1 | 0 | 0 | 0 | 2 |
| Heminotodacus | 2 | 2 | 2 | 1 | 1 | 0 | 1 | 2 |
| Hemiparatridacus | 2 | 2 | 2 | 0 | 0 | 1 | 0 | 2 |
| Javadacus | 2 | 2 | 2 | 1 | 0 | 0 | 1 | 2 |
| Niuginidacus | 2 | 2 | 2 | 0 | 0 | 1 | 0 | 2 |
| Papuadacus | 2 | 2 | 2 | 1 | 0 | 0 | 1 | 2 |
| Paradacus | 2 | 2 | 2 | 1 | 0 | 1 | 0 | 2 |
| Paratridacus | 2 | 2 | 2 | 0 | 0 | 1 | 1 | 2 |
| Sinodacus | 2 | 2 | 1 | 1 | 0 | 1 | 0 | 2 |
| Zeugodacus | 2 | 2 | 2 | 1 | 0 | 1 | 1 | 2 |

| Characters | | |
|---|---|---|
| A = abdominal terga | 1 = fused | 2 = not fused |
| B = posterior lobe of male surstylus | 1 = short | 2 = long |
| C = scuteller bristles | 1 = one pair | 2 = two pairs |
| D = pecten of cilia on tergum III | 0 = absent | 1 = present |
| E = humeral bristles | 0 = absent | 1 = present |
| F = supra-alar bristles | 0 = absent | 1 = present |
| G = prescutellar bristles | 0 = absent | 1 = present |
| H = abdominal sternum V of male | 1 = strongly concave<br>2 = weakly concave | |

indicated that *Sinodacus* and *Paradacus* form a monophyletic group. An identical analysis to that described earlier was performed on the data in Table 5. The majority rule consensus tree, derived from 377 parsimonious trees generated by this analysis (length 35, $CI = 0.657$, $RI = 0.75$) is shown in Fig. 3.

On the basis of this analysis the four species formerly assigned to the subgenus *Pacifodacus* form a monophyletic grouping within the paraphyletic taxon *Paradacus*. *Paradacus* + *Pacifodacus* form a monophyletic group within which a number of ill defined monophyletic groupings occur. There must be doubt that *Sinodacus* is monophyletic if *ortholomatus* is retained within the subgenus. Fig. 4 is the area cladogram derived by substituting areas for the taxa in Fig. 3. A number of interesting features are evident in this area cladogram. Firstly, the Outer Melanesian Rift fragments of New Britain and Halmahera (Michaux, 1995) show a sister group relationship to Asian/Indonesian areas. Secondly, that Sabah and Sulawesi are connected and form a sister group to Sulawesi + Java + Sumatra + Thailand. Lastly, the Lesser Sundas and India are sister areas.

## 4. Other cladograms involving Sulawesi

There are a number of cladograms involving Sulawesi in the literature and many of these show features similar to those discussed above. Examples that show Melanesian Rift taxa as a sister group to Asian + Indonesian taxa include *Lepidopelatum* spp. (Sapindaceae: Van Welzen et al., 1992: fig. 5), the *Besida xylinata* group (Notodontidae: Holloway, 1987: fig. 9.5), the *Narosa velutina* group (Holloway, 1987: fig. 9.6), cicadas of the subtribe Cosmopsaltriaria (Duffels, 1990: fig. 1), the danaine genus *Parantica* (in Holloway, 1984: fig. 10), and the *Halovelia malaya* group (Veliidae: Andersen, 1991: fig. 4). This last example is of special interest as African and Madagascan taxa are included with Asian and Indonesian taxa in a

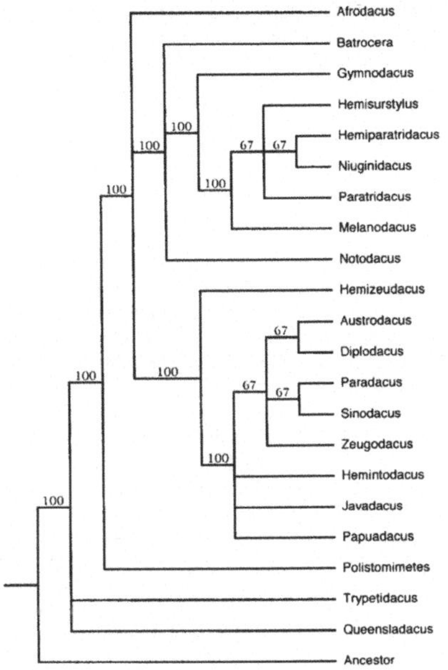


Fig. 2. Majority rule consensus tree for the *Batrocera*-group of subgenera based on the data in Table 4. Numbers refer to the percentage of trees supporting each grouping. See text for details.

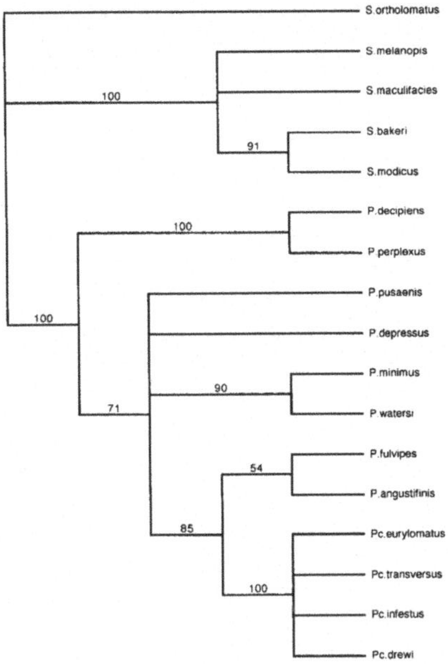


Fig. 3. Majority rule consensus tree for Sulawesian *Dacus* based on the data in Table 5. Numbers refer to the percentage of trees supporting each grouping. Pc = *Pacitodacus*. See text for details.

Table 5
Characters and character states of Sulawesian Dacus (Diptera: Tephritidae)

| | Characters | | | | | | | | | | | | | | | | | |
|---|---|---|---|---|---|---|---|---|---|---|---|---|---|---|---|---|---|---|
| | A | B | C | D | E | F | G | H | I | J | K | L | M | N | O | P | Q | R |
| Sinodcus | | | | | | | | | | | | | | | | | | |
| (Asiadacus) ortholomatus | 0 | 1 | 0 | 0 | 0 | 0 | 1 | 1 | 1 | 0 | 0 | 0 | 0 | 0 | 0 | 0 | 0 | 0 |
| (A.) melanopis | 0 | 1 | 0 | 0 | 0 | 0 | 0 | 2 | 3 | 2 | 1 | 0 | 0 | 0 | 0 | 0 | 0 | 2 |
| (A.) maculifacies | 0 | 1 | 0 | 0 | 0 | 0 | 0 | 2 | ? | 1 | 0 | 0 | ? | ? | 0 | 0 | 0 | ? |
| (A.) bakeri | 0 | 1 | 0 | 0 | 0 | 0 | 0 | 2 | 2 | 0 | 0 | 1 | 0 | 1 | 0 | 0 | 0 | 0 |
| (A.) modicus | 0 | 1 | 0 | 0 | 0 | 0 | 0 | 2 | 3 | 0 | 0 | 1 | 1 | 0 | 0 | 0 | 0 | 2 |
| (Pacifodacus) eurylomatus | 0 | 1 | 0 | 0 | 0 | 1 | 0 | 1 | 2 | 2 | 0 | 0 | 1 | 2 | 0 | 0 | 2 | 0 |
| (P.) transversus | 0 | 1 | 0 | 0 | 0 | 1 | 0 | ? | 2 | 0 | 0 | 0 | 1 | 2 | 0 | 0 | 2 | 0 |
| (P.) infestus | 0 | 1 | 0 | 0 | 0 | 1 | 0 | ? | 2 | 1 | 0 | 1 | ? | ? | 0 | 0 | 2 | 0 |
| (P.) drewi | 0 | 1 | 0 | 0 | 0 | 1 | 0 | 1 | 2 | 0 | 0 | 0 | 1 | 2 | 0 | 0 | 2 | 0 |
| Paradacus | | | | | | | | | | | | | | | | | | |
| (Paradacus) decepiens | 0 | 1 | 0 | 0 | 1 | 1 | 0 | 1 | 3 | 0 | 0 | 0 | 2 | ? | 1 | 1 | 0 | 2 |
| (P.) perplexus | 0 | 1 | 0 | 0 | 1 | 1 | 0 | ? | 3 | ? | 0 | 0 | ? | ? | 1 | 0 | 0 | 2 |
| (P.) pusaensis | 0 | 1 | 0 | 0 | 1 | 1 | 0 | ? | 3 | ? | 0 | 0 | ? | ? | 0 | 0 | 1 | 1 |
| (P.) depressus | 0 | 1 | 0 | 0 | 1 | 1 | 0 | 1 | 3 | ? | 0 | 0 | ? | ? | 0 | 0 | 2 | 2 |
| (P.) minimus | 0 | 1 | 0 | 0 | 1 | 1 | 0 | 2 | 3 | ? | 0 | 0 | ? | 1 | 0 | 0 | 2 | 2 |
| (P.) watersi | 0 | 1 | 0 | 0 | 1 | 1 | 0 | 2 | 3 | ? | 0 | 0 | ? | 2 | 0 | 0 | 2 | 2 |
| (P.) fulvipes | 0 | 1 | 0 | 0 | 1 | 1 | 0 | ? | 2 | ? | 0 | 0 | ? | 0 | 0 | 0 | 1 | 0 |
| (P.) augustifinis | 0 | 1 | 0 | 0 | 1 | 1 | 0 | 1 | 2 | 0 | 0 | 0 | 1 | 2 | 0 | 0 | 1 | 0 |
| Ancestor | ? | 0 | ? | 0 | 0 | 0 | 0 | 0 | 0 | ? | 0 | 0 | ? | ? | 0 | 0 | 0 | 0 |

Characters

| | | | | | |
|---|---|---|---|---|---|
| A | Antennae | 0 = short | 1 = long | | |
| B | Post. lobes m surstylus | 0 = short | 1 = long | | |
| C | 5th sturnum | 0 = gently concave | 1 = strongly concave | | |
| D | Prescutellar bristles | 0 = absent | 1 = present | | |
| E | Scutellar bristles | 0 = 2 | 1 = 4 | | |
| F | Anterior supraalars | 0 = absent | 1 = present | | |
| G | Abdomen | 0 = normal | 1 = narrow & straight sided | | |
| H | Costal cells | 0 = absent | 1 = haline | 2 = yellow | |
| I | Yellow vittae | 0 = absent | 1 = 1 | 2 = 2 | 3 = 3 |
| J | Face | 0 = yellow | 1 = black band | 2 = black | |
| K | m crossvein | 0 = normal | 1 = with brown mark | | |
| L | Brown spot on r3 & r5 | 0 = absent | 1 = present | | |
| M | Mesonotum | 0 = yellow | 1 = red/brown | 2 = black | |
| N | Legs | 0 = black | 1 = rufous | | |
| O | Wing crossband | 0 = absent | 1 = present | | |
| P | S-mark on wing | 0 = absent | 1 = present | | |
| Q | Costal band | 0 = absent | 1 = narrow | 2 = broad | |
| R | Median vitta | 0 = absent | 1 = reduced | 2 = normal | |

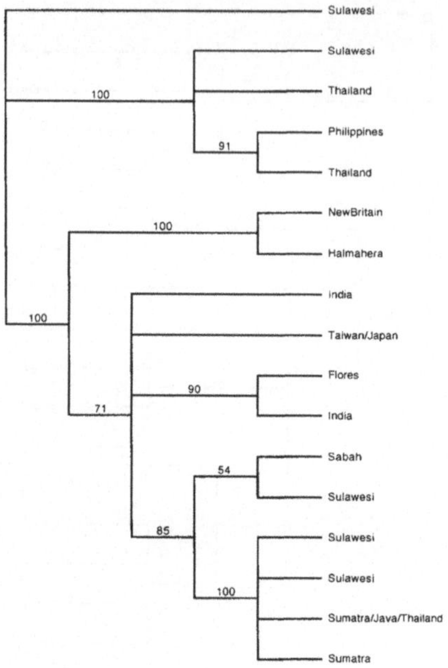


Fig. 4. Area cladogram based on Fig. 3.

monophyletic group. One of Holloway's (1990: fig. 6) general area cladograms, based on lepidopteran data, shows Sulawesi as a sister area to Melanesian Rift fragments + Australia/New Guinea foreland.

A variety of sister group relationships are shown by Sulawesian taxa. As in the case of *Dacus* discussed above, Van Welzen (1990: fig. 38) reported a sister group relationship between taxa from Sulawesi and Sabah. Sulawesi is linked with the Philippines by Holloway (1987) for the notodontid genera *Besida*, *Phalaria*, and *Erconholda* and by De Jong (1983) for the hesperid genus *Matapa*. It is of interest to note that for *Lepidopelatum* (Van Welzen et al., 1992) the Philippines also share a sister group relationship with Sumatra and the Nicobar islands. Vane-Wright (1991) linked Sulawesi to Palawan (north Borneo), Mindanao, Java, and Sumatra, on the basis of a component analysis of Sulawesian butterfly species. Finally, De Jong (1983: fig. 55) reported a sister group relationship for Sulawesi with Sundaland/Asian taxa for the genus *Matapa* (Hesperiidae).

## 5. Geological history

At the start of the Tertiary the margins of the Sunda Shelf and western parts of Southeast Asia were approximately in the positions indicated by the letters *a*, *c–f* in Fig. 1. Fig. 1: *1* is an allochthonous continental block called the Burma plate by Mitchell and Mount Victoria Land by Metcalfe (1991). The eastern margin of Mount Victoria Land (*sensu* Metcalfe) is defined by a major strike-slip fault system—the Sagaing-Namyin or Shan fault (Fig. 1: *a*)—along which there has been 450 km of relative northwards movement of the western side, and by a line of serpentinites and basalts termed the Mandalay Line (Gatinsky and Hutchison, 1986). The western margin (Fig. 1: *b*) is defined in the north by the Indo-Burma ranges and to the south by the Nias-Mentawai subduction system.

Metcalfe (1991) linked Mount Victoria Land with the Sikuleh and Natal terranes, found to the west of the Sumatran fault (Fig. 1: *c*), and western Sulawesi (Fig. 1: *5*). He envisaged Mount Victoria Land as part of an early-Cretaceous rift from Northwest Australia that collided with Southeast Asia in the late Cretaceous. Sengupta et al. (1990) regarded the Andaman terrane (Fig. 1: *2*) as a separate microcontinental block situated on a southern extension of the collision zone between India and the Eurasian continent. These authors did not recognise Mount Victoria Land as a separate terrane and regarded the northern collision zone as one between the Indian craton and the Asian margin. Sengupta et al. (1990) regard Sumatra west of the Sumatran fault and associated islands off the west coast (Fig. 1: *3*) as a southern extension of the Andaman terrane. An early-Tertiary age for the suturing of the Andaman terrane to the Southeast Asian continental margin is suggested by Sengupta et al. (1990) and Rangin et al. (1990) and for the Sikuleh and Natal terranes by Hartono and Tjokrosapoetro (1986).

The margins of the east Java, southeast and east Kalimantan, and Sabah terranes (Fig. 1: *d–f*) are defined by late Cretaceous or early Tertiary mélanges marking old collision boundaries (Hartono and Tjokrosapoetro, 1986; Bénard et al., 1990). Rangin et al. (1990) suggested that these and other areas formed a terrane of Australian affinity (the Sumba block =? Banda allochthon of Audley-Charles et al., 1988) that collided with the Sundaland margin in the early Tertiary. Southwest Sulawesi, which Hutchison (1989) identified as a microcontinental terrane, was connected to Kalimantan prior to the mid-Tertiary opening of the Macassar Straight (Gatinsky and Hutchison, 1986; McCabe and Cole, 1989) and was presumably part of Rangin et al.'s (1990) Sumba Block. Burrett et al. (1991) tentatively link parts of Sumatra and southern Kalimantan as a terrane of Palaeocene age (60 m.y. B.P.).

## 6. Conclusions

The distributional pattern evident in bird and moth data, which is also corroborated by what limited phylogenetic data are available, supports Rangin et al.'s (1990) geological hypothesis of the Sumba terrane (sensu west Mindanao + Sabah + east Kalimantan + south west Sulawesi + parts of the Lesser Sundas). It further suggests that the origin of this terrane was linked with that of Mount Victoria Land + the Andamans terrane + Sumatra west of the Sumatran fault.

The high level of endemism and often obscure phylogenetic relationships shown by elements of the Sulawesian biota suggests, with the exception of its connection to Kalimantan until the mid-Miocene, that southwest Sulawesi was isolated from other fragments. It is, therefore, highly unlikely that the areas discussed above were geographically contiguous once they had separated from the same general source area of Gondwana. It is more likely that they existed as an island group similar perhaps to today's Melanesian Rift fragments such as the Solomon or Vanuatu archipelagos. On the basis of the relationships shown by the modern biota the most likely source area was closer to India than to Australia. This is in contrast to the reconstructions of Audley-Charles et al. (1988) and Metcalfe (1991) who place these areas adjacent to Australia. The sister group relationship between taxa now living on these diverse fragments and Australasian taxa, a common feature of groups for which phylogenetic data are available, suggests that the original position of the Sumba-Andaman-Mount Victoria Land rift was adjacent to areas from which the Outer Banda Arc and the Melanesian Rifts were derived (Michaux, submitted).

## Acknowledgements

My thanks to Guy Musser and Charles Hutchison for their comments and suggestions on an earlier draft of this paper.

## References

Andersen, N.M., 1991. Cladistic biogeography of marine water-striders (Insecta, Hemiptera) in the Indo-Pacific. Aust. Syst. Bot., 4: 151–163.

Audley-Charles, M.G., 1983. Reconstruction of eastern Gondwanaland. Nature, 306: 48–50.

Audley-Charles, M.G., Ballantyne, P.D. and Hall, R., 1988. Mesozoic–Cenozoic rift-drift sequence of Asian fragments from Gondwanaland. Tectonophysics, 155: 317–330.

Beehler, B.M., Pratt, T.K. and Zimmerman, D.A., 1986. Birds of New Guinea. Princeton Univ. Press.

Belov, A.A., Gatinsky, Y.G. and Mossakovsky, A.A., 1986. A precis on pre-Alpine tectonic history of Tethyan paleooceans. Tectonophysics, 127: 197–211.

Bénard, F., Muller, C., Letouzey, J., Rangin, C. and Tahir, S., 1990. Evidence of multiphase deformation in the Rajang-Crocker Range (northern Borneo) from Landsat imagery interpretation: Geodynamic implications. Tectonophysics, 183: 321–339.

Burrett, C., Duhig, N., Berry, R. and Varne, R., 1991. Asian and south-western Pacific continental terranes derived from Gondwana, and their biogeographic significance. Aust. Syst. Bot., 4: 13–24.

Croizat, L., 1958. Panbiogeography. Vols. 1, 2a, 2b (published by author, Caracas).

Delacour, J., 1947. Birds of Malaysia. MacMillan, New York.

De Jong, R., 1983. Revision of the Oriental genus *Matapa* Moore (Lepidoptera, Hesperiidae) with discussion of its phylogeny and geographic history. Zool. Meded., 57: 243–270.

Du Pont, J.E., 1971. Philippine Birds. Delaware Mus. Nat. Hist.

Drew, R.A.I., 1989. The tropical fruit flies (Diptera: Tephritidae: Daninae) of the Australasian and Oceanian regions. Mem. Queensld. Mus., 26: 1–521.

Duffels, J.P., 1990. Biogeography of Sulawesi cicadas (Homoptera: Cicadoidea). In: W.J. Knight and J.D. Holloway (Editors), Insects and the Rain Forests of South East Asia (Wallacea). R. Entomol. Soc., London, pp. 63–72.

Gatinsky, Y.G. and Hutchison, C.S., 1986. Cathaysia, Gondwanaland, and the Paleotethys in the evolution of continental Southeast Asia. Geol. Soc. Malaysia Bull., 20: 179–199.

Hamilton, W.S., 1988. Plate tectonics and island arcs. Geol. Soc. Am. Bull., 100: 1503–1527.

Harbury, N.A., Jones, M.E., Audley-Charles, M.G., Metcalfe, I., and Mohamed, K.R., 1990. Structural evolution of Mesozoic peninsular Malaysia. J. Geol. Soc. London,. 147: 11–26.

Hall, R., 1987. Plate boundary evolution in the Halmahera region, Indonesia. Tectonophysics, 144: 337–352.

Hardy, D.E., 1982. The Dacini of Sulawesi (Diptera: Tephritidae). Treubia, 28: 173–241.

Hardy, D.E., 1983. The fruitflies of the genus *Dacus* Fabricus of Java, Sumatra and Lombok, Indonesia (Diptera: Tephritidae). Treubia, 29: 1–45.

Hartono, H.M.S. and Tjokrosapoetro, S., 1986. Geological evolution of the Indonesian Archipelago. In: GEOSEA Proc., 2. Geol. Soc. Malaysia Bull., 20: 97–136.

Hill, J.E., 1983. Bats (Mammalia: Chiroptera) from Indo-Australia. Bull. Br. Mus. Nat. Hist., 45: 103–208.

Holloway, J.D., 1976. The moths of Borneo with special reference to Mount Kinabalu. Sun U, Kuala Lumpur, 264 pp.

Holloway, J.D., 1984. Lepidoptera and the Melanesian arcs. Bishop Mus. Spec. Publ., 72: 129–169.

Holloway, J.D., 1987. Lepidoptera patterns involving Sulawesi: what do they indicate of past geography? In: T.C. Whitmore (Editor), Biogeographical Evolution of the Malay Archipelago. Clarendon Press, Oxford, pp. 103–118.

Holloway, J.D., 1990. Patterns of moth speciation in the Indo-Australian archipelago. In: E.C. Dudley (Editor), The Unity of Evolutionary Biology. Dioscorides Press, Portland, OR, pp. 340–72.

Holmes, D. and Nash, S., 1990. Birds of Sumatra and Kalimantan. Oxford Univ. Press, 83 pp.

Hutchison, C.S., 1989. Displaced terranes of the southwest Pacific. In: Z. Ben-Avrahim (Editor), The Evolution of the Pacific Ocean Margins (Oxford Monogr. Geol. Geophys., 8). Oxford Univ. Press, New York, pp. 161–175.

Hutchison, C.S., 1993. Gondwana and Cathaysian blocks, Palaeotethys sutures and Cenozoic tectonics in South-east Asia. Geol. Rundsch., 82: 388–405.

Laurie, E.M.O. and Hill, J.E., 1954. Land Mammals of New Guinea, Celebes, and Adjacent Islands. Br. Mus., London, 175 pp.

MacKinnon, J. and Phillipps, K., 1993. A Field Guide to the Birds of Borneo, Sumatra, Java, and Bali. Oxford Univ. Press, 491 pp.

McCabe, R. and Cole, J., 1989. Speculations on the late Mesozoic and Cenozoic evolution of the Southeast Asian margin. In: Z. Ben-Avrahim (Editor), The Evolution of the Pacific Ocean Margins (Oxford Monogr. Geol. Geophys., 8). Oxford Univ. Press, New York, pp. 143–160.

Metcalfe, I., 1991. Late Palaeozoic and Mesozoic palaeogeography of Southeast Asia. Palaeogeogr. Palaeoclimatol. Palaeoecol., 87: 211–221.

Michaux, B., 1991. Distributional patterns and tectonic development in Indonesia: Wallace reinterpreted. Aust. Syst. Bot., 4: 25–36.

Michaux, B., 1994. Land movements and animal distributions in east Malesia (eastern Indonesia, Papua New Guinea, and Melan esia). Palaeogeogr. Palaeoclimatol. Palaeoecol., 112: 323–343.

Mitchell, A.H., 1981. Phanerozoic plate bondaries in mainland S.E. Asia, the Himalayas and Tibet. J. Geol. Soc. London, 138: 109–122.

Musser, G.G., 1977. *Epimys benguetensis*, a composite, and one zoogeographic view of rat and mouse faunas in the Philippines and Celebes. Am. Mus. Novit., 2624: 1–15.

Musser, G.G., 1981. Results of the Archbold expeditions. No. 105. Notes on the systematics of Indo-Malay murid rodents, and descriptions of new genera and species from Ceylon, Sulawesi, and the Philippines. Bull. Am. Mus. Nat. Hist., 168: 229–334.

Musser, G.G., 1986. Sundaic *Rattus*: Definitions of *Rattus baluensis* and *Rattus korinichi*. Am. Mus. Novit., 2862: 1–24.

Musser, G.G., 1987. The mammals of Sulawesi. In: T.C. Whitmore (Editor), Biogeographical Evolution of the Malay Archipelago. Clarendon Press, Oxford, pp. 73–94.

Musser, G.G., 1990. Sulawesi rodents: species traits and chromosomes of *Haeromys minahassae* and *Echiothrix leucura* (Muridae: Murinae). Am. Mus. Novit., 2989: 1–18.

Musser, G.G., 1991. Sulawesi rodents: descriptions of new species of *Bunomys* and *Maxomys* (Muridae, Murinae). Am. Mus. Novit., 3001: 1–41.

Musser, G.G. and Califia, D., 1982. Results of the Archbold expeditions. No. 106. Identities of rats from Palau Maratua and other islands off east Borneo. Am. Mus. Novit., 2726: 1–30.

Musser, G.G. and Heaney, L.R., 1985. Philippine *Rattus*: a new species from the Sulu Archipelago. Am. Mus. Novit., 2818: 1–32.

Musser, G.G. and Heaney, L.R., 1992. Philippines rodents: Definitions of *Tarsomys* and *Limnomys* plus a preliminary assessment of phylogenetic patterns among native Philippine murines (Murinae, Muridae). Bull. Am. Mus. Nat. Hist., 211: 1–138.

Musser, G.G. and Dagosto, M., 1987. The identity of *Tarsius pumilus*, a pygmy species endemic to the montane mossy forests of central Sulawesi. Am. Mus. Novit., 2867: 1–53.

Musser, G.G. and Holden, M.E., 1991. Sulawesi rodents (Muridae: Murinae): morphological and geographic boundaries of species in the *Rattus hoffmanni* group and a new species from Palau Peleng. Bull. Am. Mus. Nat. Hist., 206: 322–413.

Pigram, C.J. and Supanjono, J.B., 1985. Origin of the Sula platform, eastern Indonesia. Geology, 13: 246–248.

Rangin, C. and Tethys Pacific Working Group, 1990. The quest for Tethys in the western Pacific. 8 paleogeodynamic maps for Cenozoic time. Bull. Soc. géol. Fr., 6: 907–913.

Sengupta, S., Ray, K.K., Acharyya, S.K. and De Smeth, J.B., 1990. Nature of ophiolite occurrences along the eastern margin of the Indian plate and their tectonic significance. Geology, 18: 439–442.

Strange, M. and Jeyarajasingham, A., 1993. Birds—a Photographic Guide to the Birds of Peninsular Malaysia and Singapore. Sun Tree, Singapore, 258 pp.

Tikader, B.K., 1984. Birds of the Andaman and Nicobar Islands. Zool. Surv. India, Calcutta, 167 pp.

Vane-Wright, R.I., 1991. Transcending the Wallace line: do the western edges of the Australian region and the Australian plate coincide? Aust. Syst. Bot., 4: 183–197.

Van Welzen, P.C., 1990. *Guioa* Cav. (Sapidaceae): Taxonomy, Phylogeny, and Historical Biogeography. Rijksherbarium, Leiden.

Van Welzen, P.C., Piskaut, P. and Windadri, F.I., 1992. *Lepidopelatum* Blume (Sapindaceae): Taxonomy, phylogeny, and historical biogeography. Blumea, 36: 439–65.

Wallace, A.R., 1880. The Malay Archipelago. MacMillan, London.

Walters, M., 1980. The Complete Birds of the World. Reed, Sydney.

# Attribution

**Michaux, B. 1996. The origin of southwest Sulawesi and other Indonesian terranes: A biological view. *Palaeogeography, Palaeoclimate and Palaeogeography,* 122:167–183.**

First published in 1996 in *Palaeogeography, Palaeoclimate and Palaeogeography,* 122:167–183. Reproduced with permission from Elsevier Publishing.

# *chapter five*

# *The Malay Archipelago*

This paper represented a further exploration of the ideas I first presented in 1990 at the Ninth Annual Willi Hennig meeting in Canberra, Australia. The main focus of this study was to describe the relationship of the Sulawesian fauna to those of adjacent and more distant areas. Comprehensive distributional data were compiled for Sulawesian birds and moths, and these were supplemented with an areagram based on a cladistic analysis of *Dacus* fruit flies. Sulawesi has long fascinated biologists who have studied it because of its strange and enigmatic biota. Wallace (Plate 5.1) was the first to give an in-depth description of its distinct animals, and published a popular account in *The Malay Archipelago.* Because Sulawesi lies at the heart of the archipelago and is surrounded by large and small islands, Wallace assumed that the Sulawesian fauna would be composed of a subset of species from the surrounding areas, perhaps with some development of endemism, because there were no obvious barriers to dispersal:

> As so often happens in nature, however, the fact turns out to be just the reverse of what we should have expected; and an examination of its animal productions, shows Celebes to be at once the poorest in number of its species and the most isolated in the character of its productions, of all the great islands in the Archipelago. (Wallace 1869)

Sulawesi (Celebes) appeared enigmatic to Wallace because its highly endemic and impoverished fauna was out of place with respect to its location. It was as though invisible barriers were stopping colonisation of the island, barriers that had been biologically isolating Sulawesi for a long time. The fauna present was characterised by high levels of endemism and consisted of a strange mixture of Asian and Australasian groups. Wallace noted that the biological relationships shown by many Sulawesian endemics to Asian relatives appeared unclear and rather distant. For example, endemic mammals such as the babirusa (Plate 5.2), the two dwarf anoas and the three squirrel genera, while clearly of Asian origin, had no obvious relatives anywhere on the Sunda Shelf. He was also surprised by the presence of Australasian groups such as the marsupial phalangers and

***Plate 5.1*** Portrait of Alfred Russel Wallace after his return to England from Indonesia.

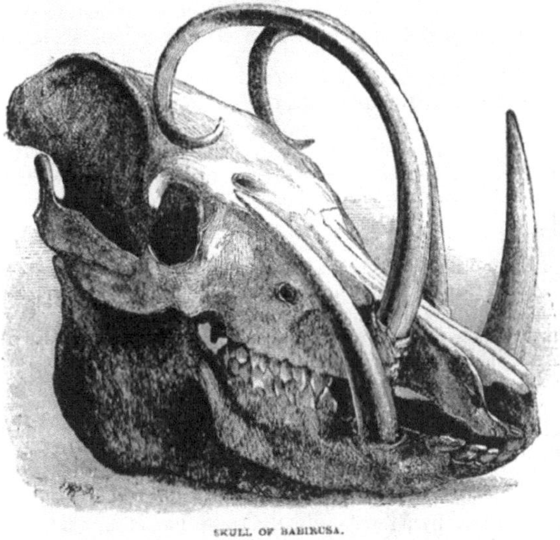


***Plate 5.2*** Skull of a Babirusa. Woodcut based on illustration by Robinson, originally published in *The Malay Archipelago.*

cockatoos so far west of the Sahul Shelf. As a consequence of these observations, Wallace regarded Sulawesi as the key to understanding the biogeography of the Malay Peninsula. Poor Wallace tried very hard to give a coherent explanation of the origin of the Sulawesian fauna. He inferred that the biota was a product of past rather than present geography, but lacked the necessary conceptual framework to explain how such a situation had arisen. It wasn't until Wallace was an old man nearing the end of his life that Wegener developed the idea of continental drift, but I'm sure he would have embraced these ideas as the key to understanding the Sulawesian biota if they had only come 50 years earlier.

## *A working-class hero*

> Hero: a person admired for achievements and noble qualities (Merriam-Webster dictionary)

Wallace's personal qualities are apparent from the account of his travels given in *The Malay Archipelago*. His energy and enthusiasm, his steadfastness in the face of hardship and difficulty, his openness to the experiences of life and above all his honesty and open mindedness shine through his writing. And the more I found out about him, the more there was to admire. In my view, Wallace was certainly the most interesting and possibly the most important of the Victorian biological theorists. The 2013 celebration of his life and work on the centennial anniversary of his death did much to rehabilitate his scientific standing and to bring him to the attention of a wider audience, not as Darwin's moon (Williams-Ellis 1966) but as an outstanding thinker in his own right.

Wallace was a man of limited formal education and no social standing in a class-obsessed society who managed to rise to a pre-eminent intellectual position within the Victorian scientific community. I regard Wallace as working class but accept van Wyhe's argument (van Wyhe 2015) that the question of class in Victorian society had as much to do with breeding as it had with wealth. For example, nouveau riche Victorian industrialists had to marry into established families before gaining entry into the upper echelons of British society. Wallace, as the son of a gentleman, was also a gentleman, albeit one without independent means. In the complex taxonomy of class structure in Victorian Britain, Wallace was not working class in the strict sense but was somebody who had to work for his living. Shermer (2002), who also regarded Wallace as working class, argued that this predisposed him to develop 'heretical' theories. He suggested that Wallace's restless intellect, unencumbered as it was by received wisdom and nurtured by exposure to the educational programmes and opportunities afforded to working-class men at Mechanics' Institutes, became receptive to radical ideas. Wallace's working background as a surveyor was also important in his development as a self-reliant, resourceful and practical man.

In the summer of 1837, at the age of fourteen, Wallace was apprenticed to his eldest brother William as a trainee surveyor. For most of the next six years, the brothers travelled extensively through rural Britain surveying canal routes (Raby 2001). Wallace enjoyed the life and became increasingly interested in the natural history of the areas he worked in. He bought himself his first identification guide, Lindley's *Elements of Botany*, which he annotated extensively from Loudon's *Encyclopaedia of Plants* to make it more useful for identifying what he was collecting, and also started reading widely on geology, including the influential *Principles of Geology* by Charles Lyell. Wallace regarded the years from 1840 to 1843 as one of the turning points in his life, a period during which he had set the course for his future. He had become fascinated with the natural world, had approached his self-education in a systematic way and had acquired a broad, practical skill base that complemented his growing theoretical outlook. What was needed now was the spark to ignite him into action.

That spark was to be provided by Henry Walter Bates. Bates was an avid collector of insects, particularly beetles, and it was he who introduced Wallace to entomology in general and beetle collecting in particular. Wallace had found a kindred spirit, somebody with similar interests with whom he could discuss ideas. It is probable that their plan to go to the Amazon was first hatched in 1846 when Bates visited Wallace in Wales. This 'rash adventure', as Williams-Ellis (1966) called it, was to be financed by the collection and sale of specimens. Wallace spent four and a half years in Amazonia; Bates would remain eleven years and go on to achieve renown as a tropical entomologist and originator of the hypothesis of Batesian mimicry. Wallace's greatest achievement during his time in the Amazon was to explore the upper reaches of the Rio Negro and the headwaters of the Orinoco, and in doing so fulfilling his ambition to follow in the footsteps of his great hero Humboldt.

Wallace, by necessity, travelled lightly and lived as the locals did. When he explored the Rio Negro, Wallace used existing trading and communication networks when they were available, or hired local guides and travelled by dugout when they weren't. A major food source for travellers was fish and one of the camp chores at the end of the day would be to go fishing. This could be done either by netting or using *timbo*, a fish poison containing the alkaloid rotenone. The collection would be inspected for new specimens before the rest went into the cooking pot. Wallace's fish drawings and Rio Negro journals were among the few documents to survive the shipwreck he suffered on his return journey to England (Raby 2001) and have recently been published (Toledo-Piza 2002; reviewed by Harold (2005)). While Wallace's scientific output from these four years was not extensive, his experiences in the Amazon did lead to personal and professional growth, transforming him from a rather gauche young man and naïve enthusiast into a confident and skilled field biologist.

It was with this background that Wallace arrived in Indonesia. He travelled extensively within the archipelago and as far east as New Guinea, all the time collecting specimens in order to finance his study of the 'species problem'. Wallace estimated that his own personal collection consisted of 3000 bird skins, 20,000 pinned beetles and butterflies and sundry other mammalian and molluscan specimens (Wallace 1869). According to Shermer (2002), his total collection from this period amounted to 125,660 specimens. Wallace was to publish 43 scientific papers during his time in Indonesia, on subjects as varied as descriptive lists, avian higher-level systematics and biogeography and evolution, and also kept up a tremendous correspondence with peers, friends and his agent, J.S. Stevens. This output included the famous Ternate paper of 1858 (Smith 2018) describing evolution by means of natural selection, which was to cause Darwin much anxiety and precipitated the writing of a hastily cobbled together joint paper read at the Linnean Society on 1 July 1858.

*The Malay Archipelago,* published seven years after his return to England, was Wallace's account of the eight years he spent exploring and collecting in the region and was his most commercially successful book. The *Malay Archipelago* is still in print today because it remains a very good read – I can recommend it to anybody interested in biogeography, natural history, Indonesia or the history of science. He was able to bring into Victorian sitting rooms vivid accounts of the life, peoples and landscapes of a region that would have seemed to his readers to be straight out of a fable. Wallace's clever interweaving of story lines, the wealth of detail so clearly presented and his obvious love for what he was writing about ensured a continuing public demand. Detailed and often amusing descriptions of incidences from his everyday life, such as staying in Dyak long houses among the famed headhunters of Borneo, or evicting giant pythons from his roof, made the extraordinary seem ordinary (Plate 5.3). What is often overlooked, however, is that the *Malay Archipelago* is also a first-hand description of a biodiversity hotspot by an experienced tropical biologist before any large-scale habitat destruction had taken place (Severin 1998). As such, it is a historically important document concerning one of the most globally important centres of biodiversity and endemicity.

Wallace's return to England and a settled life couldn't have been altogether easy for him. The English countryside must have seemed very tame in comparison to the tropics he was used to, the weather grey and dreary, the sedentary and ordered lifestyle restrictive, and the rigid social system difficult for somebody who was used to taking people as he found them. While he settled down to married life and pursued his scientific studies and career as a writer, he never quite seemed at ease. There are a number of aspects of his post-travelling life that point to a restlessness of spirit – his frequent moves of house, his financial impulsiveness and the numerous social issues he involved himself in (Williams-Ellis 1966). His radicalism didn't exactly endear him to the authorities, which probably explains his failure to achieve an appointment to any official position, and it was only through Darwin's intercession that he was eventually awarded a pension

***Plate 5.3*** Ejecting an Intruder. Woodcut based on an illustration by Baines, originally published in *The Malay Archipelago*. Wallace can be seen in the background holding a gun.

that saved him from financial ruin. His outspokenness on such issues as compulsory vaccination, land reform, the eugenics movement, the rights of ordinary men and women and spiritualism probably explains why he was quietly forgotten after his death. If it hadn't been for *Wallace's Line* – a term coined by Huxley – his name may have been completely forgotten.

## *Sulawesi revisited*

### *Geological update*

Sulawesi is a geological composite of four different components – the southwest peninsula and central Sulawesi, the southeast peninsula and Buton, the northeast peninsular and Sula/Banggai, and the northern peninsula. The relationship between southwest Sulawesi and the Sunda Shelf has been clarified following the work of Smyth et al. (2007), van Leeuwen et al. (2007) and Hall et al. (2009). Smyth et al. (2008) established that eastern Java is underlain by continental crust with an Australian zircon age profile. This continental crust is also thought to underlie the East Java Sea and southern Makassar Strait (Hall et al. 2009a) that was collectively termed Argoland by Hall and his colleagues. Similar Australian crust is thought to be at depth in western Sulawesi (van Lueewen et al.

2007), indicating that Argoland also included the southwest peninsula of Sulawesi. Whether Sumba was also part of this terrane is uncertain, but Rangin et al.'s (1990) Sumba terrane can be seen as broadly equivalent to Argoland. Argoland is thought to have docked with southwest Borneo sometime between 92 and 80 Ma, causing uplift of the Sunda Shelf, and remained attached until the opening of the Makassar Strait at 45 Ma. Although Argoland was derived from the Australian region, it is unlikely to have been the source of the Australasian element of Sulawesi's biota because Australasian groups present in Sulawesi, such as marsupials or cockatoos, are not found in Java or Borneo. Argoland may also have been derived from a deep-shelf environment (Argo Abyssal Plain) and may not have carried a terrestrial biota, but there is uncertainty about its source area, and it may have originally been adjacent to the New Guinean or northwest Australian continental shelf (Zahirovic et al. 2014). However, the Argoland terrane is very likely to have been the source for Sulawesi's palaeoendemic mammals such as the babirusa, anoas and endemic squirrel genera following its detachment from Sundaland at 45 Ma.

The southeast peninsula/Buton and northeast peninsula/Sula-Banggai are viewed as Australasian fragments (Satyana and Purwaningsih 2011; Watkinson et al. 2011) but their means of emplacement is debated. While there is a general agreement that these terranes originated close to the Bird's Head (Norvick 1979; Hill and Hall 2003; Decker et al. 2017), there are two distinct views as to their mode of emplacement. The earlier view was that the Sula-Banggai terrane had been translated westwards along the Sorong Fault (e.g. Norvick 1979), but a more recent analysis of faulting and interpretation of seismic sections to the north of Sula-Banggai led Watkinson et al. (2011) and Watkinson and Hall (2017) to question the role of the Sorong Fault in its emplacement. These authors argued that there was little evidence for substantial movement along the Sorong Fault and, despite the region being a collision zone, that it was extension that dominated its Neogene tectonics. In their alternative model, these Sulawesian terranes were originally part of a single structure – the Sula Spur – that subsequently became fragmented with the opening of a series of deep basins in the Banda Sea. The modern island of Sulawesi was assembled between 20 and 10 Ma (Nugraha and Hall 2018) following the collision of the southeast peninsula/Buton and northeast peninsula/Sula-Banggai terranes with the central/southwest Sulawesi terrane and their subsequent amalgamation with the Northern Arm Arc (Lohman et al. 2011). The northern arm of Sulawesi was an island arc linked to Eocene arcs found in the eastern Philippines, Halmahera and northern New Guinea.

## *PAE analysis of bird distributions*

There was no analysis of the bird and moth distributional data in the original paper and pattern recognition remained at a descriptive level

(Michaux 1996). The biological evidence for a Sumba terrane (north Borneo + west Mindanao + southwest Sulawesi + Lesser Sundas) and a possible connection with a Myanmar + Andaman Islands + western Sumatra block is not convincing, although the *Dacus* phylogeny does support a north Borneo and Sulawesi sister-area relationship. A PAE analysis was carried out as previously described on the original bird data shown in Michaux (1996: Table 2). The data for southeast Borneo, Europe and the Mentawai Islands were removed because there were too few species present in these areas. The analysis of the remaining 16 areas using 178 species' distributions was analysed by TNT v1.5 (Goloboff et al. 2008) using an exact search algorithm option (*ienum*). Africa was used as the outgroup area because it is not part of the Indo-Australian region. The single most parsimonious tree (length 410 steps) found is shown in Figure 5.1. A similar analysis of the moth data in Michaux (1996: Table 3) was

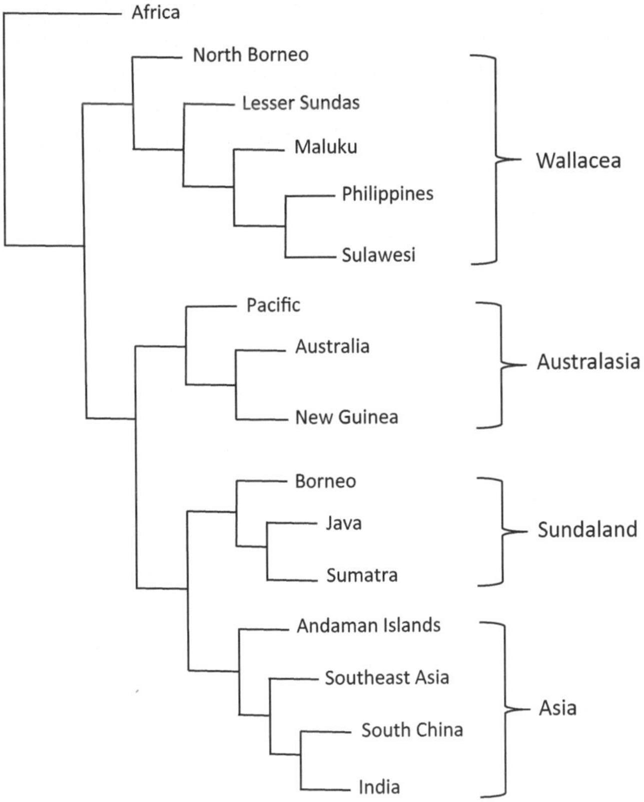


***Figure 5.1*** PAE analysis of avian distributions from Michaux (1996). Single most parsimonious tree (length = 410 steps) found by an exact search strategy (*ienum*) in TNT v 1.5 (Goloboff et al. 2008).

uninformative, a strict consensus tree identifying only three informative nodes – Australia/New Guinea +Maluku, Sulawesi + Borneo and India + north Borneo.

While I regard PAE as a preliminary form of analysis because of its lack of phylogenetic input and reliance on similarity only, there are some interesting features of the areagram shown in Figure 5.1. The most important of which is that Wallacea is a natural area if the Philippines are included as originally suggested by Dickerson et al. (1923). The presence of north Borneo (Sabah) in this clade rather than with the rest of Borneo, Java and Sumatra emphasises the importance of Sabah as an endemic area, and suggests that it has had a different biological and geological history to the rest of Borneo (Balaguru and Hall 2009). The second clade contains three natural groupings – Australasia, Sunda Shelf and Asia – with Sundaland and Asia as sister areas, which in turn are sister area to Australasia. Figure 5.1 shows that the Wallacean avifauna is distinct from those of Sundaland, Asia and Australasia, which are more similar to each other than any of them is to the Wallacean avifauna. The result of the PAE analysis of the avifauna implies that rather than being a transition zone between Asian/Sundaic and Australasian avifaunas, Wallacea might be better regarded as a separate natural biogeological area, at least as far as the avifauna is concerned.

## *The best of all possible worlds*

If you were going to design a biogeological study of Wallacea, and not just use distributional data or phylogenies derived from published studies but one designed specifically for the purpose, what might it look like? Figure 5.2 outlines one possible schema for constructing such a biogeological research programme.

### *Gathering the data*

An initial literature survey should establish what areas of endemism exist within the region of interest, what areas are likely to be geologically composite, and what groups warrant further systematic research involving targeted collection in the field or the use of museum specimens. Some phylogenies or partial phylogenies found during this initial literature search could also be included in the final data set. Widespread taxa are of no real use in biogeological studies because of the negative effect increasing numbers of terminal areas have on the resolving power of the analysis, with the resulting 'noise' generated by MASTs obscuring any cladistics signal (see the following section). More geographically restricted subordinate groups within a widespread taxon may be of use; for example, a family may be widespread within the Indo-Australian region, but constituent

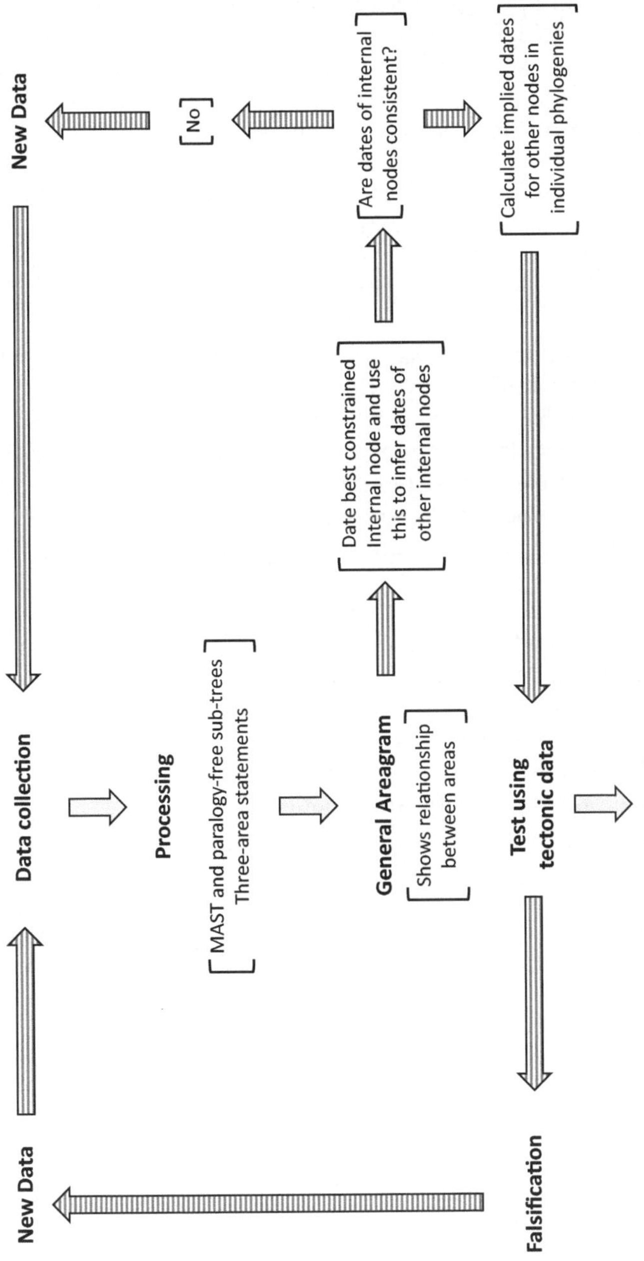


*Figure 5.2* How to do biogeology: A methodological schema. See text for explanation.

genera may have more restricted distributions. However, island endemics and restricted range species for which recent revisions and cladistics analyses or molecular phylogenies are available are the most useful in elucidating the relationship between areas. If an area can be demonstrated to be geologically composite, that is, shows different sister-area relationships in different areagrams, then separate analyses should be carried out to establish its relationship in each area clade. While I clearly have Wallacea in mind when outlining this initial step in a biogeological analysis, I believe it would be applicable to elucidating area relationships in other collision zones such as the Mediterranean or Caribbean regions.

## *Processing*

Individual trees, for the reasons discussed at the end of the previous chapter, need to be combined to produce a general areagram. The method suggested in Figure 5.2 involves producing MAST and paralogy-free subtrees from the molecular and morphological phylogenies. MASTs and redundant areas – areas that occur on multiple branches causing nodes to be paralogous – are problematic because they are a major source of incongruity between individual trees, increasing noise and obscuring any cladistic signal. All three-area statements (all implied relationships between any three areas in the data set) are then extracted from the processed trees and assembled to produce an intersection tree (general areagram). The programme LisBeth (Bagils et al. 2012) has been designed to perform these operations. The method used to combine individual phylogenies to form a consensus areagram is a matter for each researcher, but what is important is that individual trees are combined and a consensus areagram produced that summarises *all* the available information.

## *General areagram*

The general areagram is the most reliable description of area relationships. Once a general areagram has been found, it is possible to ask how many of the nodes correlate with tectonic events, for example, with the opening of the Makassar Strait or the collision of the Sula Spur with Sulawesi. One event may have a more constrained age than others and this can then be used as a calibration point enabling, for example, a choice between alternative dates for nodes that are not so well constrained. After all internal nodes have been dated, it is possible to assess the consistency of the areagram because basal nodes must be older than (or equal to) the ages of more derived nodes. Inconsistency points to the need for further phylogenies to be added to the data set or to the re-evaluation of supposed tectonic ages of inconsistent nodes. An example of this type of analysis is discussed in Chapter 9. It should now be possible to date appropriate internal nodes of *any* phylogeny using nodal ages established from the general areagram. The implied ages of the other nodes of the molecular phylogeny can then be tested using independent geological evidence.

It should be possible to compare dates for a particular geological event derived from different phylogenies and to ask questions such as do phylogenies converge on an age for particular tectonic events, do any individual phylogenies produce inconsistent dates, and what are the errors associated with each nodal date? In this way, it might be possible to reject individual phylogenies (the biological data are wrong) or proposed geological dates (the tectonic models are wrong) or the hypothesis that tectonic change and speciation are correlated for those groups of taxa for which phylogenies are available.

## *Interpretation*

Interpretation should ideally follow on from this cycle of analysis and testing, because we could then have confidence that such interpretations were more than just speculation, and the results could be used with some confidence as inputs into other research projects.

*Biological Journal of the Linnean Society*, 2005, **86**, 95–115. With 3 figures

# East meets west: biogeology of the Campbell Plateau

B. MICHAUX[1]* and R. A. B. LESCHEN[2]

[1]*Private Bag, Kaukapakapa, New Zealand*
[2]*New Zealand Arthropod Collection, Private Bag 92 170, Auckland, New Zealand*

*Received 17 August 2003; accepted for publication 7 October 2004*

The New Zealand Subantarctic Islands, emergent remnants of the Campbell Plateau, were given World Heritage status in 1998 in recognition of their importance to global biodiversity. We describe the flora and fauna of these islands and discuss the results of recent phylogenetic analyses. Part of the New Zealand Subantarctic biota appears to be relictual and to be derived from west Gondwana. The relictual element is characterized by genera endemic to the Campbell Plateau that show relationships with taxa of the southern South Island, New Zealand, southern South America, and the north Pacific. In contrast, a younger, east Gondwanan element is composed of species that are either taxonomically identical to widespread mainland species, or endemic species with close New Zealand relatives. Area cladograms support the inclusion of the southern South Island, New Zealand and Macquarie Island (although this is separate geologically) as parts of the Campbell Plateau, but suggest the Chatham Rise and Torlesse terranes of the eastern South Island, New Zealand were originally parts of east Gondwana. East and west Antarctica acted as independent plates during the breakup of Gondwana, and were separated by oceanic crust until a compressive phase sutured them along the trace of the *trans*-Antarctic mountains during the early Tertiary. The Campbell Plateau microcontinent was connected to west Antarctica until its separation at 80 Mya, contemporaneous with the separation of the southern portion of the Melanesian rift from east Gondwana. Presently the Campbell Plateau is joined to the Melanesian Rift along the Alpine Fault. Cenozoic plate tectonic reconstructions place the Campbell Plateau adjacent to the Melanesian Rift throughout the rift–drift phase, relative motion being confined to strike–slip movement over the last 20 Myr. Our synthesis of phylogenetic and plate tectonic evidence suggests that the Alpine Fault is the most recent development of a much older extensional rift/basin boundary originally separating west and east Gondwana. © 2005 The Linnean Society of London, *Biological Journal of the Linnean Society*, 2005, **86**, 95–115.

ADDITIONAL KEYWORDS: Alpine Fault – arthropods – Chatham Islands – east Gondwana – New Zealand Subantarctic Islands – west Gondwana.

## INTRODUCTION

The New Zealand Subantarctic Islands are emergent parts of the Campbell Plateau microcontinent, a roughly triangular fragment of cratonic rock derived from Gondwana during its fragmentation at 80 Mya. The rocks of the Campbell Plateau are predominantly granites of Palaeozoic or older age, with much younger basaltic shield volcanoes erupted at various centres, including Auckland and Campbell islands, during the late Tertiary. The New Zealand Subantarctic Islands were accorded World Heritage status in 1998. Five island groups – the Snares, Bounty Islands, Antipodes Islands, Auckland Islands and Campbell Island – are scattered across the Southern Ocean (Fig. 1 and Table 1), totalling only 76 485 hectares in combined area (O'Connor, 1999). Despite their restricted geographical extent and history of glaciation (McGlone *et al.*, 1997), these islands continue to support a remarkable endemic flora and fauna. The islands are major breeding centres for both endemic and circumpolar marine mammals and birds, including iconic species such as royal albatross *Diomedea epomophora* Lesson, crested penguins *Eudyptes* spp., and Hooker's sea lion *Phocarctos hookeri* (Gray). While the main islands still continue to experience the effects of introduced species, outlying islands preserve examples of pristine habitats. Biological uniqueness, importance as oceanic breeding grounds, and isolation were the

*Corresponding author. E-mail: michaux@xtra.co.nz

**Table 1.** Location and size of subantarctic islands on (or nearby*) the Campbell Plateau. Basement rocks of the Campbell Plateau are Palaeozoic and the dates for extrusive volcanics are based on Adams (1981) (Note that the Bounty and Snares Islands are not volcanic). Mesozoic and early Tertiary sedimentary rocks are present on Auckland and Campbell Islands

| Island group | Latitude/longitude | Area (ha) | Formation date (Mya) |
|---|---|---|---|
| Bounty Islands | 47°40′S/179°35′E | 135 | – |
| Snares Islands | 48°01′S/166°33′E | 328 | – |
| Antipodes Islands | 49°41′S/178°45′E | 2100 | 0.2–0.5 |
| Auckland Islands | 50°40′S/166°05′E | 61 120 | 12–25 |
| Campbell Island | 52°35′S/169°10′E | 11 330 | 6–8 |
| Macquarie Island* | 54°30′S/158°95′E | 11 800 | 27–30 |

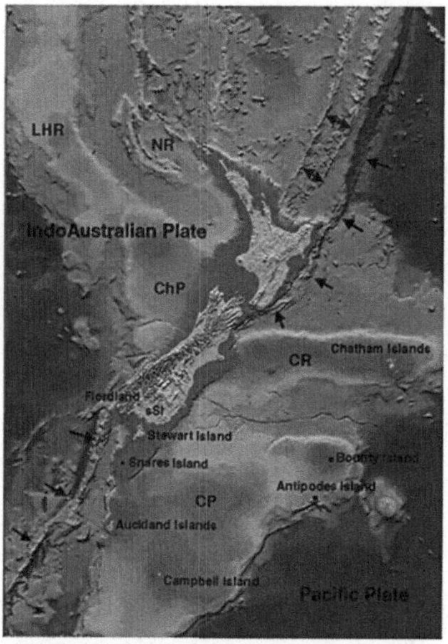


**Figure 1.** Locality map based on NIWA's *Undersea New Zealand* (CANZ, 1996). CP, Campbell Plateau; CR, Chatham Rise; ChP, Challenger Plateau; LHR, Lord Howe Rise; NR, Norfolk Ridge; sSI, Southland and Otago. Arrows indicate direction of plate motion.

criteria by which these islands were included in the World Heritage List (see http://www.wcmc.org.uk).

We start the paper with an annotated review in which we describe the biota and discuss relevant taxonomic and phylogenetic studies with the consultation of specialists (see Acknowledgements). The biota is unique mainly because much of the flora and fauna is endemic to either individual islands within groups, to island archipelagos, or to the Campbell Plateau as a whole. Endemism is apparent at many levels of the taxonomic hierarchy, from local species belonging to widespread New Zealand genera, through genera restricted to the Campbell Plateau, to monotypic genera with distant southern or even northern Pacific relatives. The mixed nature of the flora and fauna was recognized by Dugdale (1971) and Kuschel (1991), who suggested that the islands are home to both pre-Pleistocene survivors and postglacial dispersers. We note also the biogeographical relationship between the southern portions of the South Island of New Zealand (Fiordland, Otago and Southland regions) with the Campbell Plateau.

We undertake an analysis of area relationships that confirms the biota of the Campbell Plateau is composite. A summary of relationships based on phylogenetic data from the literature is shown in Table 2. Neo-endemics belonging to New Zealand, Pacific and Australasian clades coexist with palaeo-endemics of southern affinity, predominantly South American, resulting in the odd juxtaposition of 'tropical' and 'austral' groups such as parakeets (Psittacidae) and honey eaters (Meliphagidae) with Southern Ocean albatross (Diomedeidae) and penguins (Spheniscidae), or *Olearia* (Asteraceae) and *Metrosideros* (Myrtaceae), with *Stilbocarpa* (Araliaceae) and *Pleurophyllum* (Asteraceae). We interpret the neo-endemics as east Gondwanan and derived from New Zealand at ≈20 Mya, and the palaeo-endemics as remnants of an original west Gondwanan flora and fauna. While east and west Gondwana are geological terms strictly speaking, we extend their use to describe distinct areas of endemism. Geologically, we use the term east Gondwana to refer to east Antarctica + Australia + New Zealand (but not the southern South Island, New Zealand), and west Gondwana as west Antarctica + southern South America + Campbell Plateau. New Zealand Subantarctic taxa with phylogenies showing their membership of Australian and/or Pacific clades or, in the absence of phylogenetic information, whose conge-

**Table 2.** Condensed results from area cladograms including Campbell Plateau species generated from publications

| Taxon | West Gondwana | East Gondwana | Ambiguous* | Reference |
|---|---|---|---|---|
| Plants | | | | |
| *Abrotanella*† | X | – | W | Swenson & Bremer (1997) |
| *Hebe*‡ | X | X | – | Wagstaff & Garnock-Jones (1998) |
| *Lepidium*§ | – | X | – | Mitchell & Heenan (2000) |
| *Metrosideros*¶ | – | – | B | Wright *et al.* (2000) |
| *Oreobolus* | – | – | W | Seberg (1988) |
| *Pseudopanax/Raukaua* | – | X | – | Mitchell *et al.* (1997) |
| *Stilbocarpa* | X | – | – | Mitchell *et al.* (1999) |
| Beetles | | | | |
| *Stylogymnusa* | – | – | B | Ahn & Ashe (2004) |
| *Baeostethus* | – | – | B | Leschen *et al.* (2002) |
| Curculionidae (PAE)** | – | – | P | Morrone (1998) |
| *Molytini* | – | X | – | Craw (1999) |
| *Ectemnorhinchus* | – | – | W | Kuschel & Chown (1995) |
| Caddis flies | | | | |
| Hydrobiosidae†† | X | – | B | Ward *et al.* (2004) |
| Moths and butterflies | | | | |
| Lepidoptera (PAE)‡‡ | – | – | P | Emerson *et al.* (1997) |
| Birds | | | | |
| *Phalacrocorax*§§ | X | – | – | Kennedy *et al.* (2000) |
| *Cyanoramphus* | – | X | – | Chambers *et al.* (2001) |
| *Anas* | – | X | – | Kennedy & Spencer (2000) |
| Aves oceanic (PAE) | X | – | – | This study |
| Aves residents (PAE) | – | X | – | This study |

*B, basal taxa; W, widespread taxa with no apparent biogeographical pattern; P, partial analysis; †two major clades in tree; ‡two major clades; §partial analysis; ¶possible ancient east/west split; **analysis of subantarctic group; ††W distribution is scored for a derived lineage; ‡‡analysis of New Zealand taxa; §§two major clades.

ners are all found in these areas or further north are classed as east Gondwanan. Taxa with membership of southern South American clades are classed as west Gondwanan. It is not always possible to classify taxa unequivocally, e.g. genera endemic to the Campbell Plateau + New Zealand.

## BIOGEOGRAPHICAL METHODS

Although endemic Campbell Plateau taxa may indicate radiation or uniqueness, it is the more widespread taxa occurring in more than one region that are important for understanding biogeographical relationships. We use two analytical methods to determine the relationship of the Campbell Plateau to other geological areas. The paucity of data for some groups limits analyses to vascular plants, Coleoptera, Lepidoptera and birds. If phylogenetic information is available, area cladograms (Morrone & Crisci, 1995) are constructed by replacing the names of terminal taxa with the names of the areas in which they occur. In the absence of phylogenetic data areas can be classified by their shared taxa using Parsimony Analysis of Endemicity (PAE; Rosen, 1988; Cracraft, 1991). This procedure involves creating a data matrix consisting of areas and taxa which then is subjected to parsimony analysis. Taxa limited to a single distribution were excluded from the analysis because they contribute no information to area relationship. Trees were rooted with a hypothetical ancestor area coded as 0. We include PAE analyses of bird distributions using PAUP* (Swofford, 2003) for tree searching. All cladograms were produced using the heuristic search option with random addition, TBR branch swapping, and rooted using an all zero outarea. Branch support was determined by Bootstrap analyses with character re-sampling and 100 replicates and was executed in PAUP*.

## ANNOTATED REVIEW OF THE PLANTS AND ANIMALS OF THE SUBANTARCTIC

### DISCOVERY AND HISTORY OF SCIENTIFIC EXPLORATION

According to Gressitt & Wise (1971), the New Zealand Subantarctic Islands were discovered between 1788 and 1810. Commercial exploitation of their natural resources started as early as 1804 when a sealing station was established on Antipodes Island. During the age of sailing ships the archipelago lay across busy shipping lanes and many shipwrecks occurred, resulting in the establishment of shipwreck depots on both Auckland and Campbell islands in 1868. Rabbits, sheep, goats and cattle were all liberated at various times to provide relief supplies. A settlement started on Auckland Island by Charles Enderby in 1849 proved to be short lived. This early human contact resulted in modification of the natural biota on the Auckland and Campbell islands, and in the probable introduction of exotic flora and fauna including rats and mice. Sporadic scientific expeditions were carried out in the first half of the 19th Century by the Russians (Bellingshausen in 1820), Americans (Wilkes in 1840), French (D'Urville in 1840), and British (Ross in 1840). Biological collecting expeditions date from early last century when the Philosophical Institute of Canterbury expeditions of 1907 and 1909 visited all five island groups. Members of the 'Cape Expedition' made collections between 1941 and 1944 when wartime coast watchers were stationed on the Auckland and Campbell islands. These coast watchers included R. A. Falla, E. G. Turbott and C. A. Fleming, all of whom were to have distinguished scientific careers subsequently. The Bishop & Auckland Museums' 1961 expedition to the Auckland Islands was followed by two expeditions led by Falla in 1962–3 and 1966. The University of Canterbury Antipodes expedition of 1969 also visited Campbell Island, while the last general biological expedition to the Auckland Islands was in summer 1972/73, organized by the then Department of Lands & Survey (Yaldwyn, 1975). Over the past three decades more focused scientific expeditions have continued to visit the New Zealand Subantarctic Islands, particularly Auckland and Campbell islands. Scientific research effort has been across the fields of taxonomic collection (e.g. fungi and Coleoptera), ecological research (e.g. plant and animal communities) and management (e.g. pest eradication programs). In the following sections we will detail what is known of this biota.

### Fungi

McKenzie & Foggo (1989) listed the fungi of the New Zealand Subantarctic Islands. At that time only 49 named species had been reported. Thirteen species were known from the Auckland Islands, 15 from Campbell Island, 12 from both, eight from Snares, and one from Auckland and Snares islands. There are no records of fungi from either Bounty or Antipodes islands. Fifteen species were endemic and the rest were New Zealand or cosmopolitan species, some introduced along with their exotic hosts. However, recent collecting indicates there are more than 200 species on Campbell Island alone (see http://www.landcareresearch.co.nz/research/biodiversity/antarctica/index.html). Publications based on these recent collections include Aa *et al.* (2001), Mel'nik & Stephenson (2001), Johnston (2002a, b), and Golubeva & Stephenson (2003), with work on these collections continuing [for example, there are ten species of Laboulbeniales with seven known only from the Campbell Plateau (Hughs *et al.*, 2004)]. Much more research effort is required before meaningful comparisons of fungal diversity can be made for fungi from the Campbell Plateau.

### Vascular plants

General accounts of the flora of Campbell Island can be found in Gressitt (1964) and a description of plant communities and their distributions in Meurk, Foggo & Wilson (1994). Gressitt & Wise (1971) described the flora of the Auckland Islands and provided distribution details for plants of the Auckland, Snares and Antipodes groups. Plant diversity is low with only 176 species recorded from the Auckland group, 128 from Campbell, 67 from Antipodes, and a mere 22 species from Snares Island (O'Connor, 1999). No vascular plants occur on the Bounty islands. *Metrosideros* (Myrtaceae), *Olearia* (Asteraceae) and *Raukaua* (Araliaceae) forest grows on the Aucklands, while *Olearia* forest covers much of Snares Island. The more southerly Campbell Island lacks true forest cover, but a scrub association of *Hebe* (Scrophulariaceae), *Coprosma* (Rubiaceae) and *Myrsine* (Myrsinaceae) dominated by extensive stands of *Dracophyllum* (Epacridaceae), grows in sheltered conditions and at lower altitudes. Other typical plants of the New Zealand Subantarctic Islands include the 'megaherbs' [*Pleurophyllum* (Compositae), *Stilbocarpa* (Araliaceae), *Anisotome* (Apiaceae), *Bulbinella* (Liliaceae), *Gentiana* (Gentianaceae) and *Poa* tussock (Poaceae)]. The herbs of the subantarctic are generally characterized by larger leaves and more colourful flowers than congeners on mainland New Zealand.

Each island group has a small number of endemic taxa (one species on Snares, four on Antipodes, and six each on the Aucklands and Campbell) and the genus *Damnamenia* (Asteraceae) is endemic to the Auckland Islands. Overall, about a third of all vascular plants are endemic to the New Zealand Subantarctic Islands (plus Macquarie) according to the list provided by Gressitt & Wise (1971). Dicotyledons have the highest rate of endemism (almost 60%) and ferns the lowest (about 10%). One genus [*Pleurophyllum* (Asteraceae)] is endemic to the New Zealand Subantarctic Islands plus Macquarie, and another (*Stilbocarpa*) to the New

Zealand Subantarctic Islands, Macquarie and Stewart islands, and Fiordland. Floristically the islands are similar to New Zealand on the basis of shared species (about two thirds of Auckland Island's vascular plants are found in New Zealand), with some evidence that they are most similar to the southern South Island, New Zealand. This similarity is more marked in subantarctic marine algae with several shared species between Otago/Southland and the New Zealand Subantarctic Islands (Morton & Miller, 1973; Chin, Brown & Heads, 1991).

Phylogenies (Table 2) are available for *Stilbocarpa* (Mitchell, Meurk & Wagstaff, 1999), *Raukaua* (Mitchell, Frodin & Heads, 1997), *Pseudopanax* (Araliaceae) (Mitchell & Wagstaff, 1997, 2000), *Lepidium* (Brassicaceae) (Mitchell & Heenan, 2000), *Metrosideros* ( Myrtaceae) (Wright *et al.*, 2000), *Hebe* (Scrophulariaceae) (Wagstaff & Garnock-Jones, 1998; Wagstaff & Wardle, 1999; Wagstaff *et al.*, 2002), *Abrotanella* (Asteraceae) (Swenson & Bremer, 1997) and *Oreobolus* (Cyperaceae) (Seberg, 1988). Additional taxonomic papers consulted include Heads (1992, 1993, 1994a, 1994b) for *Hebe*, Heads (1996) for *Coprosma*, Heads (1998a; see also Cross, Quinn & Wagstaff, 2002) for *Olearia* and Heads (1999) for *Abrotanella*.

New Zealand Subantarctic species of *Raukaua*, *Lepidium* and *Metrosideros* show east Gondwana affinitiy. *Raukaua simplex* (Forst) Mitchell, Frodin & Heads is part of a New Zealand clade whose closest relatives are Pacific species found in Hawaii and the Marquesas. *Metrosideros umbellata* Cav. (Southern Rata) is sister taxon to all other species of an otherwise Pacific genus. *Lepidium oleraceum* Forst. (Cook's Scurvy Grass), found in New Zealand and on the Snares and Auckland Islands, is part of an endemic New Zealand clade. This clade is sister group to other New Zealand and northern hemisphere species according to Mitchell & Heenan (2000).

Subantarctic species of *Oreobolus*, *Abrotanella*, *Hebe* and *Stilbocarpa* exhibit more complex relationships. *Oreobolus pectinatus* Hook. (New Zealand + Campbell Plateau) and *O. strictus* Berggren (widespread in New Zealand) form an unresolved trichotomy with a predominantly South American clade and a Tasmanian species. Pacific and Indonesian species are more basal in Seberg's (1988) cladogram. *Abrotanella rosulata* Hook. and *A. spathulata* Hook. are endemic to the Campbell Plateau, and are placed by Swenson & Bremer (1997) with species from Fiordland and western Tasmania in a clade showing sister group relationship with species from Australia, New Zealand and New Guinea. Two other New Zealand endemics, *A. muscosa* Kirk (Stewart Island) and *A. fertilis* Swenson (widespread) are included in a southern South American clade. Two subantarctic species of *Hebe* (*H. elliptica* Pennell and *H. salicifolia* (Forst.f) Pennell) are included in a southern South American and Falkland Island clade by Wagstaff & Garnock-Jones (1998). *Hebe epacridea* (Hook.) Ckn. & Allan found in northern regions of the South Island, New Zealand, is the sister species of the third subantarctic *Hebe*, *H. odora* Ckn. None of these are species particularly closely related to each other within the genus. A similar result is shown in Wagstaff *et al.* (2002), although in that paper the subantarctic endemic species *H. benthami* is placed near to the base of a New Zealand clade. The three *Stilbocarpa* species link the New Zealand Subantarctic Islands to both Fiordland and Maquarie Island (Mitchell *et al.*, 1999). They are sister group to *Schizeilema*, a genus found in New Zealand, South America and Australia. *Azorella*, a genus that has a wider distribution including southern Indian Ocean islands, is the sister group to *Stilbocarpa* plus *Schizeilema*.

### ARACHNIDA

Forster (1955, 1964, 1970), Forster & Wilton (1973), Forster & Platnick (1985), and Taylor (2004) described and listed the spiders and harvestmen of the New Zealand Subantarctic Islands, Marris (2000) provided a list of Antipodes species and, most recently, Pugh (2004) provided a biogeographical analysis of subantarctic spiders. Atyeo (1964) and Strandtmann (1964) dealt with various families of the Prostigmata and Wallwork (1964) with the Cryptostigmata. Endemism is high with 80% of spider and all seven harvestmen species being endemic. There is also one endemic genus, *Pacificana* (Miturgidae) (Bounty) (Forster, 1964). Forster (1964) regarded the high levels of endemism as evidence of the fauna's considerable age, and not explicable by post-Pleistocene dispersal. He argued that the spider fauna of the New Zealand Subantarctic Islands was not characteristic of oceanic islands, but rather appeared to be an attenuated relict fauna from an earlier period. Forster (1964) also noted the close resemblance of the fauna to New Zealand in general and to the southern South Island in particular. The genus *Gohia* (Desidae) is endemic to the New Zealand Subantarctic Islands and the southern South Island.

Only two species of pseudoscorpions are known from the Campbell Plateau (Beier, 1964). *Systellochernes zonatus* Beier is found on Stewart, Auckland, Campbell and Antipodes islands (Beier, 1976); *Apatochernes antarcticus* Beier is found on Auckland, Campbell, Stewart, Snares, Antipodes and the South islands (Beier, 1976).

Nine of the 14 indigenous species of prostigmatid mites are endemic (64%). *Bdellodes petila* Atyeo (Bdellidae) is found in Otago and Campbell Island. Two cryptostigmatid genera, *Globoppia* (Oppiidae) and

*Oppia* (Oppiidae), link the New Zealand Subantarctic Islands to southern South America. All five oribatid species are endemic (Wallwork, 1964).

### TERRESTRIAL CRUSTACEA

At least 16 species of amphipods (Ceinidae, 1 sp.; Talitridae, 7 spp.) and isopods (Oniscidae, 4 spp.; Scyphacidae, 1 sp.; Styloniscidae, 4 spp.) are known from the Campbell Plateau (Bousfield, 1964; Pugh, Dartnall & McInnes, 2002). The New Zealand Talitridae were revised by Duncan (1994) and there are five species endemic to the New Zealand Subantarctic Islands and one, *Puhuruhuru patersoni* (Stephensen), links the Campbell Plateau with Otago and Southland (Duncan, 1994).

### COLLEMBOLA

Wise (1964a) listed 32 species indigenous to the New Zealand Subantarctic Islands of which 27 are endemic (84%). Wise (1964a) argued that the Collembola fauna strongly resembles that of New Zealand, whereas Salmon (1949) stressed its uniqueness. Greenslade (1990) suggested that the Collembola of Macquarie have closer affinities to the west, rather than to the New Zealand Subantarctic Islands.

### PLECOPTERA

There are seven species of stoneflies, all endemic to the Campbell Plateau (Illies, 1964; McLellan, 1977). The genus *Aucklandobius* (four species) is endemic to the Auckland Islands and *Rungaperla* (two species) is endemic to Campbell Island. The single species from the Snares, *Nesoperla johnsi* McLellan is endemic, but with an outlying species of *Nesoperla* found throughout New Zealand. *Aucklandobius* and *Nesoperla* are members of Acroperlini and *Rungaperla* of Zelandoperlini tribes of the subfamily Zelandoperlinae in family Gripopterygidae (McLellan, 1977). This southern family of 47 genera and some 250 species is found only in Australia, New Zealand, South America, Falkland Islands, Snares, Auckland Islands and Campbell Island. Illies (1964) considered that the Auckland and Campbell islands genera originated in the Mesozoic, and that cool, wet conditions and the presence of running water have prevailed since connections with New Zealand were broken. However, it now appears that a later wind-borne winged species from New Zealand colonized the Snares and, therefore, there is also a possibility that wind-borne winged New Zealand stonefly species colonized Auckland and Campbell islands too (see McLellan, 2001, who postulates a similar derivation for the Falkland Islands stonefly fauna).

The species on Auckland and Campbell Islands are wingless apart from winged *Aucklandobius complimentarius* Enderlein (Auckland Islands) and the micropterous *N. johnsi* (Snares). *Aucklandobius complimentarius* and *A. gressitti* Illies have aquatic nymphs but the remaining New Zealand Subantarctic species have terrestrial or semiterrestrial nymphs, a common occurrence in South American, Falkland and New Zealand species which live in alpine zones, or at lower altitudes in higher latitudes, where they are exposed to colder temperatures and high winds (McLellan, 2001).

### ORTHOPTERA

Each subantarctic island group has its own endemic genus of cave weta (Rhaphidophoridae). *Ischyroplectron* occurs on Bounty Island, *Dendroplectron* on the Auckland Islands, *Notoplectron* on Campbell Island, and *Insulanoplectron* on Snares and Antipodes islands (Richards, 1964, 1971; Marris, 2000). Gwynne (1995) placed the Rhaphidophoridae as the basal taxon to all other tettigonioids. There is a ground weta (Anostostomatidae) on Snares Island, the endemic *Hemiandrus subantarcticus* (Salmon). The genus is found otherwise in New Zealand.

### HEMIPTERA

Beardsley (1964) discussed the Coccoidea from Campbell Island and Cox (1987) reviewed the Pseudococcidae. Three species of mealybug occur: *Balanococcus danthoniae* (Morrison) is also found on Antipodes (Marris, 2000) and Stewart islands, and is common on grasses throughout New Zealand; *Chryseococcus longispinus* (Beardsley) is also found on Campbell Island; and *Ventrispina otagoensis* (Brittin) found on Campbell Island and widespread in the South Island, New Zealand. There are two more *Ventrispina* species in the South Island, New Zealand and another three species found only on Mt Kosciusko, Australia. A fourth species, *Phenacoleachia australis* Beardsley is one of only two species in the genus, the other, *P. zealandica* (Maskell), is found in New Zealand. This genus has been considered to constitute a monotypic family (Phenacoleachiidae; see Cook, Gullan & Trueman, 2002), although Beardsley (1964) suggested it was best placed as a primitive subfamily of the Pseudococcidae.

### COLEOPTERA

A checklist of the Coleoptera (about 140 spp.) known from the New Zealand Subantarctic Islands is available online (Johnston, Leschen & Stephenson, 2001). Gourlay (1950), Brookes (1951), and Marris (2000) detailed the Coleoptera of Auckland, Campbell, and Antipodes islands respectively. Taxonomic papers consulted included: Kuschel (1964, 1971), May (1971), Kuschel & Chown (1995), Chown (1994), and Craw (1999) (Curculionidae); Spilman (1967) (Salpingidae); Townsend (1971) and Johns (1974) (Carabidae); Steel (1964), Ahn & Ashe (1996), Ashe (2000), Carlton & Leschen (2001), and Leschen *et al.* (2002) (Staphylinidae); Samuelson (1964), Gressitt & Samuelson (1964), Ordish (1971, 1974), Watt (1971), and Hansen (1998) (various families). Results of phylogenetic studies are summarized in Table 2.

Coleoptera is the most diverse arthropod group found in the New Zealand Subantarctic region with 122 described indigenous species, although many more are undescribed. Of these, 102 (82%) are endemic to the New Zealand Subantarctic Islands and many are flightless. Endemism at the generic level is also high. Of the 65 genera present on the archipelago, 12 (18%) are endemic to it. The similarity of the beetle fauna to the southern South Island, as noted for plants, is apparent also with nine genera restricted to the New Zealand Subantarctic Islands and Fiordland/Southland/Otago [*Ptilodes* (Ptiliidae), *Prodontria* (Scarabaeidae), *Notacalles*, *Pachyderris*, *Peristoreus*, *Notinus*, *Exeiratus* (Curculionidae), *Crymus* (Staphylinidae) and an undescribed genus of chrysomelid beetle].

While New Zealand Subantarctic Coleoptera show a marked similarity to New Zealand as a whole, with a further 17 genera endemic to New Zealand Subantarctic Islands + New Zealand + Chatham (26%), seven genera also link the New Zealand Subantarctic Islands to the Patagonian Region of southern South America [*Bountya*, *Kenodactylus*, *Oopterus* (Carabidae), *Crymus*, *Stenomalium*, *Leptusa* (Halmeusa) (Staphylinidae) and *Nestrius* (Curculionidae)]. Kuschel (1971) noted that eight out of nine New Zealand Subantarctic curculionid subfamilies (recognized at that time) also occur in Patagonia. Both faunas are rich in endemic species, are exclusively ground dwelling, and have an unusually large ratio of ectophytic to endophytic larvae. In both areas the dominant subfamily is Rhytirhinini and the second most abundant is Adelognatha. Macquarie Island Coleoptera shared with the New Zealand Subantarctic archipelago include *Omaliomimus albipennis* (Kiesenwetter), *O. venator* (Broun), *Stenomalium sulcithorax* (Broun), *Leptusa antarctica* Kiesenwetter (Staphylinidae) and *Epichorius sorensoni* (Brookes) (Byrrhidae) (Greenslade, 1990).

A feature of the fauna is the presence of what may be phylogenetically isolated monotypic genera. Three carabids, *Bountya* (Bounty), *Calathosoma* (Auckland) and *Synteratus* (Snares) have no close New Zealand relatives according to Johns (1974). *Antarcticodomus fallai* Brooks (Salpingidae: Aegialitinae) of Auckland and Campbell islands is the sole described member of a monotypic genus, one of two genera in the isolated subfamily Aegialitinae. Four other undescribed species are known from Antipodes, Auckland, Bounty, Campbell, Chatham, Snares and Stewart islands (Emberson, 1998; J. Marris, pers. comm.). The other members of this subfamily, four species of *Aegialites*, are all confined to the northern hemisphere Pacific coast from California to the Kuriles (Spilman, 1967). This odd distribution pattern is mirrored by two staphylinid genera *Baeostethus* (Antipodes, Auckland, Campbell) and *Stylogymnusa* (Auckland). *Baeostethus* may be the sister taxon to all other Liparocephalini found exclusively in the Northern Hemisphere (Leschen *et al.*, 2002), while *Stylogymnusa* is the most primitive member of the aleocharine tribe Gymnusini, which is otherwise widespread in the Holarctic (Hammond, 1975; Klimaszewski, 1979; Ashe, 2000; Ahn & Ashe, 2004). The weevil *Notonesius aucklandicus* Kuschel of Auckland Island, is unique in that its larvae burrow in wet peat (rather than inside plant tissue) and possess a considerably elongated 8th abdominal segment, 'holding a partially retractile long tube that carries paired spiracles on a terminal disc fringed with 16 long setae of two kinds' (Kuschel, 1991). May (1971) called this appendage a 'rat tail'.

The presence of four endemic, monotypic genera and diversity shown within several genera is odd, and is not consistent with the present land area of the Auckland Islands. For example, there are six endemic species of *Gromilus*, three endemic species of *Loxomerus*, and two endemic species in nine other genera present in the Auckland Islands. Similar examples can be provided for other islands within the archipelago. It is difficult to imagine either this degree of speciation taking place in such a restricted area or that these species groups represent independent lineages that have colonized the islands at different times. In our view these data strongly suggest that part of the beetle fauna is both older than the present island, and derived from a more extensive land area (phylogenetic data will help test this hypothesis).

Kuschel & Chown (1995) provided a phylogeny of the Ectemnorhincus group of weevils. While this study confirmed the close relationship of the two subantarctic genera *Heterexis* and *Oclandius*, the area cladogram effectively reduces to an uninformative two-area statement. Similar problems arise with the area cladograms based on some of the available phylogenies for the Liparocephalini (Leschen *et al.*, 2002) and Gymnusini (Ashe, 2000). In both the area cladogram is effectively reduced to a two-area statement linking the New Zealand Subantarctic Islands to the northern Pacific. However, these studies do confirm the basal position of these genera within their respective groups and their extreme geographical and phylogenetic isolation.

The area cladogram based on the Molytini phylogeny provided in Craw (1999) links the New Zealand Subantarctic Islands to both the north (Chatham Islands and the Poor Knights) and to Fiordland. The northern group consists of four species in the genus *Hadramphus*. *Hadramphus stilbocarpae* Kuschel, present on Stewart Island, Snares Island and Fiordland, is most closely related to *H. spinipennis* Broun of the Chathams, and these are sister group to *H. pittospori* (Kuschel) of the Poor Knights Islands. The second group consists of 11 species in the genus *Lyperobius*. The Snares species *L. nesidiotes* Kuschel is most closely related to *L. australis* Craw of Fiordland. *Lyperobius nesidiotes* is a member of a clade that also includes species from the southern South Island and alpine regions.

Morrone's (1998) PAE of subantarctic weevils indicated that each subantarctic island group formed a 'monophyletic' clade, and that subantarctic islands share a closer relationship to the nearest continental land mass than to each other. Udvardy's (1987) hypothesis that the subantarctic islands as a whole were an area of endemism is not borne out by Morrone's (1998) study. Of particular relevance here is that Chatham Islands + New Zealand is sister taxon to the New Zealand Subantarctic Islands. The resolution of the positions of Antipodes and Bounty islands is poor owing to the low number, or absence of weevils on these islands.

### Diptera

Diptera are the second most diverse group of arthropods on the New Zealand Subantarctic archipelago. The following sources were consulted: Hutton (1902), Miller (1950), Harrison (1955, 1964), Alexander (1955, 1964), Richards (1955), Oldroyd (1955), Quate (1964), Tokunaga (1964), Steffan (1964), Yukawa (1964), Smith (1964), Duckhouse (1971), Sublette & Worth (1980), Greenslade (1990), and Evenhuis (2003). Names were also verified in Macfarlane *et al.* (2004).

Gressitt (1964) listed 81 species for Campbell Island (about half endemic) and O'Connor (1999) listed 110 for the Auckland Islands. Flightlessness is a feature of dipteran evolution on these islands. Although diversity is low, some genera have undergone local diversification. For example, seven endemic species of *Spilogona* (Muscidae) are found on Campbell Island. Twelve endemic monotypic genera are found in the archipelago; seven on Auckland Island [*Kuschelius*, *Maryella*, *Mecaorus*, *Nakataia*, *Nesiocladius* (Chironomidae), *Howickia* (Sphaeroceridae) and *Paradoxa* (Mycetophilidae)] one on Campbell Island [*Gynnidocladius* (Chironomidae)], three on both Auckland and Campbell islands [*Hevelius*, *Pterosis* (Chironomidae) and *Apterachalcus* (Dolichopodidae)], and one genus is restricted to the Bounty Islands [*Aphrosylopsis* (Dolichopodidae)]. At least five genera [*Gressittius*, *Maoridiamesa* (Chironomidae), *Baeopterus* (Coelopidae), *Xenocalliphora* (Calliphoridae) and *Dipsomyia* (Empidae)] and one endemic family (Australimyzidae) are endemic to the New Zealand Subantarctic Islands plus New Zealand.

*Apetaenus* (Tethinidae) is a subantarctic endemic found on Campbell, Antipodes, Bounty and Macquarie islands, and the southern Indian Ocean islands of Kerguelen and Crozet. Other taxa that link the archipelago to other southern areas include: *Smittia* (Chironomidae) and *Coelopella* (Coelopidae) (Campbell and Macquarie islands, south Australia, New Zealand), *Icaridion* (Coelopidae) (Campbell, Auckland, Antipodes, Snares and Macquarie islands, New Zealand), *Schoenophilus* (Dolichopodidae) (Campbell, Auckland and Macquarie islands), *Parochlus* (Chironomidae) (Campbell, Auckland and Antipodes islands, New Zealand, southern South America, South Shetlands, South Georgia, Crozet), and *Scatella* (Ephydridae) (Campbell and Falklands islands).

### Hymenoptera

Twenty-five species of New Zealand Subantarctic Island Hymenoptera are known and details for most of these can be found in Kerrich (1964), Kerrich & Yoshimoto (1964), Muesebeck (1964), Townes (1964), Yoshimoto (1964a, 1964b), Valentine (1971), Early (1978, 1980) and Greenslade (1990). Nineteen of the 25 species are endemic to the New Zealand Subantarctic Islands (76%), and 13 of the 25 are entirely flightless or have flightless morphs. There are two endemic genera, *Nesopatasson* (Mymaridae) (Auckland Islands) and *Nesomymar* (Mymaridae) (Campbell Island). *Antarctopria latigaster* Brues is essentially a true subantarctic species but has turned up on Stewart Island and on the South Island though in very low numbers. Valentine (1971) suggested that *Nesopatasson* is most closely related to a genus found in South America, while Noyes & Valentine (1989) allied the genus with the genera *Paracmotemnus* and *Nesomymar*. The species *Kleidotoma subantarcticana* Yoshimoto may be conspecific with coastal *Kleidotoma* from the South Island and Stewart Island (J. Early, pers. comm.). The other genera are found in New Zealand.

### Lepidoptera

Dugdale (1971; Dugdale (1975) and Patrick (1994) provided distribution details for Lepidoptera. Campbell Island species are discussed in Yano (1964), Munroe

(1964), and Dugdale (1964). Flightlessness is prominent also in this group. Of the 42 species then known from the archipelago, Dugdale (1971) reported that two thirds had reduced wings. Thirty-six (46%) of the 78 indigenous species now known from the New Zealand Subantarctic Islands are endemic (Patrick, 1994). Three genera are endemic to Campbell Island [*Campbellana* (Carposinidae), *Exsilirarcha* (Crambidae) and *Sorensenata* (Tortricidae)] and one is endemic to Auckland Islands [*Protyparcha* (Crambidae)]. A close similarity with New Zealand (particularly the southern South Island) is apparent also in Lepidoptera (e.g. the Snares fauna is most similar to that present on Stewart Island and Fiordland; J. S. Dugdale, pers. comm.). There are 13 genera endemic to the New Zealand Subantarctic Islands plus New Zealand, eight of which are widespread within New Zealand, including the Chatham Islands. The other five genera [*Pyrgotis* (Tortricidae), *Mallobathra* (Psychidae), *Izatha* (Oecophoridae), *Epiphryne* (Geometridae), and *Grypotheca* (Psychidae)] are absent from the Chathams.

Dugdale (1964) commented on the 'curiously incomplete' nature of the geometrid fauna of Campbell Island, noting the absence of New Zealand alpine species and members of the distinctive New Zealand geometrid assemblage associated with *Dracophyllum* (other than *Epiphryne charidema* Meyrick), *Coprosma* [other than *Austrocidaria similata* (Walker)] and *Hebe* which are present on Campbell Island. Patrick (1994) noted also the absence of tortricids from Antipodes Island and the paucity of geometrids. Geometrids are uncommon also among the Falkland Islands' Lepidoptera (Dugdale, 1971). Dugdale (1971) interpreted the lepidopteran fauna of the New Zealand Subantarctic Islands as a mixture of pre-Pleistocene survivors and post-Pleistocene waifs. Survivors are *Campbellana* (Carposinidae), *Protyparcha* and *Exsilirarcha* (Crambidae), *Tinearupa* (Oecophoridae), *Reductoderces* spp. (Psychidae), *Antipodesma turbotti* Salmon and Bradley (Tineidae), and *Sorensenata* (Tortricidae). Post-Pleistocene waifs are *Eudonia*, *Scoparia* (Crambidae), *Elachista* (as *Irenicodes*, Elachistidae), *Austrocidaria*, *Epiphryne*, *Pasiphila* (Geometridae), *Graphania* (Noctuidae), *Proterodesma* s.l. (Tineidae), *Apoctena*, *Ericodesma* and *Pyrgotis* (Tortricidae). The above remarks exclude the Snares Islands, which has only one genus in common with the other subantarctic islands (J. S. Dugdale, pers. comm.).

Emerson, Wallis & Patrick (1997) investigated the relationship between the southern South Island, Snares and Auckland islands using a PAE based on 532 Lepidoptera distributions. In this analysis, the New Zealand Subantarctic Islands have a sister group relationship with Southland and Otago, and a more distant relationship with other South Island, New Zealand regions. This result supports the hypothesis that the New Zealand Subantarctic Islands have a greater compositional similarity to the southern South Island than to other parts of the South Island, New Zealand (Table 2).

## TRICHOPTERA

Three species of caddis fly belonging to two families are known from the New Zealand Subantarctic Islands (Wise, 1964b; Schmid, 1989). A single species, *Oxyethira albiceps* (McLachlan), of the family Hydroptiidae is found throughout the New Zealand Subantarctic Islands apart from the Bounty Islands, and is also widespread on the New Zealand mainland and the Chatham Islands.

The Hydrobiosidae are represented by two species, *Tiphobiosis kuscheli* Wise and *Costachorema notopterum* Wise, that are endemic to Auckland Island (Schmid, 1989). Henderson (1991) undertook a phylogenetic analysis of the Hydrobiosidae using the data provided in Schmid (1989). A recent reanalysis of the family by Ward *et al.* (2004) placed *Tiphobiosis*, a genus restricted to the New Zealand Subantarctic Islands and New Zealand, at the base of a New Zealand clade that is sister taxon to *Clavichorema* from Chile. *Costachorema*, also restricted to the New Zealand Subantarctic Islands and New Zealand, is nested within the same clade of New Zealand genera (Table 2).

## FISH

Coastal rockpool fish from the New Zealand Subantarctic Islands are described in Parrott (1958) and Paulin & Roberts (1992). Diversity is low, with only eight species known from the Auckland Islands and 29 species from Snares Island. The New Zealand Subantarctic Islands rockpool fauna is dominated by benthic carnivores, in contrast to the dominant planktivores and herbivores of the mainland New Zealand fauna. There are only two described endemic species, the clingfish *Gastrocymba quadriradiata* (Rendahl) (Auckland, Bounty, Antipodes and Campbell islands) and the triplefin '*Tripterygian*' *jenningsi* (Auckland and Antipodes Islands). According to Paulin & Roberts (1992) the rockpool fauna of Auckland Island shows greatest similarity to that of the South Island, New Zealand, with six shared species. There is only one species of freshwater fish found in the New Zealand Subantarctic Islands, the widespread New Zealand galaxid, *Galaxias brevipinnis* Günther. This species is restricted to Auckland and Campbell islands where suitable, permanent habitat occurs (O'Connor, 1999).

### Birds

'The initial and most lasting impression of New Zealand's subantarctic islands and their surrounding oceans is one of seabirds wheeling and soaring against stark, misty cliffs or effortlessly skimming over turbulent seas on long, narrow wings.' (O'Connor, 1999)

There are two types of birds found on these islands: pelagic birds that come to these remote places to breed and whose normal habitat is the open ocean, and species that live there permanently. The pelagic birds [albatross, mollymawks (Diomedeidae), shearwaters, petrels and prions (Procellariidae)] are found in great numbers only during the breeding season. Outside the breeding season these birds wander the open ocean. The taxonomy of the Diomedeidae is far from settled (Robertson & Nunn, 1998) but approximately 40 species nest on the New Zealand Subantarctic Islands. Many of these species are essentially circumpolar in distribution, whereas others are endemic to single islands or island groups.

Permanent residents include both terrestrial and marine birds. The diversity of terrestrial birds on each island is related to island size. Auckland Islands, the largest of the archipelagoes, has 14 species, including the New Zealand falcon [*Falco novaeseelandiae* Gmelin], two honeyeaters [*Anthornis melanura* (Sparrman) and *Prosthemadera novaeseelandiae* (Gmelin)], two parakeets [*Cyanoramphus novaezelandiae* (Sparrman) and *C. auriceps* (Kuhl)], a tomtit [*Petroica macocephala* (Gmelin)], two rails [*Lewinia muelleri* (Rothschild) and *Gallirallus philippensis* Linnaeus], a dotterel (*Charadrius bicinctus* Jardine et Selby) and a snipe (*Coenocorypha aucklandica* Gray). Turbott (1990) recognized 19 endemic taxa (including marine shag and penguin species) resident in the New Zealand Subantarctic Islands. The majority of these are treated as subspecies, although recent molecular studies on teals (Kennedy & Spencer, 2000) and parakeets (Chambers *et al.*, 2001) suggest that many may have to be reclassified as full species. An endemic snipe species was discovered on one of the predator-free islands off Campbell Island in 1997 (O'Connor, 1999), bringing the total number of endemics to 20. Eight of these are found on Auckland Islands, five on the Antipodes, four on the Snares, three on Campbell Island and two on Bounty Island. Terrestrial resident birds can be characterized as New Zealand-derived, depauperate, and with probable high levels of endemism at the species level.

Cladograms for *Cyanoramphus* parakeets (Psittacidae) (Chambers *et al.*, 2001) and Australasian teals (Anatidae) (Kennedy & Spencer, 2000) connect the New Zealand Subantarctic Islands to the Pacific, New Guinea and Australia. New Zealand Subantarctic and New Zealand *Cyanoramphus* group in a clade whose closest relative is found in Norfolk Island. The subantarctic and New Zealand teal species form a clade that is a sister group to an Australian/New Zealand/New Guinea clade.

The area cladogram based on *Phalacrocorax* (Phalacrocoracidae) (Kennedy, Gray & Spencer, 2000) is potentially more informative biogeographically as the genus is geographically widespread. Unfortunately, the molecular phylogeny of Kennedy *et al.* (2000) is in disagreement with the morphologically based phylogeny of Siegel-Causey (1988). The problem is compounded by the paucity of New Zealand Subantarctic species included in the molecular analysis, and the possible effect that inclusion of more species might have on the topology of the molecular-based phylogeny. The single New Zealand Subantarctic Islands species included in Kennedy *et al.* (2000), *Phalacrocorax campbelli* (Filhol), has a sister group relationship to the southern South Island/Stewart Island species [*P. chalconotus* (Gray)] plus the Chatham species *P. onslowi* (Forbes). These three species form a derived group within a larger South American plus Macquarie clade. Interestingly, other New Zealand and Chatham Islands species occur in a widespread northern clade that is phylogenetically distant from the subantarctic species (see Table 2).

A PAE analysis of 79 subantarctic birds is shown in Figure 2. Distribution data came from Harper & Kinsky (1978) and Turbott (1990). The analysis produced two most parsimonious trees (Fig. 2A: length = 153, CI = 0.52, RI = 0.59). They differ in the position of the Campbell Plateau (New Zealand Subantarctic Islands), which can be a member either of a northern or southern clade. In order to try to resolve the position of the Campbell Plateau the data were divided into oceanic ($N = 53$) and resident species ($N = 26$) and a second analysis performed. When oceanic species are treated separately five trees were produced. The strict consensus tree is shown in Figure 2B (length = 100, CI = 0.53, RI = 0.63). In this tree Campbell Plateau links to other subantarctic Islands and southern South America. The Chathams and New Zealand form a sister group to Australia. Pacific islands form a third clade. There is no resolution of relationships among these groups or Antarctica. If the distributions of resident birds are analysed, a single most parsimonious tree (Fig. 2C: length = 40, CI = 0.65, RI = 0.75) placed the Campbell Plateau as sister area to New Zealand, and forming a clade with Australia and Pacific islands.

Oceanic birds appear to have a separate history to that of resident birds. The former indicate that the Campbell Plateau is related most closely with Macquarie and other subantarctic islands plus South America, the latter that it is most closely related to New Zealand and Chatham Islands (see Table 2). The analysis highlights the problem associated with com-

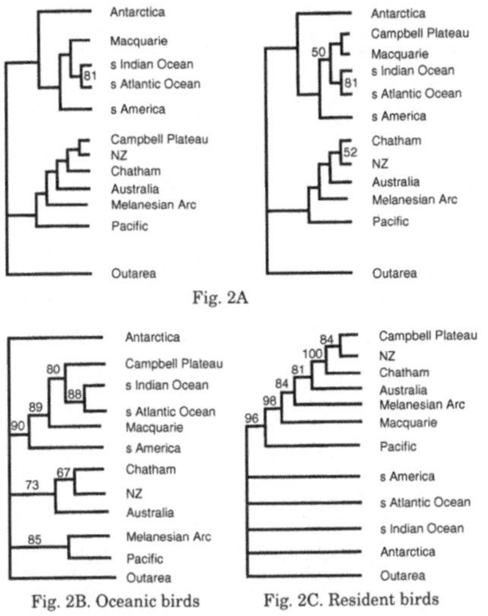


**Figure 2.** PAE results from subantarctic bird distributions. Details in text and raw data available from the authors. A, 79 subantarctic bird distributions. Two minimal trees (length = 153, CI = 0.52, RI = 0.59). Bootstrap values are indicated by each branch. B, 53 pelagic species. Consensus tree (length = 100, CI = 0.53, RI = 0.63). Bootstrap values are indicated by each branch. C, 26 resident species. Single tree (length = 40, CI = 0.65, RI = 0.75). Bootstrap values are indicated by each branch.

posite data sets, because the results of the combined analysis may differ from those achieved by analysing the parts independently.

## CHARACTERISTICS OF THE SUBANTARCTIC BIOTA

New Zealand Subantarctic taxa exhibit significant levels of endemism, from a relative low of about 30% for vascular plants, to 50% for Lepidoptera and Diptera, and in excess of 80% for other arthropods. Even accepting that such estimations are likely to change as future taxonomic and sampling works are undertaken, evidently the biota of the New Zealand Subantarctic Islands has been isolated for a significant period of time allowing for the evolution of endemic groups to occur. The degree of endemism exhibited appears taxonomically bimodal. Groups such as resident birds and vascular plants exhibit endemism at the species or subspecies level, whereas arthropods also exhibit endemism at the generic level. Many of these endemic genera are monotypic and some may be phylogenetically isolated from other members of their groups on the Campbell Plateau or on the mainland of New Zealand. Assuming that different levels of endemism reflect isolation rather than differing evolutionary rates, some arthropods appear to have been isolated for considerably longer than other groups such as resident birds or many vascular plants. Forster (1964), for example, considered the spider fauna to be of considerable geological age (although this statement refers to the knowledge of the fauna in the 1960s).

As expected, diversity on the subantarctic islands is generally low with relatively few taxa compared with similar sized mainland areas or islands, such as the Chatham Islands, (Emberson, 1998) located at lower latitudes. A few groups have radiated extensively to produce high numbers of species. Examples would include weevils (Coleoptera: Curculionidae), some dipteran genera such as Limnophora, and the subantarctic shags (Aves: Phalacrocoracidae). The disharmonic nature of the subantarctic biota is in part a consequence of this low diversity. Forster (1964) termed the spider fauna 'attenuated', and Dugdale (1964) commented on 'the curiously incomplete' nature of the geometrid fauna. Even shallow marine fish communities have an odd ecological balance in comparison with similar environments in the southern South Island (Paulin & Roberts, 1992).

The phenomena described above are not unique to the New Zealand Subantarctic Islands, but seem to be exhibited by many isolated islands. Gillespie & Roderick (2002) classified islands into two types, Darwinian or de novo islands, and fragment or continental islands. We will refer to these latter as Wallacean islands in honour of A. R. Wallace, who championed the role of geological change as an explanation of distributions within the islands of the Malay Archipelago, now Indonesia. According to Gillespie & Roderick (2002) the biological history of these two types of islands are different. On Darwinian islands the original 'ecological space' is empty and is exploited through the adaptive radiation of waifs and strays, provided isolation is high or long enough to allow for speciation. This results in the evolution of endemic species that these authors term neo-endemics. Disharmony on isolated Darwinian islands is a consequence of the lottery of long-distance dispersal. Isolated Wallacean islands have a history of extinction as land area decreases ('relaxation') perhaps followed by speciation amongst the survivors, leading to what Gillespie & Roderick (2002) termed palaeo-endemics (endemic genera). Disharmony on Wallacean islands results initially from distributional sampling (or possibly a disharmonic source area), but subsequent extinction will accentuate it. The New Zealand Subantarctic Islands clearly show characteristics of both type of islands as defined by Gillespie & Roderick (2002).

## RELATIONSHIP OF THE NEW ZEALAND SUBANTARCTIC ISLANDS TO OTHER AREAS

Relationships between areas can be deduced from the relationships shown by taxa present in those areas. Evidence ranges from the descriptive (hypotheses about taxa relationships without phylogenetic data, or similarity of taxa), through PAE (analysis of species presence/absence) to area cladograms based on phylogenetic data. These lines of evidence form a hierarchy, with descriptive evidence (qualitative) being the weakest and phylogenetic evidence (empirical) the strongest.

### GEOGRAPHICAL LIMITS OF THE CAMPBELL PLATEAU

We include the southern South Island (Fiordland, Southland and Otago) as part of the Campbell Plateau microcontinent both biologically and geologically. Taxa endemic to the New Zealand Subantarctic Islands and the southern South Island include the arachnid genus *Gohia*, one amphipod species (*Puhuruhuru patersoni*), and nine coleopteran genera (see text). A PAE analysis of Lepidoptera linked the New Zealand Subantarctic Islands to the southern South Island in a clade that is sister area to other South Island regions. Phylogenetic support comes from the weevil subfamily Molytini and the vascular plant genera *Abrotanella* and *Stilbocarpa*. Descriptive evidence to support this hypothesis includes greater (subjective) similarity of subantarctic vascular plants, marine algae, and fish to the southern South Island than to New Zealand in general.

We have not analysed the flora and fauna of Macquarie Island in total, only those parts that relate to the New Zealand Subantarctic Islands, but we would suggest that Macquarie Island is biologically, but not geologically, related to the Campbell Plateau. Macquarie Island is a slab of oceanic crust formed at a triple junction that was originally adjacent to the Campbell Plateau and which subsequently became isolated through sea floor spreading. Descriptive evidence suggesting that the Campbell Plateau and Macquarie Island were once adjacent to each other includes the sharing of six endemic coleopteran species, an endemic fly genus '*Schoenophilus*' and one endemic vascular plant genus (*Pleurophyllum*). One of two minimal trees from a PAE of birds links Macquarie Island with the Campbell Plateau (Fig. 2). Phylogenetic support comes from *Stilbocarpa*, in which Macquarie Island is part of a clade including southern South Island and the New Zealand Subantarctic Islands, and to a lesser extent by *Phalacrocorax*, where Macquarie Island has an unresolved position within a clade that includes Campbell Plateau.

We can find little evidence to link the Chatham Islands exclusively with the Campbell Plateau (see Craw, 1989) because shared taxa between the Campbell Plateau and Chatham Islands also occur in New Zealand (and sometimes further a field). From the limited data we have available for this three-area statement, the Chatham Islands are sister area to New Zealand, not the Campbell Plateau, even though these areas share widespread taxa. Similarity based on the sharing of widespread species is analogous to similarity based on primitive characters in phylogeny, and suffers the same problem of potential paraphyly.

### MELANESIAN RIFT AND PACIFIC CONNECTION

The Melanesian rift is a complex of continental fragments (Fig. 1) that includes the Challenger Plateau and west coast of the South Island, New Zealand, the Lord Howe Rise, New Caledonia Ridge, and Three Kings Rise (Coleman, 1997; Gaina *et al.*, 1998; Michaux, 1998). New Zealand, Lord Howe Island, Norfolk Island, New Caledonia, and the central Highlands of New Guinea are emergent parts of this rift structure. Michaux & White (1999) discussed the area relationships between south-west Pacific islands and the Melanesian Rift on the basis of a phylogeny for *Bactrocera* (Diptera).

Although the New Zealand Subantarctic Islands show similarity with New Zealand on the basis of shared species, we have argued that the southern parts of the South Island, New Zealand are part of the Campbell Plateau. Therefore, similarity between the New Zealand Subantarctic Islands and mainland New Zealand may not necessarily indicate a Melanesian Rift or Pacific connection, unless supported by phylogenetic evidence. The PAE of resident birds (Fig. 2) showed a clear northern relationship for Campbell Plateau species. The avian families Meliphagidae, Eopsaltriidae and Sylviidae, all of which have representatives on the Campbell Plateau, are distributed widely in the Pacific, Australia, New Guinea, and Indonesia. Area cladograms for *Cyanoramphus* (parakeets), *Anas* (teal), *Molytini* (Coleoptera), *Raukaua*, *Lepidium* and *Metrosideros* (vascular plants) all connect the Campbell Plateau with the Melanesian Rift and/or the Pacific (see Table 2). As we will discuss below, these elements of the subantarctic biota, usually neo-endemics, are interpreted as east Gondwanan.

### SOUTHERN SOUTH AMERICAN CONNECTION

A southern South American connection is evident particularly within New Zealand Subantarctic

Coleoptera, with eight genera shared between the two areas and a general similarity between the coleopteran fauna of Patagonia and the New Zealand Subantarctic Islands. Other genera shared between the two regions include *Globoppia* and *Oppia* (cryptostigmatid mites), *Parachlus* and *Scatella* (Diptera), *Nesopatasson* (Hymenoptera), and the plecopteran family Gripopterygidae. The PAE of oceanic birds (Fig. 2B) linked all the subantarctic islands into a clade with southern South America as sister area. Phylogenetic evidence for a southern connection includes the cladograms of *Phalacrocorax* (shags), *Oreobolus*, *Abrotanella*, *Hebe*, *Stilbocarpa* (vascular plants, see Wardle *et al.*, 2001 for other examples) and the Hydrobiosidae (Trichoptera). The southern element of the Campbell Plateau biota includes palaeoendemics that we identify as west Gondwanan in origin (Table 2).

## IMPLICATIONS FOR CRETACEOUS–CENOZOIC PALAEO-RECONSTRUCTIONS OF NEW ZEALAND

Reconstructing the Pacific margin of Gondwana is difficult because the boundary between the Indo-Australian and Pacific plates is complex. There have been many models suggested for the development of New Zealand. Kamp (1986) provided a number of reconstructions. Our analysis of the subantarctic biota suggests that, contrary to most models, the Melanesian Rift and Campbell Plateau should not be treated as a single entity. In our view these units, although historically proximate, had separate histories from ≈100 Mya until united in the latter part of the Tertiary. Although lithological similarities exist between the Palaeozoic foreland blocks in north-west Nelson and Fiordland (e.g. Spörli, 1987), detailed analyses show differences in lithology, grade of metamorphism and tectonic history, which raise questions about any proposed contiguity. Cooper (1975) suggested that the Nelson foreland is most similar to rocks of south-east Australia, and Cooper *et al.* (1982) compared Fiordland with Marie Byrd Land of west Antarctica. We interpret the disjunct distributions across the Alpine Fault discussed in Heads (1998b) as a result of Oligocene collision and floral and faunal mixing, and not as evidence of geographical contiguity between Nelson and the southern South Island, New Zealand prior to 80 Mya break-up. We see little biological or geological connection between the Campbell Plateau and the Chatham Rise, and would place the latter with other Torlesse terranes forming the east coast of the South Island, New Zealand, and presently part of the Pacific Plate, on the Indo-Australian plate prior to the Oligocene.

### CAMPBELL PLATEAU AND WEST GONDWANA

The Campbell Plateau was originally adjacent to Marie Byrd Land in west Antarctica. First based on the fit of 2000 m contours (Griffiths, 1971) and similarity in geology (e.g. Bradshaw *et al.*, 1997), this has now been confirmed by satellite altimetry data reported in McAdoo & Laxon (1997). Unlike east Antarctica, which has behaved as a single block, west Antarctica is composite (reconstructions in Kamp, 1986, and Spörli, 1987). McLoughlin (2001) has provided a recent review of the history of west Antarctica and its relationship to west Gondwana. McLoughlin (2001) identified four microplates in present-day west Antarctica: Ellsworth–Whitmore Mountains, Antarctic Peninsula, Thurston Island and Marie Byrd Land. Although resolution of the geology and history of these blocks is poor owing to extensive ice cover, all are continental blocks with Protozoic/Palaeozoic metamorphic and igneous basements intruded by Mesozoic igneous rocks. According to McLoughlin (2001) there has been little relative movement between them since the early Mesozoic, with the possible exception of Marie Byrd Land.

West Antarctica, South America and Africa remained in contact after the separation of Madagascar/India from Africa and east Antarctica between 165 and 162 Mya. Sea floor spreading in the Weddell Sea also dates from 162 Mya. Africa started to separate from southern South America at 135 Mya, while southern South America remained attached to the Antarctic Peninsula microplate (part of west Antarctica) until the opening of the Drake Passage between 32 and 28 Mya. Rifting between east and west Antarctica was probably active from early Cretaceous along the Byrd Sub-polar basin (Grunow, Kent & Dalziel, 1991), and along the West Antarctic Rift System during the Late Cretaceous and Cenozoic (Cande *et al.*, 2000). According to these authors new ocean crust was formed from 43 Mya, resulting in the separation of east and west Antarctica by some 180 km. The Cretaceous disruptive phase resulted in the development of early Cretaceous floral provincialism between east and west Gondwana. Southern South America and the Antarctic Peninsula had a diverse fauna with unique endemic genera, and characterized by bennettitaleans with large leaves (McLoughlin, 2001). We suggest that the palaeo-endemics of the New Zealand Subantarctic Islands (or more likely their ancestors) and the southern South American connections discussed previously also originated in the early Cretaceous at 100 Mya. The separation of the Campbell Plateau from Marie Byrd Land is well constrained to 84 Mya by identification of C34 magnetic anomaly in oceanic crust adjacent to Campbell Plateau. A reconstruction for Anomaly 32A (71 Mya) is presented in Figure 3.

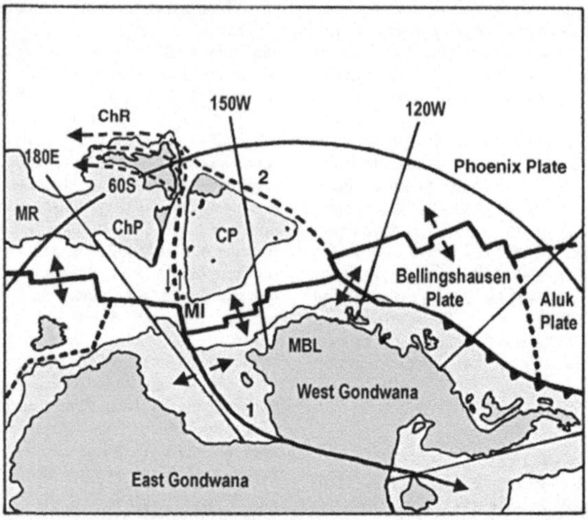


**Figure 3.** 71 Ma (Anomaly 32 A) Reconstruction. CP, Campbell Plateau microcontinent; ChR, possible position of Chatham Rise; ChP, Challenger Plateau; MBL, Marie Byrd Land; MI, Macquarie Island; MR, Melanesian Rift. Heavy line delineates active tectonic boundary, arrows indicate plate movements. 1, Proto east–west Gondwana boundary of the South Atlantic with the Pacific Ocean (dashed lines indicate possible plate boundary configuration in the North Island, New Zealand). 2, Possible plate boundary. Bellingshasen and Aluk Plates from Stock *et al.* (1996).

## Assembly of New Zealand during the Cenozoic

In Figure 3 we suggest that a plate boundary ran between the Challenger Plateau, situated at the southern end of the Melanesian Rift, and the Campbell Plateau microcontinent. The nature of this boundary is uncertain, but was likely to have been either a transform fault or rift, or even a combination of the two. Kamp (1986) located a triple junction (ridge–ridge–transform) adjacent to the Campbell Plateau at 72 Mya, allowing for some 330 km of displacement along a transform fault within the Campbell Plateau to accommodate differential spreading rates in the south Tasman Sea and Pacific–Antarctic Ridge. He placed a rift system separating the southern edge of the Melanesian Rift from the Campbell Plateau. Laird (1993) also suggested a rift existed between the Campbell Plateau and the Melanesian Rift at about 72 Mya. This tectonic structure may relate to the Tauru Fault, a major crustal and upper mantle structure in the south of the South Island, New Zealand described by Sutherland & Melhuish (2000), which is probably pre-Eocene in age. In Figure 3 we suggest that this rift or transform boundary is an extension of the plate boundary between east and west Gondwana, connecting the South Atlantic to the Pacific. By 43 Mya new ocean crust was propagated along the line of this proposed transform in the Adare Trough (Cande *et al.*, 2000), the Emerald Basin and Solander Trough (Sutherland, 2000). Rifting gave way to oblique compression by 25 Mya.

Adams (1981) has described a linear arrangement of volcanic centres in the South Island, New Zealand and Campbell Plateau that show an eastward decrease in age from about 23 Mya to 0.5 Mya. Eruption isochrons have rotated 60° anticlockwise between 20 Mya and 2 Mya, paralleling the rotation of New Zealand continental crust with respect to the Antarctic–Pacific Rise. According to Adams (1981) this pattern is consistent with the movement of the Pacific plate over a linear mantle source. We suggest that it was a consequence of the Campbell Plateau overriding an earlier spreading centre between the Melanesian Rift and Campbell Plateau.

Collision between the Campbell Plateau and the Melanesian Rift completed the compressive phase of the Indian/Pacific plate margin, transforming the boundary within New Zealand to one of strike–slip along the Alpine Fault. The plate boundary marked by the Alpine Fault is thus considerably older than 20 million years. While the data that Adams (1981) presents are sparse, they do indicate that by 10 Mya the Campbell Plateau and Canterbury/Chatham Rise were acting as a single unit. We would speculate that the transfer of the Torlesse terranes from the Indian plate to their present position on the Pacific plate was a consequence of major reorganization of plate boundaries in the north of New Zealand between 27 Mya and 20 Mya (e.g. Ballance, Pettinga & Webb, 1981).

## BIOGEOLOGY OF NEW ZEALAND'S SUBANTARCTIC ISLANDS

Michaux (2001) discussed the hierarchical nature of processes that affect distributions. He identified a number of these processes that operate over different timescales: geological, geomorphological and ecological. Because longer timescale processes provide the context in which the shorter timescale processes operate, a logical hierarchy exists between them. We term this hierarchical approach to understanding plant and animal distributions 'biogeology'.

The Campbell Plateau originated as a microcontinent rifted from the Marie Byrd Land region of West Gondwana at 84 Mya. We know from Late Cretaceous sediments preserved on the Campbell Plateau (Dettmann *et al.*, 1990; Beggs, 1993) and Southland (Bowen, 1978) that there are extensive coal deposits of Late Cretaceous age, indicating that the microcontinent was forested prior to time of rifting (Mildenhall, 1980). Pollen recovered from these sediments showed that *Nothofagus* (Dettmann *et al.*, 1990), podocarps and *Araucaria* (Bowen, 1978) were present. We can reasonably assume that forest and other environments supported a high-latitude biota of west Gondwanan origin that rafted away to evolve in isolation. Michaux (2001) termed this vicariant event, operating over a geological timescale, as 'splitting'.

We know relatively little about what changes occurred in the 60 million years that the Campbell Plateau was isolated in the Southern Ocean, because so little of the plateau is available for exploration. However, some general principles can be used to deduce what may have happened. We do know that in the Great South Basin topography was eroded and subsidence resulted in marine transgressions by the start of the Tertiary. Subsidence was greatest along the southern and eastern margins, the west and north remaining high (Beggs, 1993; Sutherland, 2000). Peneplanation and subsidence, processes operating in geomorphological time, would have had a profound effect on the composition of the original biota. We can imagine that the Campbell Plateau became a mosaic of low-lying land, lake and swamp environments, surrounded by marginal seas. From an evolutionary perspective the initial rift–drift phase, when the microcontinent was formed, would have led to 'relaxation' and an overall decrease in diversity (Gillespie & Roderick, 2002). Subsequent fragmentation of the landscape would have further promoted extinction, but also increased speciation rates amongst survivors as their populations became fragmented. The effect of this extinction/speciation cycle on an isolated biota would have been to reduce overall diversity further and increase endemism. As the Campbell Plateau continued to subside, and land area became more restricted and fragmented, these effects would have intensified. Climatic conditions in the Southern Ocean deteriorated from 35 Mya as Australia finally completed its separation from east Antarctica (Norvick & Smith, 2001) and South America detached from west Gondwana, causing initiation of the circum-Antarctic current and subsequent glaciation in Antarctica. Deteriorating climate was likely to have reduced diversity further by increasing extinction rates.

Towards the end of the Oligocene (*c.* 25 Mya) the Campbell Plateau biota's long isolation came to an end. From the preceding discussion we can surmise that by this time the biota would have been characterized by low overall diversity (due to extinction and habitat-poor environments), high levels and extreme degrees of endemism (due to its long isolation), of austral affinity (because of its west Gondwanan source area), and cool-adapted (due to its long history at high latitudes and the influence of deteriorating Antarctic climate). In many ways the survival of this biota was an unlikely event. We have no way of knowing how much of the microcontinent was above water by 25 Mya: some must have done so to account for the persistence of palaeo-endemics.

The period 25–20 Mya, the end of the Oligocene to the earliest Miocene, was one of significant plate boundary rearrangements in the New Zealand region. To the north an active subduction system propagated south with the opening of the South Fiji Basin. The Northland allochthon was also emplaced at this time. Large shield volcanoes erupted along the continental margin of present-day Northland. Significant volcanic activity was associated also with active subduction. We include the suturing of the Campbell Plateau as part of the boundary rearrangements that occurred at this time.

We suggest that one consequence of amalgamation was an interchange of organisms between the Campbell Plateau & Melanesian Rift. Michaux (2001) referred to this process of dispersal in geological time as 'juxtaposition', and the exchange of elements of distinct biotas, as new environments were created through changing geomorphology, as 'mixing'. We date the origin of the New Zealand Subantarctic neo-endemics to this event. Although it may seem strange that essentially northern, tropical groups would be able to establish in cool, temperate environments, it should be remembered that climatic conditions were warmer in the early Miocene and that many empty niches were presumably present because of the depauperate nature of the Campbell Plateau biota. We suggest also that the enigma of the New Zealand alpine fauna (see Trewick & Wallis, 2001) might have its solution in these processes of juxtaposition and mixing. It has long been a mystery how alpine forms could have evolved so quickly from a biota living in a land of minimal relief and warm climate. Indeed, some have suggested at least part of the alpine biota of New Zealand had to be derived from other places, such as the Holarctic (e.g. Fleming, 1980). In our view, a proportion of the Alpine biota is of Campbell Plateau origin, which was 'preadapted' to cool, wet and windy conditions created as the Southern Alps rose over the last 10 or so million years.

The New Zealand Subantarctic Islands, Fiordland, Southland, Otago, and southern South America preserve a unique collection of west Gondwanan flora and fauna. Much of the original biodiversity of west Gondwana would have been lost with the glaciation of west Antarctica. In southern South America and mainland New Zealand there has been substantial intermixing and dilution of the original west Gondwanan element, but on the New Zealand Subantarctic Islands this mixing has been minimal. For this reason alone, the New Zealand Subantarctic Islands deserve their World Heritage status.

## ACKNOWLEDGEMENTS

We would like to thank Adrian Paterson (Lincoln University, New Zealand) and an anonymous reviewer for their comments, which greatly improved the paper. Roger Brand (Austral-Pacific) has discussed New Zealand geology over many years with the senior author and Mark Bullians (Ministry of Agriculture and Forestry) helped with references. We would like to thank the following people for reading the manuscript and providing comments: John Dugdale (Landcare Research, general and Lepidoptera), John Early (Auckland War Memorial Museum, Hymenoptera), Mark Harvey (Western Australian Museum, pseudoscorpions), Michael Heads (University of the South Pacific, general), Rosa Henderson (Landcare Research, Coccidae), Peter Johnston (Landcare Research, fungi), John Marris (Lincoln University, New Zealand, general), Ian McLellan (Westport, New Zealand, Plecoptera), Cor Vink (San Diego State University, spiders), Steve Wagstaff (Landcare Research, plants), John Ward (Canterbury Museum, Trichoptera), and Zhi-Qiang Zhang (Landcare Research, mites). The views expressed in the paper, and the errors, are of course ours. This work was made possible in part by FRST (contract C09X0002) and National Science Foundation (grant 9971695).

## REFERENCES

**Aa HA, Vanev SG, Mel'nik VA, Stephenson SL. 2001.** A new species of *Cytosporella* (Coelomycetes, Deuteromycotina) from subantarctic Campbell Island (New Zealand). *New Zealand Journal of Botany* **39:** 543–545.

**Adams CJ. 1981.** Migration of late Cenozoic volcanism in the South Island of New Zealand and the Campbell Plateau. *Nature* **294:** 153–155.

**Ahn K-J, Ashe JS. 1996.** Phylogeny of the intertidal aleocharine tribe Liparocephalini (Coleoptera: Staphylinidae). *Systematic Entomology* **21:** 99–114.

**Ahn K-J, Ashe JS. 2004.** Phylogeny of the Myllaenini and related taxa (Coleoptera: Staphylinidae: Aleocharinae). *Cladistics* **20:** 123–138.

**Alexander CP. 1955.** The crane-flies of the subantarctic islands of New Zealand (Diptera). *Records of the Dominion Museum* **2:** 233–239.

**Alexander CP. 1964.** Insects of the Campbell Island. Diptera: Trichoceridae and Tipulidae. *Pacific Insects Monograph* **7:** 272–279.

**Ashe JS. 2000.** Mouthpart structure of *Stylogymnusa subantarctica* Hammond, 1975 (Coleoptera: Staphylinidae: Aleocharinae) with a reanalysis of the phylogenetic position of the genus. *Zoological Journal of the Linnean Society* **130:** 471–498.

**Atyeo WT. 1964.** Insects of Campbell Island. Prostigmata: Bdellidae. *Pacific Insects Monograph* **7:** 166–169.

**Ballance PF, Pettinga JR, Webb C. 1981.** A model of the Cenozoic evolution of northern New Zealand and adjacent areas of the southwest Pacific. *Tectonophysics* **87:** 37–48.

**Beardsley JW. 1964.** Insects of Campbell Island. Homoptera: Coccoidea. *Pacific Insects Monograph* **7:** 238–252.

**Beggs JM. 1993.** Depositional and tectonic history of the Great South Basin. In: Hsü KJ, Ballance PF, eds. *South Pacific sedimentary basins. Sedimentary basins of the World* 2. Amsterdam: Elsevier, 365–373.

**Beier M. 1964.** Insects of Campbell Island. Pseudoscorpionidae. *Pacific Insects Monograph* **7:** 116–120.

**Beier M. 1976.** The pseudoscorpions of New Zealand, Norfolk and Lord Howe. *New Zealand Journal of Zoology* **3:** 199–246.

**Bousfield EL. 1964.** Insects of Campbell Island. Talitrid amphipod crustaceans. *Pacific Insects Monograph* **7:** 45–57.

**Bowen FE. 1978.** Ohai coalfield, Southland. In: Suggate RP, Stevens GR, Te Punga MT, eds. *The geology of New Zealand* (2 volumes) Wellington: New Zealand Geological Survey, 397–398.

**Bradshaw JD, Pankhurst RJ, Weaver SD, Storey BC, Muir RJ, Ireland TR. 1997.** New Zealand superterranes recognized in Marie Byrd Land and Thurston Island. In: Ricci CA, ed. *The Antarctic region: geological evolution and processes. Proceedings of the VII International Symposium on Antarctic Earth Sciences.* Siena: Terra Antarctica Publications, 429–436.

**Brookes AE. 1951.** *The Coleoptera of the Auckland and Camp-*

*bell Islands*. Cape Expedition Series Bulletin 5. Wellington, DSIR.

**Cande SC, Stock JM, Müller RD, Ishihara T. 2000.** Cenozoic motion between East and West Antarctica. *Nature* **404:** 145–150.

**CANZ. 1996.** *Undersea New Zealand (New Zealand region physiography)*, scale 1: 4 000 000, 2nd edn. New Zealand Oceanographic Institute Chart, Miscellaneous Series No. 74. Wellington, New Zealand: NIWA.

**Carlton CE, Leschen RAB. 2001.** Species in sympatry. *Pselaphotheseus* of Campbell Island (Coleoptera: Staphylinidae: Pselaphinae). *New Zealand Journal of Zoology* **29:** 287–393.

**Chambers GK, Boon WM, Buckley TR, Hitchmough RA. 2001.** Using molecular methods to understand the Gondwanan affinities of the New Zealand biota: three case studies. *Australian Journal of Botany* **49:** 377–387.

**Chin NKM, Brown MT, Heads MJ. 1991.** The biogeography of Lessoniaceae, with special reference to *Macrocystis* C. Agardh (Phaeophyta: Laminariales). *Hydrobiologia* **215:** 1–11.

**Chown SL. 1994.** Historical ecology of sub-Antarctic weevils (Coleoptera: Curculionidae): patterns and processes on isolated islands. *Journal of Natural History* **28:** 411–433.

**Coleman PJ. 1997.** Australia and the Melanesian arcs: a review of tectonic settings. *AGSO Journal of Australian Geology and Geophysics* **17:** 113–125.

**Cook LG, Gullan PJ, Trueman HE. 2002.** A preliminary phylogeny of the scale insects (Hemiptera: Sternorrhyncha: Coccoidea) based on nuclear small-subunit ribosomal DNA. *Molecular Phylogenetics and Evolution* **25:** 43–52.

**Cooper RA. 1975.** New Zealand and south east Australia in the early Paleozoic. *New Zealand Journal of Geology and Geophysics* **18:** 1–20.

**Cooper RA, Landis CA, Le Masurier WC, Spenden IG. 1982.** Geological history and regional patterns in New Zealand and west Antarctica, their paleotectonic and paleogeographic significance. In: Craddock C, ed. *Antarctic geoscience*. USA: University of Wisconsin Press, 43–53.

**Cox JM. 1987.** Pseudococcidae (Insecta: Hemiptera). *Fauna of New Zealand* **11:** 1–232.

**Cracraft J. 1991.** Patterns of diversification within continental biotas: hierarchical congruence among areas of endemism of Australian vertebrates. *Australian Systematic Botany* **4:** 211–227.

**Craw RC. 1989.** Continuing the synthesis between panbiogeography, phylogenetic systematics and geology as illustrated by empirical studies on the biogeography of New Zealand and the Chatham Islands. *Systematic Zoology* **37:** 291–310.

**Craw RC. 1999.** Molytini (Isecta: Coleoptera: Curculionidae: Molytinae). *Fauna of New Zealand* **39:** 1–68.

**Cross EW, Quinn CJSJ, Wagstaff SJ. 2002.** Molecular evidence for the polyphyly of *Olearia* (Astereae: Asteraceae). *Plant Systematics and Evolution* **235:** 99–120.

**Dettmann ME, Pocknall DT, Romero EJ, Zamaloa MC. 1990.** *Nothofagidtes* Erdtman ex Potonié 1960: a catalogue of species with notes on the palaeogeographic distribution of *Nothofagus* Bl. (Southern Beech). *New Zealand Geological Survey Palaeontological Bulletin* **60:** 1–79.

**Duckhouse DA. 1971.** Entomology of the Aucklands and other islands south of New Zealand. Diptera: Psychodidae. *Pacific Insects Monograph* **27:** 317–325.

**Dugdale JS. 1964.** Insects of Campbell Island. Appendix Lepidoptera: Geometridae. *Pacific Insects Monograph* **7:** 607–623.

**Dugdale JS. 1971.** Entomology of the Aucklands and other islands south of New Zealand: Lepidoptera, excluding non-crambine Pyralidae. *Pacific Insects Monograph* **27:** 55–172.

**Dugdale JS. 1975.** Report on entomological work carried out during the Auckland Islands Expedition 1972–73. In: Yaldwyn JC, eds. *Preliminary results of the Auckland Islands Expedition 1972–73*. New Zealand: Department of Lands & Survey, 74–83.

**Duncan KW. 1994.** Terrestrial Talitridae (Crustacea: Amphipoda). *Fauna of New Zealand* **31:** 1–128.

**Early JW. 1978.** New Diapriinae (Hymenoptera: Diapriidae) from the South Island and Subantarctic Islands of New Zealand. *Journal of the Royal Society of New Zealand* **8:** 207–228.

**Early JW. 1980.** The Diapriidae (Hymenoptera) of the Southern Islands of New Zealand. *Journal of the Royal Society of New Zealand* **10:** 153–171.

**Emberson RM. 1998.** The beetle (Coleoptera) fauna of the Chatham Islands. *New Zealand Entomologist* **21:** 25–64.

**Emerson BC, Wallis GP, Patrick BH. 1997.** Biogeographic area relationships in southern New Zealand: a cladistic analysis of Lepidoptera distributions. *Journal of Biogeography* **24:** 89–99.

**Evenhuis NL. 2003.** Catalog of the Diptera of the Australasian and Oceanian Regions. http://hbs.bishopmuseum.org/aocat/aocathome.html

**Fleming CA. 1980.** *The geological history of New Zealand and its life*. Auckland: Auckland University Press.

**Forster RR. 1955.** Spiders from the subantarctic islands of New Zealand. *Records of the Dominion Museum* **2:** 167–203.

**Forster RR. 1964.** The Araneae and Opiliones of the subantarctic islands of New Zealand. *Pacific Insects Monograph* **7:** 58–115.

**Forster RR. 1970.** The spiders of New Zealand. Part III. *Otago Museum Bulletin* **3:** 1–184.

**Forster RR, Platnick NI. 1985.** A review of the austral spider family Orsolobidae (Arachnida, Araneae), with notes on the superfamily Dysderoidea. *Bulletin of the American Museum of Natural History* **181:** 1–230.

**Forster RR, Wilton CL. 1973.** The spiders of New Zealand. Part IV. *Otago Museum Bulletin* **4:** 1–309.

**Gaina C, Roest WR, Müller RD, Symonds P. 1998.** The opening of the Tasman Sea: a gravity anomaly animation. *Earth Interactions* **2–004** (Electronic Journal).

**Gillespie RG, Roderick GK. 2002.** Arthropods on islands: colonisation, speciation, and conservation. *Annual Review of Entomology* **47:** 595–632.

**Golubeva OG, Stephenson SL. 2003.** Zoosporic fungi from subantarctic Campbell Island. *New Zealand Journal of Botany* **41:** 319–324.

**Gourlay ES. 1950.** Auckland Island Coleoptera. *Transactions of the Royal Society of New Zealand* **78:** 171–202.

**Greenslade P. 1990.** Notes on the biogeography of the free-living terrestrial invertebrate fauna of Macquarie Island with an annotated checklist. *Papers and Proceedings of the Royal Society of Tasmania* **124:** 35–51.

**Gressitt JL. 1964.** Insects of Campbell Island. Introduction. *Pacific Insects Monograph* **7:** 3–33.

**Gressitt JL, Samuelson GA. 1964.** Insects of the Campbell Island. Coleoptera: Hydraenidae, Ptiliidae, Leptodiridae, Byrrhidae, Lathridiidae, Melandryidae. *Pacific Insects Monograph* **7:** 376–390.

**Gressitt JL, Wise KAJ. 1971.** Entomology of the Aucklands and other islands south of New Zealand. Introduction. *Pacific Insects Monograph* **27:** 1–45.

**Griffiths JR. 1971.** Reconstructions of the south-west Pacific margin of Gondwanaland. *Nature* **234:** 203–207.

**Grunow AM, Kent DV, Dalziel IWD. 1991.** New paleomagnetic data from Thurston Island; implications for the tectonics of west Antarctica and Weddell Sea opening. *Journal of Geophysical Research, B, Solid Earth and Planets* **96:** 17,935–17,954.

**Gwynne DT. 1995.** Phylogeny of the *Ensifera* (Orthoptera): a hypothesis supporting multiple origins of acoustical signalling, complex spermatophores and maternal care in crickets, katydids, and weta. *Journal of Orthoptera Research* **4:** 203–218.

**Hammond PM. 1975.** The phylogeny of a remarkable new genus and species of gymnusine staphylinid (Coleoptera) from the Auckland Islands. *Journal of Entomology* **44:** 153–173.

**Hansen M. 1998.** *World catalogue of insects: Hydraenidae (Coleoptera) 1.* Stenstrup: Apollo Books.

**Harper PC, Kinsky FC. 1978.** *Southern albatrosses and petrels.* Wellington: Victoria University Press.

**Harrison RA. 1955.** Report on the Diptera of Auckland and Campbell islands. *Records of the Dominion Museum* **2:** 205–231.

**Harrison RA. 1964.** Insects of Campbell Island. Diptera. *Pacific Islands Monograph* **7:** 304–324.

**Heads M. 1992.** Taxonomic notes on the *Hebe* complex (Scrophulariaceae) in the New Zealand mountains. *Candollea* **47:** 583–595.

**Heads M. 1993.** Biogeography and biodiversity in *Hebe*, a South Pacific genus of Scrophulariaceae. *Candollea* **48:** 19–59.

**Heads M. 1994a.** Biogeography and evolution in the *Hebe* complex (Scrophulariaceae) *Leonohebe* and *Chionohebe*. *Candollea* **49:** 81–119.

**Heads M. 1994b.** A biogeographic review of *Parahebe* (Scrophulariaceae). *Botanical Journal of the Linnean Society* **115:** 65–89.

**Heads M. 1996.** Biogeography, taxonomy and evolution in the Pacific genus *Coprosma* (Rubiaceae). *Candollea* **51:** 381–405.

**Heads M. 1998a.** Biodiversity in the New Zealand divaricating tree daisies: *Olearia* sect. nov. (Compositae). *Botanical Journal of the Linnean Society* **127:** 239–285.

**Heads M. 1998b.** Biogeographic disjunction along the Alpine fault, New Zealand. *Biological Journal of the Linnean Society* **63:** 161–176.

**Heads M. 1999.** Vicariance biogeography and terrane tectonics in the South Pacific: analysis of the genus *Abrotanella* (Compositae). *Biological Journal of the Linnean Society* **67:** 391–432.

**Henderson IM. 1991.** Biogeography without area? *Australian Systematic Botany* **4:** 59–71.

**Hughes M, Weir A, Leschen R, Judd C, Gillen B. 2004.** New species and records of Laboulbeniales from the subantarctic islands of New Zealand. *Mycologia* **96:** 1355–1369.

**Hutton FW. 1902.** On a small collection of Diptera from the southern islands of New Zealand. *Transactions of the New Zealand Institute* **34:** 169–175.

**Illies J. 1964.** Insects of Campbell Island. Plecoptera. *Pacific Insects Monograph* **7:** 208–215.

**Johns PM. 1974.** Arthropoda of the subantarctic islands of New Zealand (1) Coleoptera: Carabidae. Southern New Zealand, Patagonia, and Falkland Islands insular Carabidae. *Journal of the Royal Society of New Zealand* **4:** 288–302.

**Johnston PR. 2002a.** *Biscogniauxia*, Campbell Island, rats, and beetles. *The Mycologist* **16:** 172–174.

**Johnston PR. 2002b.** *Clypeostroma spilomeum* and *C. hemisphaericum* reassessed. *New Zealand Journal of Botany* **40:** 265–268.

**Johnston P, Leschen RAB, Stephenson S. 2001**. Subantarctic Fungi, slime moulds and beetles. http://www.landcareresearch.co.nz/research/biodiversity/antarctica/

**Kamp PJJ. 1986.** Late Cretaceou-Cenozoic tectonic development of the southwest Pacific region. *Tectonophysics* **121:** 225–251.

**Kennedy M, Gray RD, Spencer HG. 2000.** The phylogenetic relationships of the shags and cormorants: can sequence data resolve a disagreement between behavior and morphology? *Molecular Phylogenetics and Evolution* **17:** 345–359.

**Kennedy M, Spencer HG. 2000.** Phylogeny, biogeography, and taxonomy of Australasian teals. *The Auk* **117:** 154–163.

**Kerrich CG. 1964.** Insects of Campbell Island. Hymenoptera: Encyrtidae. *Pacific Insects Monograph* **7:** 504–506.

**Kerrich CG, Yoshimoto CM. 1964.** Insects of Campbell Island. Hymenoptera: Eulophidae. *Pacific Insects Monograph* **7:** 501–503.

**Klimaszewski J. 1979.** A revision of the Gymnusini and Deinopsini of the world (Coleoptera: Staphylinidae, Aleocharinae). *Agriculture Canada Monograph* **7:** 1–169.

**Kuschel G. 1964.** Insects of Campbell Island. Coleoptera: Curculionidae of the subantarctic islands of New Zealand. *Pacific Insects Monograph* **7:** 416–493.

**Kuschel G. 1971.** Entomology of the Aucklands and other islands south of New Zealand. Coleoptera: Curculionidae. *Pacific Insects Monograph* **27:** 225–259.

**Kuschel G. 1991.** Biogeographic aspects of the subantarctic islands. *Atti dei Convegni Lincei* **85:** 575–591.

**Kuschel G, Chown SL. 1995.** Phylogeny and systematics of the *Ectemnorhinus*-group of genera (Insecta: Coleoptera). *Invertebrate Taxonomy* **9:** 841–863.

**Laird MG. 1993.** Cretaceous continental rifts: New Zealand region. In: Hsü KJ, Ballance PF, eds. *South Pacific sedimentary basins* Sedimentary Basins of the World 2. Amsterdam: Elsevier, 37–49.

**Leschen RAB, Bullians M, Michaux B, Ahn K-J. 2002.** Systematics of *Baeostethus chiltoni*, a subantarctic liparocephaline (Coleoptera: Staphylinidae: Aleocharinae): a pangean relic or a more recent immigrant? *Journal of the Royal Society of New Zealand* **32:** 189–201.

**Macfarlane RP, Andrew IG, Sinclair BJ, Harrison RA, Dugdale JS, Boothroyd IKG, Toft R, Marshall SA, Johns PM, Forsyth DJ, Matile L, Holder P, Oliver HA. 2004.** Checklist of New Zealand Diptera. http://www.ento.org.nz/diptera.htm

**Marris JWM. 2000.** The beetle (Coleoptera) fauna of the Antipodes Islands, with comments on the impact of mice; and an annotated checklist of the insect and arachnid fauna. *Journal of the Royal Society of New Zealand* **30:** 169–195.

**May BM. 1971.** Entomology of the Aucklands and other islands south of New Zealand: Immature stages of Curculionidae. *Pacific Insects Monograph* **27:** 271–316.

**McAdoo D, Laxon S. 1997.** Antarctic tectonics: constraints from an ERS-1 satellite marine gravity field. *Science* **276:** 556–560.

**McGlone MS, Moar NT, Wardle P, Meurk CD. 1997.** Lateglacial and Holocene vegetation and environment of Campbell Island, far southern New Zealand. *The Holocene* **7:** 1–12.

**McKenzie EHC, Foggo MN. 1989.** Fungi of New Zealand subantarctic islands. *New Zealand Journal of Botany* **27:** 91–100.

**McLellan ID. 1977.** New alpine and southern Plecoptera from New Zealand and a classification of the Gripopterygidae. *New Zealand Journal of Zoology* **4:** 119–147.

**McLellan ID. 2001.** *Falklandoperla kelper* new genus and species of Gripopterygidae (Plecoptera) from the Falkland Islands. *Aquatic Insects* **23:** 153–160.

**McLoughlin S. 2001.** The breakup history of Gondwana and its impact on pre-Cenozoic floristic provincialism. *Australian Systematic Botany* **49:** 271–300.

**Mel'nik VA, Stephenson SL. 2001.** Deuteromycetes associated with the litter of a planted *Picea sitchensis* on subantarctic Campbell Island (New Zealand). *Mikologiya I Fitopatologiya* **35:** 36–40.

**Meurk CD, Foggo MN, Wilson JB. 1994.** The vegetation of subantarctic Campbell Island. *New Zealand Journal of Ecology* **18:** 123–168.

**Michaux B. 1998.** Terrestrial birds of the Indo-Pacific. In: Hall R, Holloway JD, eds. *Biogeography and geological evolution of SE Asia*. Leiden: Backhuys, 361–391.

**Michaux B. 2001.** Dispersal versus vicariance, artifice rather than contest. In: Metcalfe I, Smith JMB, Morwood M, Davidson I, eds. *Faunal and floral migrations and evolution in SE Asia-Australasia*. Lisse: Balkema, 311–318.

**Michaux B, White IM. 1999.** Systematics and biogeography of southwest Pacific *Bactrocera* (Diptera: Tephritidae: Dacini). *Palaeogeography, Palaeoclimatology, Palaeoecology* **153:** 337–351.

**Mildenhall DC. 1980.** New Zealand late Cretaceous and Cenozoic plant biogeography: a contribution. *Palaeogeography, Palaeoclimatology, Palaeoecology* **31:** 197–233.

**Miller D. 1950.** *Catalogue of the Diptera of the New Zealand Sub-Region*. Department of Science and Industrial Research Bulletin 100. Wellington, DSIR.

**Mitchell AD, Frodin DG, Heads MJ. 1997.** Reinstatement of *Raukaua*, a genus of the Araliaceae centred in New Zealand. *New Zealand Journal of Botany* **35:** 309–315.

**Mitchell AD, Heenan PB. 2000.** Systematic relationships of New Zealand endemic Brassicaceae inferred from rDNA sequence data. *Systematic Botany* **25:** 98–105.

**Mitchell AD, Meurk CD, Wagstaff SJ. 1999.** Evolution of *Stilbocarpa*, a megaherb from New Zealand's sub-antarctic islands. *New Zealand Journal of Botany* **37:** 205–211.

**Mitchell AD, Wagstaff SJ. 1997.** Phylogenetic relationships of *Pseudopanax* species (Araliaceae) inferred from parsimony analysis of rDNA sequence data and morphology. *Plant Systematics and Evolution* **208:** 121–138.

**Mitchell AD, Wagstaff SJ. 2000.** Phylogeny and biogeography of the Chilean *Pseudopanax laetevirens*. *New Zealand Journal of Botany* **38:** 409–414.

**Morrone JJ. 1998.** On Urvardy's Insulantarctica province: a test from weevils (Coleoptera: Curculionidae). *Journal of Biogeography* **25:** 947–955.

**Morrone JJ, Crisci JV. 1995.** Historical biogeography: introduction to methods. *Annual Review of Ecology and Systematics* **26:** 373–401.

**Morton J, Miller M. 1973.** *The New Zealand sea shore*, 2nd edn. Auckland: Collins.

**Muesebeck CFW. 1964.** Insects of Campbell Island. Hymenoptera: Braconidae. *Pacific Insects Monograph* **7:** 494–495.

**Munroe E. 1964.** Insects of Campbell Island. Lepidoptera: Pyralidae. *Pacific Insects Monograph* **7:** 260–271.

**Norvick MS, Smith MA. 2001.** Mapping the plate tectonic reconstruction of southern and southeastern Australia and implications for petroleum systems. *Australian Petroleum Production and Exploration Association Journal* **2001:** 15–35.

**Noyes JS, Valentine EW. 1989.** Mymaridae (Insecta: Hymenoptera): introduction and review of genera. *Fauna of New Zealand* **17:** 1–95.

**O'Connor T, ed. 1999.** *New Zealand's Subantarctic Islands*. Auckland: Reed.

**Oldroyd H. 1955.** A wingless dolichopodid (Diptera) from Campbell Island. *Records of the Dominion Museum* **2:** 243–246.

**Ordish RG. 1971.** The entomology of the Aucklands and other islands south of New Zealand. Coleoptera: Hydranidae. *Pacific Insects Monograph* **27:** 185–192.

**Ordish RG. 1974.** Arthropoda of the subantarctic islands of New Zealand (3) Coleoptera. Hydrophilidae. *Journal of the Royal Society of New Zealand* **4:** 307–314.

**Parrott AW. 1958.** Fishes from the Auckland and Campbell islands. *Records of the Dominion Museum* **3:** 109–119.

**Patrick B. 1994.** Antipodes Island Lepidoptera. *Journal of the Royal Society of New Zealand* **24:** 91–116.

**Paulin CD, Roberts CD. 1992.** *The rockpool fishes of New Zealand.* Wellington: Museum of New Zealand.

**Pugh PJA. 2004.** Biogeography of spiders (Araneae: Arachnida) on the islands of the Southern Ocean. *Journal of Natural History* **38:** 1461–1487.

**Pugh PJA, Dartnall HJG, McInnes SJ. 2002.** The non-marine Crustacea of Antarctica and the Islands of the southern ocean: biodiversity and biogeography. *Journal of Natural History* **36:** 1047–1103.

**Quate LW. 1964.** Insects of Campbell Island. Diptera: Psychodidae. *Pacific Insects Monograph* **7:** 280–288.

**Richards AM. 1964.** Insects of Campbell Island. Orthoptera: Rhaphidophoridae of Auckland and Campbell Is. *Pacific Insects Monograph* **7:** 216–225.

**Richards AM. 1971.** Revision of the Rhaphidophoridae (Orthoptera) of New Zealand. Part XIII. A new genus from the Snares Island. *Pacific Insects* **12:** 865–869.

**Richards OW. 1955.** A species of Sphaeoceridae (Diptera) from Campbell Island. *Records of the Dominion Museum* **2:** 241.

**Robertson CJR, Nunn GB. 1998.** Toward a new taxonomy for albatrosses. In: Robertson G, Gales R, eds. *Albatross biology and conservation.* Chipping Norton, NSW: Surrey Beatty & Sons, 13–19.

**Rosen BR. 1988.** From fossils to earth history: applied historical biogeography. In: Myers AA, Giller PS, eds. *Analytical biogeography.* London: Chapman & Hall, 437–481.

**Salmon JT. 1949.** *New Sub-Antarctic Collembola.* Cape Expedition Series Bulletin **4:** 1–56.

**Samuelson GA. 1964.** Insects of the Campbell Island. Appendix. Coleoptera: Hydraenidae, Leptodiridae (larvae). *Pacific Insects Monograph* **7:** 624–627.

**Schmid F. 1989.** Les Hydrobiosides (Trichoptera, Annulpalpia). *Bulletin de L'institut Royal des Sciences Naturelles de Belgique, Entomologie* **59** (Suppl.): 1–154.

**Seberg O. 1988.** Taxonomy, phylogeny, and biogeography of the genus *Oreobolus* R.Br (Cyperaceae), with comments on the biogeography of the South Pacific continents. *Botanical Journal of the Linnean Society* **96:** 119–195.

**Siegel-Causey D. 1988.** Phylogeny of the Phalacrocoracidae. *The Condor* **90:** 885–905.

**Smith KGV. 1964.** Insects of Campbell Island. Diptera: Empididae. *Pacific Insects Monograph* **7:** 325–328.

**Spilman TJ. 1967.** The Heteromerous intertidal beetles (Coleoptera: Salpingidae: Aegialitinae). *Pacific Insects Monograph* **9:** 1–21.

**Spörli KB. 1987.** Development of the New Zealand microcontinent. In: Monger JWH, Francheteau J, eds. *Circum-Pacific orogenic belts and evolution of the Pacific Ocean Basin.* Washington, DC: American Geophysical Union, 115–132.

**Steel WO. 1964.** Insects of Campbell Island. Coleoptera: Staphylinidae. *Pacific Insects Monograph* **7:** 340–375.

**Steffan WA. 1964.** Insects of Campbell Island. Diptera: Sciaridae. *Pacific Insects Monograph* **7:** 292–299.

**Stock JM, Cande SC, Raymond CA. 1996.** Updated history of the Bellingshausen Plate. Fall Meeting, American Geophysical Union: San Francisco, CA, December 15–19, 1996.

**Strandtmann RW. 1964.** Insects of Campbell Island. Prostigmata: Eupodidae, Penthalodidae, Rhagidiidae, Nanorchestidae, Tydeidae, Ereynetidae. *Pacific Insects Monograph* **7:** 148–165.

**Sublette JE, Worth WW. 1980.** The Chironomidae and Ceratopogonidae (Diptera) of New Zealand's subantarctic islands. *New Zealand Journal of Zoology* **7:** 299–378.

**Sutherland R. 2000.** Recent progress in understanding the structure and petroleum potential of southern South Island's frontier basins. *Proceedings 2000 New Zealand Petroleum Conference, 19–22 March 2000*, 153–162.

**Sutherland R, Melhuish A. 2000.** Formation and evolution of the Solander Basin, southwestern South Island, New Zealand, controlled by a major fault in continental crust and upper mantle. *Tectonics* **19:** 44–61.

**Swenson U, Bremer K. 1997.** Pacific biogeography of the Asteraceae genus *Abrotanella* Senecioneae. Blennospermatinae. *Systematic Botany* **22:** 493–508.

**Swofford DL. 2003.** *Phylogenetic Analysis Using Parsimony (PAUP*), v. 4 0.*, Sunderland, MA: Sinauer Associates.

**Taylor CK. 2004.** New Zealand harvestmen of the subfamily Megalopsalidinae (Opilones: Monoscutidae) – the genus *Pantopsalis. Tuhinga* **15:** 53–76.

**Tokunaga M. 1964.** Insects of Campbell Island. Diptera: Ceratopogonidae. *Pacific Insects Monograph* **7:** 289–291.

**Townes H. 1964.** Insects of Campbell Island. Hymenoptera: Ichneumonidae. *Pacific Insects Monograph* **7:** 496–500.

**Townsend JL. 1971.** The entomology of the Aucklands and other islands south of New Zealand. Coleoptera: Caribidae: Broscini. *Pacific Insects Monograph* **27:** 173–184.

**Trewick SA, Wallis GP. 2001.** Bridging the 'Beech-gap': New Zealand invertebrate phylogeography implicates Pleistocene glaciation and Pliocene isolation. *Evolution* **55:** 2170–2180.

**Turbott EG. 1990.** *Checklist of the birds of New Zealand*, 3rd edn. Wellington: Ornithological Society of New Zealand.

**Udvardy MDF. 1987.** The biogeographical realm Antarctica: a proposal. *Journal of the Royal Society of New Zealand* **17:** 187–194.

**Valentine EW. 1971.** Entomology of the Aucklands and other islands south of New Zealand. Hymenoptera: Mymaridae. *Pacific Insects Monograph* **7:** 327–333.

**Wagstaff SJ, Bayly MJ, Garnock-Jones PJ, Albach DC. 2002.** Classification, origin and diversification of the New Zealand Hebes (Scrophulariaceae). *Annals of the Missouri Botanical Garden* **89:** 38–63.

**Wagstaff SJ, Garnock-Jones PJ. 1998.** Evolution and biogeography of the *Hebe* complex (Scrophulariaceae) inferred from ITS sequences. *New Zealand Journal of Botany* **36:** 425–437.

**Wagstaff SJ, Wardle P. 1999.** Whipcord Hebes – systematics, distribution, ecology and evolution. *New Zealand Journal of Botany* **37:** 17–39.

**Wallwork JA. 1964.** Insects of Campbell Island. Cryptostigmata (Oribatei) Oppiidae. *Pacific Insects Monograph* **7:** 138–147.

**Ward JB, Leschen RAB, Smith BJ, Dean JC. 2004.** Phylogeny of the caddisfly (Trichoptera) family Hydrobiosidae using larval and adult morphology, with the description of a new genus and species from Fiordland, New Zealand. *Records of the Canterbury Museum* **18:** 23–43.

**Wardle P, Ezcurra C, Ramírez C, Wagstaff SJ. 2001.** Comparison of the flora and vegetation of the southern Andes and New Zealand. *New Zealand Journal of Botany* **39:** 69–108.

**Watt JC. 1971.** Entomology of the Aucklands and other islands to the south of New Zealand. Coleoptera: Scarabaeidae, Byrrhidae, Ptinidae, Tenebrionidae. *Pacific Insects Monograph* **27:** 193–224.

**Wise KAJ. 1964a.** Insects of Campbell Island. *Collembola Pacific Insects Monograph* **7:** 178–207.

**Wise KAJ. 1964b.** Insects of Campbell Island. Trichoptera. *Pacific Insects Monograph* **7:** 253–254.

**Wright SD, Yong CG, Dawson JW, Whittaker DJ, Gardner RC. 2000.** Riding the ice age El Niño? Pacific biogeography and evolution of *Metrosideros* subg. *Metrosideros* (Myrtaceae) inferred from nuclear ribosomal DNA. *Proceedings of the National Academy of Sciences, USA* **10:** 4118–4123.

**Yaldwyn JC. 1975.** *Preliminary results of the Auckland Islands expedition 1972–73*, vol. 3. New Zealand: Department of Lands and Survey.

**Yano K. 1964.** Insects of Campbell Island. Lepidoptera: Tineidae, Elachistidae, Hyponomeutidae, Oecophoridae, Tortricidae, Pterophoridae, Noctuidae. *Pacific Insects Monograph* **7:** 255–259.

**Yoshimoto CM. 1964a.** Insects of Campbell Island. Hymenoptera: Cynipoidea: Eucoilinae. *Pacific Insects Monograph* **7:** 509–512.

**Yoshimoto CM. 1964b.** Insects of Campbell Island. Hymenoptera: Diapriidae. *Pacific Insects Monograph* **7:** 507–508.

**Yukawa J. 1964.** Insects of Campbell Island. Diptera: Cecidomydiidae. *Pacific Insects Monograph* **7:** 300–303.

## Attribution

**Michaux, B., Leschen, R.A.B. 2005. East meets west: Biogeology of the Campbell Plateau. *Biological Journal of the Linnean Society*, 86:95–115.**

First published in 2005 in *Biological Journal of the Linnean Society*, 86(1):95–115, doi.org/10.1111/j.1095-8312.2005.00511.x. Reprinted with permission from Oxford University Press.

# chapter six

# The furious fifties

I'm not quite sure how Rich Leschen and I decided to write about the biota of New Zealand Subantarctic Islands, which are exposed parts of the Campbell Plateau microcontinent, but decide we did and this paper was the result. For me, it represented a change of focus away from Wallacea, which I felt I'd covered as well as I could, to regions closer to home – it was to be my first look at a part of New Zealand in any real detail. It's also the first time the term biogeology was used in one of my publications.

## Plunder, pillage and propitiation

The New Zealand Subantarctic Islands are a collection of archipelagos scattered across the Southern Ocean to the south and southeast of Stewart Island, New Zealand. Located within the Roaring Forties and Furious Fifties, their climate is cool, wet and exceedingly windy. For example, the southernmost island – Campbell Island – experiences winds in excess of 35 knots for an average of 280 days a year and in excess of 50 knots for 100 days, has only 660 hours of sunshine a year and temperatures that rarely exceed 12°C (O'Connor 1999). Yet despite the inclement conditions – or perhaps even because of them – the islands are home to a remarkable biota. Dominated by almost unimaginable numbers of seabirds that use the islands as breeding grounds – an estimated six million birds nest on the Snares Group alone – there are also substantial numbers of seals (including the endemic Hooker's sea lion (*Phocarctos hookeri*) and New Zealand fur seal (*Arctocephalus forsteri*)), most southerly forests of rata (*Metrosideros umbellate*) and fields of colourful megaherbs. Accorded UNESCO World Heritage status in 1998, these islands are among the most important parts of the New Zealand Conservation estate. Their history can be thought of as being in three acts: discovery, exploitation and restoration (Plate 6.1).

All the New Zealand Subantarctic Islands were known to Maori and named by them (NZ Gazetteer, LINZ, www.linz.govt.nz/): *Moutere Mahue* (Antipodes), *Motu Ihupuku* (Campbell Island), *Motu Maha* (Auckland Island), *Moutere Hauriri* (Bounty Island) and *Tini Heke* (Snares Island). Clashes with sealing gangs in the early years of the nineteenth century (Smith 2002) indicated that they represented important food resources for pre-nineteenth century Maori. The two largest islands – Auckland and

***Plate 6.1*** Col-Lyall Saddle, Campbell Island.

**Source: Sharon Kast.**

Campbell – were discovered by Europeans in 1806 and 1810 respectively during searches for new sealing grounds, encouraged perhaps by the earlier discovery of the smaller islands. The Bounty Islands, named by the infamous Captain Bligh after his ship, had been the first to be found in 1788, followed by the Snares in 1791 and the Antipodes (originally called the Penantipodes) in 1800 (O'Connor 1999).

Discovery was quickly followed by unchecked exploitation that caused considerable damage to the habitats of these mostly unmodified environments, a pattern repeated across all the Subantarctic archipelagos (Russ 2007). Sealers were the first to commercially exploit the natural resources of the islands by slaughtering seals to process their skins and extract oil from their blubber for the London market. There were two peaks in harvesting: the years around 1810 following the islands' discovery and in the mid-1820s. Smith (2002) attributed the intervening hiatus partly to a slump in prices but mainly to clashes with Maori, who presumably were incensed at the wholesale destruction they witnessed of their nga waahi tuku iho tuku iho. By 1840, the seal colonies were gone and the industry had collapsed. Some sporadic harvesting continued during the second half of the nineteenth century with occasional culls in response to fishing industry pressure until 1946 (Smith 2002). To give some idea of the scale of this destruction, the present population of New Zealand Fur Seals is estimated to be 50,000, which can be compared to the 60,000 skins of this species harvested from just the Snares Group in a single season by one vessel (O'Connor, 1999). Modern trawling in subantarctic waters is still responsible for seal and seabird deaths as bycatch, which is of particular concern for the endangered Hooker's sea lion (http://archive.stats.govt.nz/) (Plate 6.2).

Whalers hunted the Southern Right Whale or Tahura (*Eubalaena australis)* in the waters around the Subantarctic Islands and mainland New Zealand from 1827 until 1980, with peak numbers killed in the mid-1840s. Hunting was conducted from shore-based whaling stations because these whales used the sheltered inland waters around New Zealand and the New Zealand Subantarctic Islands as winter breeding grounds. Jackson et al. (2016) estimated that somewhere between 35,000 and 41,000 Southern Right Whales were killed during this period, reducing the population to perhaps 200 individuals by 1900. In 1992, a breeding population of Southern Right Whales was discovered at the Auckland Islands (Rayment et al. 2012), and an increasing number of animals are now being seen around mainland New Zealand again following a complete absence between 1928 and 1963. While these are pleasing developments, the recovery in numbers of Southern Right Whales has been slow with the

***Plate 6.2*** Hooker Sealions (*Phorcarctos hookerii*), Sandy Bay, Enderby Island.

**Source: Sharon Kast.**

present population still only 12% of an estimated pre-whaling population of 28,000 (Jackson et al. 2016).

The removal of so many seals and whales would have had a direct impact on the functioning of the marine ecosystems around the Subantarctic Islands, but the accidental introduction of rats and mice, which probably dates from the first European contact, had profound effects on nesting sea birds, invertebrates and the vegetation. Sorensen reported that Norway rats (*Rattus norvegicus*) had reached plague proportions by 1942 on Campbell Island and were even eating wax candles, soap and the putty used for waterproofing wooden dinghies (Bailey and Sorensen 1962). Feral cats simply added to the negative impact that the rats and mice were having on bird populations. These effects were compounded by the liberation of pigs, rabbits and goats as food animals for castaways on Auckland, Campbell and Enderby islands, and the introduction of sheep and cattle when farming was established on the larger islands. Campbell Island pastoral leases were advertised in 1895 and the island was stocked with sheep and cattle (Russ 2007). At its peak, there were 4000 to 5000 head of stock, but falling wool prices and increased shipping costs led to the initiative's abandonment in 1932. Auckland Island was also farmed for a time and it too was abandoned in the early 1930s. Feral animals continued to be present following the collapse of these industries.

The effect of grazing animals on the flora was disastrous. Bailey and Sorensen (1962), in a report by a Denver Museum expedition to Campbell Island in January and February 1958, commented on the almost complete disappearance of palatable species such as tall tussock, *Danthonia* (now *Chionochloa*) and its replacement by *Poa litorosa*, the confinement of the megaherbs *Pleurophyllum, Anisotome* and *Stilbocarpa* (Plate 6.3) to inaccessible sites such as steep slopes and ledges, the increase in abundance of unpalatable species and the erosion of upland areas caused by overgrazing. Meurk et al. (1994) described the recovery of the Campbell Island flora following the progressive removal of feral sheep and cattle, particularly on the more fertile sites and those closest to seed sources. The effects of grazing were still apparent in the complex vegetation patterns these authors observed and documented, and which they attributed to progressive recolonisation of overgrazed areas. The result was a complex mosaic of near pristine, variably altered and degraded patches juxtaposed in a seemingly random way. Meurk et al. (1994) suggested that the vegetation would recover over the coming decades, which seems to be the case for the tall tussock *Chionochloa* (Keey 2004).

Introduced plants were not thought to present a serious ecological threat by Meurk et al. (1994), their being generally restricted to areas of previous human habitation such as Beeman Point wharf and the (now unmanned) meteorological station (Keey 2004), but some adventive species are already widespread within tussock and swamps, and *Poa pratensis*

***Plate 6.3*** Megaherbs: Campbell Island carrot (*Anisotome latifolia*) and Ross lily (*Bulbinella rossii*).

**Source: Sharon Kast.**

forms dense swards along the littoral fringe of Perseverance Harbour. Meurk et al. (1994) argued that native plants will normally be able to outcompete adventive species under the exacting conditions that exist in the Subantarctic. Whether this turns out to be true or not will be settled in time, but the effects of adventive plants – actual or potential – on the functioning of pre-human ecosystems is generally given less prominence than the effects of introduced mammals.

Two things saved the situation in the 1940s from being a complete ecological disaster. First, there was a slow change in attitude about the value of the flora and fauna of these islands among politicians and law makers; second, even the most degraded island groups had pest-free and unmodified satellite islands that acted as refugia that preserved parts of the original biota. For example, the Campbell Island teal (*Anas nesiotis*) was able to survive on Dent Island, and the Campbell Island snipe (*Coenocorypha aucklandica perseverance*) on Jacquemart Island, which were both rat-free. Pest and feral animal eradication programmes over the past few decades have promoted the recovery of the flora and fauna on the main island groups. Cattle and sheep were completely removed from Campbell Island by 1991 and rats were eradicated in 2001 (McClelland and Tyree 2002; McClelland 2011). Cats died out as an unexpected consequence of the removal of the cattle and sheep, possibly as a result of tussock reinvading grazed areas. Goats were also removed from Auckland Island and rabbits, cattle and mice from Enderby Island by the early 1990s. The Antipodes were declared mice-free in 2018 following an eradication programme undertaken in June of 2016. Interestingly, this project was largely funded through public donations and dollar-for-dollar matching by the Morgan Foundation. As of 2018, the only introduced animals present in the New Zealand Subantarctic Islands are pigs, cats and mice on Auckland Island. The effect on the flora and fauna following the removal of these introduced animals was immediate, with the return of teal and snipe to 'mainland' Campbell Island and the spread of vegetation from inaccessible sites that palatable endemic species had survived on. The spectacular recovery of the natural vegetation and mammal and bird populations is the basis for a growing ecotourism industry. This industry, strictly controlled to minimise possible risks, generates revenue, but above all, builds a political constituency to help ensure their protection in perpetuity. Presumably other less iconic species, such as invertebrates, have also started to recover and expand. While the pre-human state has gone forever, pest eradication has recreated some of the least-modified environments in the world.

## *Charles Fleming*

I met Charles Fleming at a conference in Wellington during the early-1980s, not long before his untimely death. We shared an interest in Tertiary fossil Mollusca and he was interested in a talk I'd given at a Systematic Society of New Zealand conference in Wellington reporting the preliminary results of genetic studies of the New Zealand marine gastropod *Amalda*. It was evidently typical of the man, who was one of New Zealand's foremost scientists, to spend time chatting with a graduate student and taking an interest in their work. He was rather old-school even then – smartly suited with a spotted bow tie – an independently wealthy man who had no great need to conform to anyone's expectations.

Early in 1942, shortly after graduating, Fleming had been recruited as part of the 'Cape Expedition', a New Zealand government wartime initiative to set up three coast-watching stations on the Auckland and Campbell Islands in response to worries that German naval raiders were using the harbours at Auckland and Campbell Islands as bases. Fleming was part of the first expedition and spent a year (February 1942–February 1943) stationed at Carnley Harbour, Auckland Island, where, in addition to performing his coast-watching duties, he also studied the island's flora, fauna and geology. Reading the diaries he kept while stationed on Auckland Island (McEwen 2006) gives an intimate insight into the daily lives of the expedition members, and details his numerous studies and observations of the natural world around him. Fleming was not the only eminent New Zealand scientist to be part of the Cape Expedition over the five years of its operation; distinguished members also included Graham Turbott and Robert Falla (who not only served in the field but also acted as a government adviser and made sure each new batch of recruits included scientists). The research of these scientific pioneers to the Subantarctic Islands was published after the war by the Department of Scientific and Industrial Research as the *Cape Expedition – Scientific Results of the New Zealand Sub Antarctic Expedition 1941–1945* (https://collections.tepapa.govt.nz/object/659). Fleming also visited the Snares Island shortly after the war in 1947, when he collected specimens and described the geology, and was to return to Auckland Island one last time at the end of 1972 as a member of a joint Lands and Survey and National Museum expedition. There was an unscheduled stop at the Snares on the return trip when members, including Fleming, landed on a number of islands of the Western Chain (McKewen 2005).

While I can find no direct evidence that Fleming (or other scientists who served on Campbell or Auckland Islands) actively lobbied the post-war New Zealand government to protect and conserve the flora and fauna of the Subantarctic Islands in the immediate post-war years, it would be surprising if he didn't, especially in the light of his later environmental activism (McKewen 2005). We do know that Fleming argued that access to the islands should be strictly controlled and limited to scientific purposes during his time as convener of the Scientific Subcommittee of Lands and Survey's Subantarctic Reserves Committee in the early 1970s. For example, the committee rejected proposals for mineral exploration on Auckland and Campbell Islands, and a request to use Auckland Island as a base by treasure hunters wishing to salvage gold from the wreck of the *General Grant* that foundered there in 1866. As far as Fleming was concerned, 'Scientists in New Zealand have always had particular regard for islands where human modification has been kept to a minimum' (McKewen 2005). Perhaps the change in official attitude from one of benign neglect to active conservation was a direct consequence of these scientists' wartime experience and realisation of how special the islands were. In any event, the government's decision to include scientists rather than sending purely military detachments to these remote outposts is indicative that such a change was already nascent by 1940.

Dell (2000) suggested that his stay on Auckland Island as a young man sparked Fleming's interest in biogeography. Fleming's first biogeographic essay appeared in *Tuatara* in 1949 (Fleming 1949) in which he classified the New Zealand biota in terms of old endemics, and older and younger dispersers from various places such as Australia, Antarctica and the Indo-Pacific. McKewen (2005) pointed out that there is no mention of continental drift in Fleming (1949), which was a popular and widely discussed theory among biologists at the time but anathema to New Zealand geologists. McKewen suggested that this reflected Fleming's geological affiliations and she is undoubtedly correct in this, but even his biogeographical opus magnum published in 1979 – *The Geological History of New Zealand and Its Life* – had little reference to plate tectonics apart from using the breakup of Gondwana and the formation of ocean basins to explain the presence of palaeo-endemics in New Zealand.

I'm surprised Fleming wasn't more enthusiastic about tectonic theory as he was aware of its development from its earliest days. His visit to the Scripps Institution of Oceanography in 1960 and his attendance at the 10th Pacific Science Congress held in Honolulu exposed him to the radical new ideas being developed by Scripps Institution scientists such as Ronald Mason and Arthur Raff, who discovered magnetic lineations on the ocean floor (Mason and Raff 1961), and Robert Dietz who developed the key plate tectonic concept of sea-floor spreading (Dietz 1961). This was plate tectonic theory at the ground level. According to his daughter (McKewen 2005), he returned from Honolulu 'greatly excited'. Fleming also had a long correspondence with Tuzo Wilson, another early plate tectonics pioneer, and must surely have been familiar with Wilson's work on transform faults and their role in plate tectonic theory. There was clearly a difference between Fleming's personal view, which appears positive and enthusiastic, and his professional reticence to promote plate tectonics in his writings. His caution was undoubtedly a result of the conservatism of the New Zealand geological community of which he was part, and one can understand why he opted for a 'don't rock the boat' approach, but I also think it shows Fleming's dislike of being overly theoretical. He was much more at home with the 'facts'.

*The Geological History of New Zealand and Its Life* (Fleming 1979) was certainly influential in my own development as a biogeologist. It is unique in the biogeographic literature because it was written from the perspective of a palaeontologist, and thus deals with New Zealand biogeography from a geological perspective. Craw (1978) rejected Fleming's biogeographical and evolutionary views because they were derived from a Darwinian perspective, a view I also share with Craw, but what I liked about the book was Fleming's ability to synthesise a wealth of information and present it in a coherent form. He was able to give a panoramic sweep taking in 500 million years of history, and in the process brought together material that would have otherwise remained scattered across numerous technical works.

The picture Fleming painted was of an ever-changing and dynamic biota over this time span. His analysis of these changes was a refinement and update of the concept of biogeographical elements first developed in Fleming (1949), and how the relative importance of each element changed through time. Fleming interpreted these changes in terms of dispersal pathways or – when discussing the development of endemic elements – lack of dispersal pathways. I was particularly interested in what he had to say about the Cenozoic and especially the Lower Miocene influx of tropical Indo-Malay elements such as reef-building corals, large foraminifera and coconuts into northern New Zealand. My own research into New Zealand fossil and extant ancillid gastropods (Michaux 1991) included species that were part of this influx. While Fleming interpreted this event in terms of northern dispersal pathways, I was much more interested in its association with the emplacement of the Northland Allochthon (Ballance and Spörli 1979). To me the two events were linked as I've previously alluded to in Chapter 2 and to which I will return to discuss in detail in Chapter 9.

I think Fleming's strength as a scientist was his great knowledge and practical experience over a broad range of disciplines – palaeontology, systematics, ornithology, zoology, ecology, to name a few – that gave his work a breadth that demanded respect. As I have said, he was not really a theoretician and I suspect he was not greatly interested in theoretical 'speculation'. In the epilogue to *The Geological History of New Zealand and Its Life* he wrote:

> the gulf between raw data and paleogoegraphic conclusions has often been crossed by a rather flimsy bridge of projection and extrapolation, and conclusions must always be subject to modification or abandoned in the light of new data.

This I rather think sums up his preference for the empirical.

Charles Fleming was probably the last great generalist of New Zealand science, and biogeography was the means by which he was able to integrate these interests. How then should we assess his contribution to New Zealand biogeography? In the present fashion for neo-dispersalism, one would have thought his reputation as a leading dispersalist biogeographer would be assured, but I suspect the scant attention paid to history by this movement would have grated on him. The panbiogeographers dismissed him as a relic of the past and out of touch with modern thought. I valued his empirical approach in detailing the distribution of 'biotic elements' in time and space, which I think is enough to confirm his place as an important figure of his time and a significant contributor to a growing understanding of New Zealand's biogeographic history. I also valued the fact that he was a generalist at a time of increasing specialisation when the task of the scientist was becoming more prescribed and narrow. While being an expert in a specialist field has its uses, specialisation inevitably leads to a narrowing of vision and loss of any possibility of synthesis and the development of a holistic view.

## Geology of the Campbell Plateau

The geology of the Campbell Plateau is mostly unknown because it's almost entirely submerged. All we know directly of the geology, apart from some dredged samples and a few exploration well logs, comes from the islands themselves. Figure 6.1 shows the basement rocks of the North and South Islands, New Zealand and their Provincial affiliations and should be referred to in the following section and succeeding chapters.

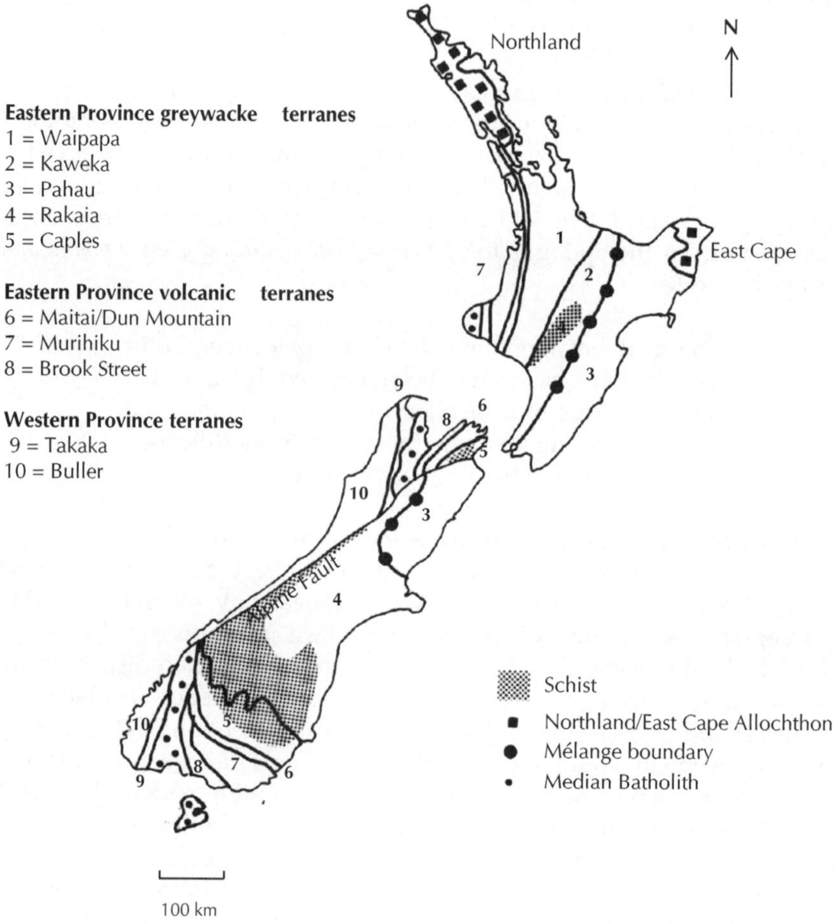


*Figure 6.1* Basement terranes of New Zealand. Western Province terranes are Palaeozoic in age and formed part of the Gondwanan margin prior to the suturing of Eastern Province terranes in the Early Cretaceous. The Median Batholith marks the suture zone between these two provinces. The Northland and East Cape Allochthon is a Neogene obduction suite that was contiguous before later plate boundary readjustments translated the East Cape region southwards.

## *Basement rocks*

Campbell Plateau basement rocks are composed of silicic to intermediate plutonic rocks and quartzoze metasedimentary rocks (Beggs et al. 1990). Granites exposed on the Bounty Islands were emplaced at 188 Ma (Adams and Gullen 1978). This age is somewhat enigmatic because it is too young to be associated with the Devonian Tuhua orogeny and too old for the Cretaceous Rangitata orogeny. Granitic rocks are also exposed on Auckland Island (Beggs et al. 1990) and Snares Island (Scott et al. 2015). The I-type (subduction related) Snare's Granite is dated at 109 Ma and intrudes the Broughten Granodiorite (dated at 114 Ma). Both were deformed at 95 Ma, possibly related to the broadly contemporaneous formation of an extensional ductile shear zone on Stewart Island. Unmetamorphosed basaltic dykes are found on Snares Island, recording a change from subduction to extensional tectonics. The granites of the Snares Group and Auckland Island have been compared to granitic rocks from the west coast of the South Island, Stewart Island and in exploration wells in the Great South Basin on the basis of petrology and age (Beggs et al. 1990). However, there are also Cretaceous I-type granites and evidence for an abrupt change to extensional tectonics in the mid-Cretaceous in Marie Byrd Land, West Antarctica (Di Venere et al. 1994; Weaver et al. 1994), so the issue of any potential correlation of the Campbell Plateau granites is unresolved by these data.

Metasedimentary basement rocks are exposed on Campbell Island (Beggs 1978), and have been recovered from drill holes in the Great South Basin (Beggs et al. 1990) and from dredge samples around the Bounty Islands (Adams and Gullen 1978). Beggs (1978) suggested that the metasedimentary rocks are equivalent to the Western Province Greenland Group from the west coast of the South Island, New Zealand – although again there are similar Lower Palaeozoic metamorphosed turbidites of the Swanson Formation in Marie Byrd Land (Di Venere et al. 1994). An early Triassic metamorphic age reported by Adams and Gullen (1978) for dredged argillites from around the Bounty Islands is anomalous if these rocks are correlates of Western Province sediments. There is no evidence of any Eastern Province terrane (Mortimer et al. 2014) extending onto the Campbell Plateau, and the presence of a broad band of magnetic anomalies – the Campbell Magnetic Zone – is not, as previously suggested (Davey and Christoffel 1978), equivalent to the Stokes Magnetic Anomaly (Beggs et al. 1990). The lack of Eastern Province rocks along the northern edge of the Campbell Plateau suggests a possible closer relationship to Marie Byrd Land, which also lacks equivalents to Eastern Province terranes, than to the Western Province of the South Island.

## *Cenozoic geology*

Cenozoic volcanic rocks are found on Auckland Island, Campbell Island and the Antipodes. The Antipodes are composed of an eroded Pleistocene volcano less than 500,000 years old (Scott et al. 2013). Two episodes of volcanic activity are recorded on both Auckland and Campbell Islands (Hoernle et al. 2006). The oldest activity occurred on Auckland Island when the Carnley Volcano erupted between 37 and 19 Ma (25–21 Ma according to Ritchie and Turnbull (1985)). Further volcanic activity occurred on Auckland Island between 17 and 15 Ma, contemporaneous with the eruption of the Dunedin volcanics and the emplacement of the Menhir Gabbro on Campbell Island (Adams et al. 1979). The youngest activity on Campbell Island was between 7.5 and 6.6 Ma (Hoernle et al. 2006). Adams (1981) claimed that the extensive Cenozoic volcanism on the Campbell Plateau, Chatham Islands and the South Island, New Zealand, showed a linear pattern of eruption ages, becoming younger towards the southeast. He explained this linear pattern by a northwest movement of the Campbell Plateau over a linear mantle source. However, Hoernle et al. (2006) reported 'no correlation among age, location or composition' of the volcanic centres and suggested that magmas had been produced by decompression melting. The implications of this view are discussed here.

The oldest sediments are found on Campbell Island and have been dated as latest Cretaceous to Palaeogene (Beggs 1978; Hollis et al. 1997). The basal Garden Cove Formation consists of up to 100 m of clastics that are non-marine in the lower part and fine upwards. These sediments contain pollen of Proteaceae, podocarps and *Nothofagus*. The overlying Tucker Cove Formation consists of up to 200 m of micritic limestone with a basal sand layer and dates from the Early Eocene to the Oligocene. Hollis et al. (1997) described a mid-Eocene unconformity within the formation. Sedimentary rocks from Auckland Island (Ritchie and Turnbull 1985) consist of a basal volcanic debris flow (Camp Cove Conglomerate) dated as Late Oligocene to earliest Miocene, which is overlain by the mid-Miocene Musgrave Formation that contains pollen of *Nothofagus matauraensis* Couper.

The Campbell Plateau separated from Marie Byrd Land, West Antarctica, at 84 Ma (McAdoo and Laxon 1997). The present thickness of the Campbell Plateau continental crust is about 27 km in the central region and 13 km under the Great South Basin (Grobys et al. 2008). This suggested to Grobys et al. (2008) that stretching occurred in the Great South Basin prior to separation from Marie Byrd Land. Eagles et al. (2004) favoured a formation date prior to 83 Ma, Carter (1988) between 100 and 90 Ma and Cook et al. (1999) at 105 Ma based on the oldest sediments. There are some 8 km of sediments within the basin.

## *Old taxa on young islands*

The phenomenon of islands containing taxa that are older than the islands themselves has been documented in such places as Hawaii (Lewin 1985), the Mascarenes (Le Péchon et al. 2015), New Caledonia (Heads 2008) and the Kermadec Islands (Bronstein et al. 2017). While molecular dating evidence isn't available for any New Zealand Subantarctic endemic generus, there are lines of evidence that indicate that this phenomenon can be seen here too (Kuschel 1975).

The annotated review of all groups from the New Zealand Subantarctic Islands presented in Michaux and Leschen (2005) showed that overall endemism is high (over 80% for arthropods) with many endemic genera and basal species also present. While taxonomic rank does not necessarily equate with age of origin – for example, an individual taxon may have many autapomorphies that could mislead a taxonomist as to the taxon's placement – when many groups occupy basal or isolated phylogenetic positions the simplest hypothesis is that they represent palaeo-endemics rather than misidentified neo-endemics.

Another feature highlighted by Michaux and Leschen (2005) was the high numbers of endemic species in some genera. For example, there are six endemic species of *Gromilus* (Coleoptera) on Auckland Island and seven endemic species of *Spilogona* (Diptera) on Campbell Island. This is clearly an unusual pattern of diversity on what are very small islands. Neo-dispersalists would favour multiple, independent invasions and those who favour sympatric speciation would point to this as a good example of such a process. The question is, did the plateau subside completely and extirpate all terrestrial organisms allowing colonists to disperse on repeated occasions after volcanic eruptions had formed new land, or was there a single colonisation event followed by sympatric speciation, or were these taxa derived from elsewhere on the Campbell Plateau as available land shrank following general subsidence and marine transgression?

The Campbell Plateau was forested at the time of its detachment from West Antarctica and this cool-temperate assemblage was still present on what is now Campbell Island during the Palaeocene. *Nothofagus* pollen was also present during the mid-Miocene of Auckland Island. This implies continuity of a cool-temperate forest on parts of the Campbell Plateau during this time interval. Localised land of some relief can also be inferred from the thickness of clastic Palaeocene sediments found on Campbell Island, and the more general presence of land by the 8 km or so of sediments within the Great South Basin. While these observations are suggestive of land continuity, the key to understanding the subsidence history of the plateau lies, I maintain, in its volcanic record.

Hoernle et al. (2006) argued that the extensive Cenozoic, intra-plate volcanism in the South Island, New Zealand and Campbell Plateau cannot be explained by movement over either a linear mantle source or hot spot, but is best explained as a result of decompression melting. In their model, a loss of higher-density lower-continental lithosphere allows asthenospheric material to rise and partially melt. The amount of lower crustal material lost governs the height of the melting column and hence the composition and volumes of the magmas generated. The large shield volcanoes erupted on Auckland and Campbell Islands were formed above regions of greater lower-lithospheric crustal loss. If this model is correct, then Campbell Plateau crust would have had a lower average density relative to normal continental crust because of the loss of higher-density lower lithosphere. In other words, it would have increased buoyancy. A combination of a crustal thickness of up to 27 km, which is comparable with much of the crust underlying the east coast of the South Island (Mortimer et al. 2002), and a decrease in density with concomitant isostatic rebound could account for a long subaerial history of at least parts of the plateau. Volcanism declined after the Miocene allowing the Campbell Plateau to subside to its present depths and stranding a remnant biota on these volcanic life rafts.

## *The biology of the Chatham Islands*

Chatham Island and other members of the group are emergent parts of the Chatham Rise, which is separated from the Campbell Plateau by the Bounty Trough. These three geological units are usually, indeed universally, treated as a single unit detached from Marie Byrd Land, West Antarctica, in the Late Cretaceous. The Chatham Islands are also usually included as part of the New Zealand Subantarctic Islands from a biological viewpoint too (cf. the parallel arcs model of Robin Craw (Craw 1988)). At first glance, this appears to be reasonable because the Chathams and Subantarctic Islands are small, isolated outcrops within the Southern Ocean that are covered in peat and nutrient-deficient soils, are home to many oceanic seabirds that breed there, have a flora with a prominent 'megaherb' component, and have biotas that contain many endemics. However, the analysis of the avifauna in Michaux and Leschen (2005) indicated that no New Zealand Subantarctic Island, or the Subantarctic Islands as a whole, show a sister-group relationship to the Chatham Islands. The terrestrial birds grouped the Subantarctic Islands in an Australasian/Pacific clade that included both the Chatham Islands and New Zealand. The oceanic seabirds grouped the New Zealand Subantarctic Islands in a clade containing other Subantarctic archipelagos of the southern Indian and Atlantic Oceans and southern South America. The Chatham Islands, on the other hand, were linked to New Zealand, Australia and Melanesia in a sister clade.

The biological link between the Chatham Island's biota and New Zealand is also apparent in other groups. Heenan et al. (2010) investigated the phylogenetic relationships of 35 endemic plants from the Chathams and concluded that the overwhelming majority were sister species to common and widespread New Zealand species, and that only four taxa found in the Chatham Islands were sister species to southern South Island and/or Subantarctic taxa. Linse et al. (2006) examined the Mollusca of Antarctica and the various Subantarctic archipelagos and grouped southern New Zealand with the Auckland and Campbell Islands on the basis of overall biotic similarity, while placing the Chatham Islands at some distance and more basally in the phenogram. An analysis of the beetle fauna described in Emberson (2002) showed that this fauna has high levels of endemism and an overwhelming similarity with New Zealand. Of the 318 Coleoptera found on the Chathams that were detailed in Emberson (2002), approximately 30% – mostly undescribed taxa – had no extra-limital details, 30% were endemic and 40% were also found in New Zealand. Only two species were sister taxa to southern South Island species and a further two to New Zealand plus the Subantarctic Islands.

There is very little biological similarity between the Chatham Islands and the Subantarctic Islands other than through species that are also shared with mainland New Zealand. While the analysis presented here is far from exhaustive and, with the exception of Heenan et al. (2010), is not based on phylogenetic evidence, a provisional hypothesis must be that the Chatham Islands are not biologically related to the Subantarctic archipelagos. On the contrary, the Chatham Islands are clearly related to former east Gondwanan, Australasian areas. If they are not biologically related to Campbell Plateau/West Antarctica/southern South America, then the hypothesis of geological relatedness should be tested, because it is reasonable to assume that geological relatedness should be reflected to some degree in the biology.

Gondwana Research 16 (2009) 655–668

Contents lists available at ScienceDirect

Gondwana Research

journal homepage: www.elsevier.com/locate/gr

# Reciprocality between biology and geology: Reconstructing polar Gondwana

B. Michaux

*Private Bag, Kaukapakapa, New Zealand*

ARTICLE INFO

*Article history:*
Received 11 March 2009
Received in revised form 31 May 2009
Accepted 9 June 2009
Available online 21 June 2009

*Keywords:*
Reconstruction models
Phylogenies
Wegener
East Gondwana
West Gondwana
Chatham Islands
Reciprocal illumination

ABSTRACT

The reciprocal nature of the relationship between historical geology (reconstruction models) and biology (constructing phylogenies) is discussed and the conceptual basis of such a relationship is examined through its historical development. Examples to illustrate aspects of the relationship are drawn from the Cretaceous breakup of polar Gondwana and the Cenozoic history of some of the resultant microcontinental fragments. A new mid-Cretaceous (circa 100 Ma) rift zone, separating the west Gondwanan Campbell Plateau, southern New Zealand, from the east Gondwanan Melanesian Rift is proposed, and biological and geological evidence for it is presented and discussed. It is also suggested that the Bounty Trough, Chatham Rise, and Hikurangi Plateau unit is incorrectly placed in reconstruction models, and it should be fitted outboard of the Melanesian Rift until its mid-Cenozoic attachment to the Campbell Plateau. It is concluded that both reconstruction modelling and phylogenetic analyses have much to gain through 'reciprocal illumination'.



## 1. Introduction

Fig. 1 shows an Early Cretaceous reconstruction of polar Gondwana (southern South America, west Antarctica, Campbell Plateau, southeast Australia, New Zealand, and adjacent parts of east Antarctica) that was tectonically active, and had been for some time. Earlier mid-Jurassic emplacement of the Ferrar tholeiites into an active rift system in Antarctica (which extended into Tasmania and southeast Australia), and linked to the eruption of the Karoo basalts in southern Africa, was the earliest sign of the breakup of the super-continent of Gondwana (Elliot and Fleming, 2000; Riley and Knight, 2001; Veevers, 2006). Elliot and Fleming (2000) suggested that a triple junction existed in the proto-Weddell Sea region linking these spreading centres. During the Cretaceous, active rifting resumed outboard of the Ferrar volcanics and became centred on the West Antarctic Rift System (WARS). This major continental rift, comparable to today's Great Rift Valley of Africa, never developed to the stage of sea-floor spreading, except briefly in the mid-Cenozoic (43 Ma) when the rift became re-activated in the Adare Trough (Cande et al., 2000). However, it was sporadically active for much of the Cretaceous and resulted in extensive crustal thinning. WARS was a major feature of the Cretaceous polar landscape (Fitzgerald, 2002) and separated low relief, cratonic east Gondwana (east Antarctica/Australia) from the higher, often mountainous relief of the geologically composite west Gondwana (southern South America/west Antarctica/Campbell Plateau) (Adams et al., 2005; Vaughan et al., 2005; Torsvik et al., 2008).

The mountains and coastal plains were probably wet and temperate even at high latitudes. The combination of warmth and moisture allowed forests dominated by *Nothofagus* (Dettman and Pocknall, 1990), podocarps and araucarians (Mildenhall, 1980) to grow over much of west Gondwana and coastal east Gondwana, despite the long dark of the polar winter (McLoughlin, 2001).

The Cretaceous was an important time in the evolution of modern life as many major biological groups, including flowering plants (Bremer and Gustafsson, 1997; McLoughlin, 2001), birds and mammals (Hedges et al., 1996), were in the process of diversifying. It was also a period of accelerating fragmentation of Gondwana as a whole, and of the polar region in particular, as rifts opened and new oceans were born (Fig. 1). What happened to the flora and fauna of polar Gondwana as these tectonic events unfolded? Is there any evidence of a 'tectonic imprint' on the distribution patterns of their modern descendants? Wegener was the first to provide a possible answer to the latter question in the fourth and final edition of *Die Entstehung der Kontinente und Ozeane* (Wegener, 1929; English translation Biram, 1966). While Wegener's arguments will be discussed more fully later, he not only used modern and fossil distributions to corroborate his reconstruction models (geology→biology), but he also used distributions of extant species from South America, Australia, and New Zealand to suggest novel pre-breakup reconstructions for these areas (geology←biology). The arrow indicates the directionality of the hypothesis-test cycle. Wegener's use of phylogenies (hypotheses of biological relationships) as both test of geological hypotheses and generator of reconstruction hypotheses highlights an important aspect of the conceptual relationship between biology and geology – reciprocality. While aspects of reciprocality have continued to be explored in the modern geological

*E-mail address:* michaux@xtra.co.nz.

1342-937X/$ – see front matter © 2009 International Association for Gondwana Research. Published by Elsevier B.V. All rights reserved.
doi:10.1016/j.gr.2009.06.002

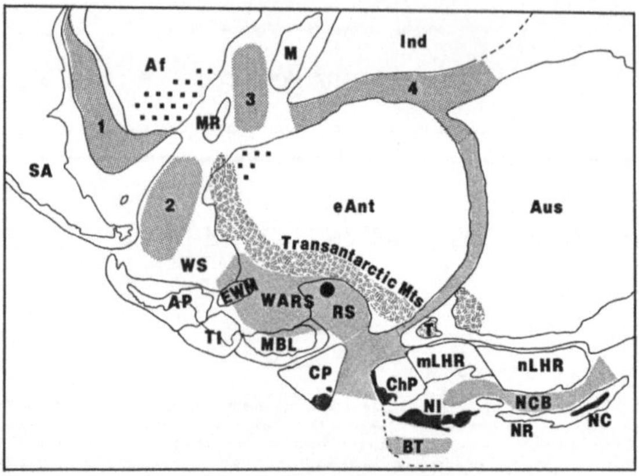


**Fig. 1.** A schematic reconstruction of Cretaceous Gondwana (circa 120–100 Ma). Af = Africa, AP = Antarctic Peninsula, Aus = Australia, BT = Bounty Trough, CP = Campbell Plateau, ChP = Challenger Plateau, EWM = Ellsworth/Whitmore Mountains, mLHR = middle Lord Howe Rise, nLHR = northern Lord Howe Rise, MBL = Marie Byrd Land, MR = Mozambique Ridge, NC = New Caledonia, NI = North Island, New Zealand, NR = Norfolk Ridge, RS = Ross Sea, T = Tasmania, TI = Thurston Island, WARS = West Antarctic Rift System, WS = Weddell Sea. Heavy stipple = oceanic crust: 1 = South Atlantic, 2 = Weddell Sea, 3 = Mozambique Basin, 4 = Enderby Basin/south Indian Ocean. Light stipple = rifted continental crust. Irregular hatching = Ferrar tholeiites. Squares = Karoo volcanics. Filled circle = South Pole.

(e.g. Spalletti et al., 2003; Benedetto, 2004; Isozaki, 2009) and biological literatures (e.g. Craw, 1988), a re-evaluation of Wegener's synthetic approach is overdue.

This paper is divided into three parts. In the first part I will outline the development of the reciprocal nature of biology and geology. Surprisingly reciprocality had been proposed more than 50 years before Wegener by the biogeographer Alfred Russel Wallace, co-author of the theory of evolution by means of natural selection (Wallace, 1858). I will briefly outline Wallace's biogeographic ideas as they pertain to reciprocality, before examining Wegener's reconstruction hypothesis of the Cretaceous polar region in more detail. In part two, I will describe modern views of the mid-Cretaceous (circa 100 Ma) tectonic history of polar Gondwana, with particular reference to west Gondwana, and examine the biological implications for the modern descendents of this polar biota. In the third part, I will describe the Upper Cretaceous (circa 70 Ma) breakup of east Gondwana and illustrate, using published phylogenies, how these biological data can be used to test aspects of the breakup model and, perhaps more importantly, to suggest refinements to that model. I will discuss an alternative reconstruction of the Chatham Rise, New Zealand to illustrate this latter point.

## 2. Reciprocality

Wallace first became aware that biology and geology were connected during his four years collecting specimens in Amazonia (1848–52). Because he was among the first to record accurately where specimens were collected he noticed that some species' distributions are bounded by major geographical features such as rivers, and that often a similar species was to be found the other side of the boundary. For Wallace it was clear that a major physical boundary could act not only as a barrier to dispersal (his term) but that the barrier was in some way *causal* in promoting speciation either side of the boundary. While congruence between biological and geological patterns led Wallace to an understanding of one aspect of the relationship between biology and geology (geology→biology), it was a *lack* of congruence that was to finally lead him to understand the full reciprocal nature of the relationship.

Wallace was to develop his ideas during his time in Indonesia (1854–62), where he spent over eight years travelling and collecting. Two examples can be used to illustrate this development. The first concerns his analysis of the faunas of the Greater Sunda Islands (Borneo, Java, Sumatra). On the basis of their faunal composition and hypothesised species' relationships, he argued that the islands were once connected to the Asian mainland and even suggested the order in which they became isolated as sea level rose at the end of the last Ice Age. His discussion relied on the argument that undifferentiated faunas meant there were no barriers to dispersal, but that once sea levels rose and the islands became progressively isolated, speciation was promoted and faunas became differentiated. The more differentiated those faunas, the longer was their period of isolation. What he had done here was to propose testable geological hypotheses on the basis of biological data (geology←biology). However, his real leap of intellectual imagination came when he studied the fauna of Sulawesi. Situated in the centre of the Indonesian archipelago, and only thirty or so kilometres from Borneo, the fauna of Sulawesi is unique. It is unique because many species found in Borneo were absent, a distinctly Australasian element was present, and there were many endemic species and genera of often obscure phylogenetic affinity. The fauna of Sulawesi had the characteristics of a geographically isolated island, despite Sulawesi's central position within the archipelago, therefore its fauna, Wallace argued, must be a result of a *past* geography very different from its present one. Wallace's effort to understand the origin of Sulawesi and its fauna was ultimately futile because he lived at a time when large scale horizontal movements were simply not part of the accepted geological framework.

Wegener was to provide the appropriate geological framework some five decades later with the publication of *Die Enfstehung der Kontinente und Ozeane* in 1915. Chapter 6 of the English edition (Biram, 1966) of *The Origin of Continents and Oceans* – Palaeontological and Biological Arguments – starts with a summary of Wegener's views of the reciprocal nature between geology and biology, which he regarded as *fundamental*:

> "Palaeontology, zoogeography and phytogeography also have a significant contribution to make in unravelling the problem of how the earth appeared in prehistoric times; the geophysicist easily finds himself on the wrong track if he does not keep constantly in mind the results provided by these branches of science, in order to check his own." (Wegener, 1929; Biram, 1966: p97)

Wegener was clearly advocating for the utility of a biological approach to reconstruction (geology←biology). He continued:

> "On the other hand, if biologists are concerned at all with the question of drift, they should use the facts of geology and geophysics in forming their own judgements. If they do not, unprofitable mistakes are a constant threat to them." (Wegener, 1929; Biram, 1966: p97)

Again, this is a clear statement of a position that advocates geological input into explanations for biological (distributional) patterns (geology→biology). Conceptually, Wegener's and Wallace's ideas are almost identical, and the question must be asked whether Wallace's ideas directly influenced Wegener. Certainly Wegener knew of Wallace as he quoted *The Geographical Distribution of Animals* (German edition published in 1876), and even discussed (favourably) Wallace's analysis of the Australian fauna. However, *The Geographical Distribution of Animals* does not contain any real discussion of reciprocality, and is more a work of compilation than ideas (Michaux, 2008). While direct influence cannot be ruled out entirely, on the balance of probabilities I would suggest that Wegener developed his ideas independently.

Wegener discussed examples illustrating both aspects of the reciprocal nature of the relationship between geology and biology in chapter six of *The Origin of Continents and Oceans*. As an example of the importance of geological reconstructions to biological thinking (geology→biology) he quoted at length a work on earthworms byMichaelson (1922), in which this author had plotted the distribution of modern Oligochaeta on past reconstructions, showing how modern distributions are intelligible in the light of past, rather than modern, geographies. In the second example, Wegener used a (then) recent revision of the Microlepidoptera of Australia, New Zealand, and southern South America to suggest a novel reconstruction of these areas. He interpreted the microlepidopteran distributions as indicating two distinct patterns of relationship between these three areas. One connected southern South America to Australia via east Antarctica, the other connected southern South America to New Zealand via west Antarctica (see also Brundin (1966)). Unfortunately, a paucity of geological data then available from Antarctica precluded any possibility of testing these ideas, but these biological data have obvious implications for the reconstruction of polar Gondwana.

## 3. Mid-Cretaceous west Gondwanan tectonics

### 3.1. West Antarctic Rift System (WARS)

Antarctica was to remain a scientific *terra incognito*, geologically speaking, until relatively recently when modern technologies, such as satellite-based radar altimetry and airborne gravity and magnetic surveys, overcame the twin difficulties of isolation and ice cover. Together with targeted field work, these technologies have provided a wealth of information that has transformed our geological understanding of the Antarctic continent. One area of intensive research has been the West Antarctic Rift System (WARS, Fig. 1). WARS is a major zone of continental extension stretching some 3000 km from the Ross Sea to the base of the Antarctic Peninsula (Behrendt et al., 1991). The southern margin of the rift is well defined by the Transantarctic Mountains; the northern margin is less well defined, but Luyendyk et al. (2003) placed it at the northern boundary of the Ford Ranges in Marie Byrd Land. This would make the western section of the rift about 1200 km wide, a figure broadly similar to those of Bialas et al. (2007) and Siddoway (2008). According to Winberry and Anandakishnan (2004) crustal thickness within the Ross Sea sector of WARS averaged 25 km (range 21–31 km), while Siddoway (2008) reported a crustal thickness of 17–19 km in the western and central Ross Basin (23–24 km over basement highs), and 23 km for the Ross Province of MBL (as opposed to 30–34 km for unaffected MBL crust). These studies indicate that WARS is underlain by thinned continental crust.

There is broad agreement about the timing of Cretaceous extension in the western (Ross Sea) section of WARS. Fitzgerald (2002) estimated rifting occurred over a 20 million year period between 105 and 85 Ma. Luyendyk et al. (2003) also suggested rifting started at 105 Ma (initially in a NNE-SSW direction, then a NE-SW direction), but that rifting was only active until 96 Ma. Siddoway (2008) dated the main rifting phase between 100 and 90 Ma, and suggested that the underlying mechanism was back-arc stretching/transtension related to oblique subduction along the outboard margin. Siddoway (2008) estimated that some $1.2 \times 10^6$ km$^2$ of continental crust was affected during this mid-Cretaceous extension. Rift-related granites from Edward VII Peninsula, Marie Byrd Land, emplaced about 100–90 Ma (Bradshaw et al., 1997) further support a mid-Cretaceous age of rifting in WARS.

### 3.2. West Gondwana as a geological entity

During the mid-Cretaceous, WARS effectively divided Gondwana into two distinct geological provinces — east and west Gondwana. East Gondwana (east Antarctica, Australia, and parts of New Zealand) was an ancient craton, some of which dated back to the Neoprotozoic (1000–800 Ma) supercontinent of Rodinia (Torsvik et al., 2008). In contrast, west Gondwana (southern South America, west Antarctica and parts New Zealand) consisted of a collage of terranes. These terranes (Ellsworth/Whitmore Mountains (EWM), Antarctic Peninsula (AP), Thurston Island (TI), and Marie Byrd Land (MBL)/Campbell Plateau (CP) (McAdoo and Laxon, 1997; Fitzgerald, 2002)) had probably amalgamated by the start of the Cretaceous, or shortly thereafter. In the reconstruction of König and Jokat (2006) the Antarctic Peninsula was originally attached to Patagonia and adjacent to East Antarctica. Seafloor spreading in the Weddell Sea and rifting between Patagonia and the Antarctic Peninsula during the Jurassic resulted in the westwards translation and clockwise rotation of the Antarctic Peninsula (McAdoo and Laxon, 1997), and eventual amalgamation with the Thurston Island block between 130 and 110 Ma (Torsvik et al., 2008). The Ellsworth/Whitmore Mountain block is interpreted as a detached east Gondwanan cratonic fragment derived from the Natal embayment region during the initial breakup stage, and rotated anticlockwise to its mid-Cretaceous position (Randall and Niocaill, 2004; Torsvik et al., 2008).

New Zealand terranes have long been classified into three distinct suites (Coombs et al., 1976): a western province composed of Palaeozoic metasedimentary rocks which are intruded by Palaeozoic to Early Cretaceous granites and interpreted as an autochthonous palaeomargin of Gondwana; a median tectonic zone (MTZ = Median Batholith of Mortimer et al., 1999) composed of Palaeozoic to Early Cretaceous subduction-related volcanics (e.g. Brook Street and Dun Mountain terranes) and sediments (e.g. Murihiku and Maitai terranes); allochthonous eastern province terranes interpreted as arc or forearc accretionary prisms (Caples, Waipapa and Torlesse terranes). Wandres and Bradshaw (2005) suggested that the emplacement of the eastern province terranes was a consequence of subduction of the Phoenix Plate under the Gondwanan margin. While these authors regarded the eastern province terranes as allochthonous, they placed the source area for these rocks locally, adjacent to Marie Byrd Land. In contrast, Adams et al. (1998) suggested the source region was the New England Fold Belt of eastern Australia, and that they were transported some 2000 km southwards as a result of oblique subduction. The entire palaeo-Pacific margin was a long-lived orogenic belt resulting from subduction and plate convergence beginning in the Lower Palaeozoic and continuing to the mid-Cretaceous (Wandres and Bradshaw, 2005; Vaughan et al., 2005; Vaughan and Pankhurst, 2008). The change from convergent to extensional tectonics occurred in the Early Jurassic in northeast Australia, in the Cretaceous in New Zealand/west Antarctica, and in the Cenozoic in the Antarctic Peninsula (Adams et al., 1998).

A number of authors have proposed correlations between the various terranes found from New Zealand with those of west Antarctica and southern South America. Adams et al. (2005) traced western province terranes (which they regarded as part of the Lachlan fold belt) from southeastern Australia into New Zealand, the Campbell Plateau, western Marie Byrd Land, the Antarctic Peninsula, and southern South America. Bradshaw (2008) also traced the Lachlan fold belt into west Antarctica. Possible correlations of MTZ terranes of New Zealand are unresolved. Wandres and Bradshaw (2005) suggested that the MTZ could be traced across the Campbell Plateau (Bounty Island Granite) and into the Amundsen Province of eastern Marie Byrd Land (Bradshaw et al., 1997). However, Adams (2007) suggested that the Bounty Island granites were not part of the MTZ, relating them instead to similar aged granites on Thurston Island, west Antarctica. The magnetic data reported in Sutherland (1999a) suggests that the MTZ is only present in western Campbell Plateau. Finally, eastern province terranes have been recognised in the Antarctic Peninsula (the Western Domain of Vaughan and Storey (2000)) and the western archipelago of southern Chile (Adams et al., 2005). No eastern province terranes have been identified in Marie Byrd Land or the Campbell Plateau.

### 3.3. Campbell Rift

West Gondwana was bounded by the Weddell Sea and WARS, which separated southern South America and west Antarctica respectively from east Antarctica (Fig. 1). The mid-Cretaceous (100 Ma) reconstruction shown in Fig. 1 is abstracted in Fig. 2A (note that the view is reversed in Fig. 2). In Fig. 2A a boundary, termed the Campbell Rift, is shown separating the Campbell Plateau/Marie Byrd Land from the Australasian sector of east Gondwana. It is proposed that a mid-Cretaceous triple junction (Fig. 2A: proto-MTJ) existed within the Ross Sea sector connecting a rift–rift–rift system. WARS and the Australian–east Antarctica rift formed two of these arms. The Australian–east Antarctica rift was similar in a number of ways to WARS. Early rifting in the late Jurassic or earliest Cretaceous produced oceanic crust only in the west as India/Madagascar rifted from east Gondwana (Powell et al., 1988; Veevers, 2006). Slow and sporadic spreading during the Cretaceous produced a rift valley geomorphology, but no oceanic crust, in the central and eastern sectors (Powell et al., 1988; Norvick and Smith, 2001; Palmowski et al., 2001). Unlike WARS, the major spreading phase occurred in the Upper Cretaceous and Cenozoic, as opposed to mid-Cretaceous, and its development culminated in the formation of the Southern Ocean as Australia separated from east Antarctica (Fig. 2C).

A third arm of the triple junction is a geometric necessity and there are a number of lines of evidence suggesting that this rift should be placed between the west Gondwanan Campbell Plateau and the Challenger Plateau/Norfolk Ridge sector of east Gondwana. Firstly, the western margin of the Campbell Plateau is straight and has no continental slope to speak of — typical characteristics of rifted margins in general. While this modern margin is the product of mid-Cenozoic rifting following the opening of the Emerald Basin, restoring the Resolution Ridge, a detached continental sliver (Wood et al., 1996; Lebrun et al., 2003), to its pre-rift position along the western margin of the Campbell Plateau does not materially affect this observation. Secondly, it is known that the Campbell Plateau has undergone extension (Sutherland and Melhuish, 2000; Grobys et al., 2008a). Grobys et al. (2008a) argued that the extension phase responsible for crustal thinning preceded the opening of the Great South Basin which probably occurred in the mid-Cretaceous (100 Ma) (Beggs, 1993; Laird, 1993; Cook et al., 1999; Sutherland, 2000, Grobys et al., 2008a) as a precursor to the tectonic events leading to the eventual detachment of the Campbell Plateau from Marie Byrd Land. Siddoway (2008: Fig. 9) showed NW-SE oriented extension at circa 100 Ma between the Campbell and Challenger Plateaux.

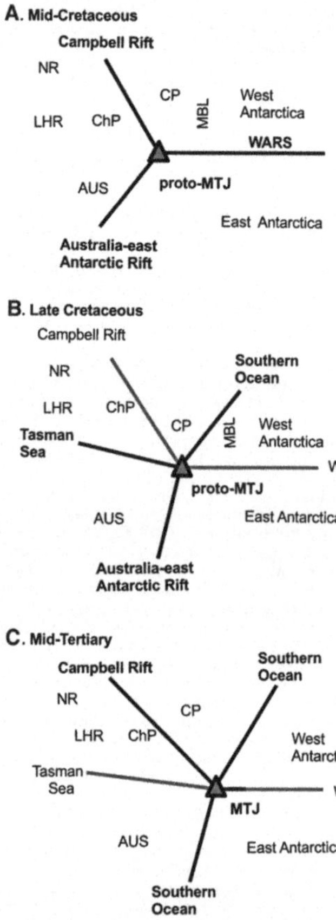


**Fig. 2.** Tectonic development in the polar Gondwana region at three time intervals. Dark lines = active tectonic boundaries, grey lines = inactive tectonic boundaries, triangle = triple junction, MTJ = Macquarie Triple Junction. Aus = Australian craton, CP = Campbell Plateau, ChP = Challenger Plateau, LHR = Lord Howe Rise, MBL = Marie Byrd Land, NR = Norfolk Ridge, WARS = West Antarctic Rift System.

Faulting parallel to the southern margin of the Challenger Plateau (Cape Foulwind Fault Zone and Paparoa Tectonic Zone) is present in both on- and offshore Cretaceous sequences of the west coast of the South Island, New Zealand (Bishop, 1992; Laird, 1993). These faults are thought to have been active in the Late Cretaceous and have been interpreted as the onshore extension of transform faults associated with the Tasman Sea spreading centre (Laird, 1993). However, coarse, Middle Cretaceous clastics underlie these late Cretaceous and Cenozoic rocks in the onshore section of the Cape Foulwind Fault zone, indicating that the Cape Foulwind faults date from at least the mid-Cretaceous and that they were reactivated in the Late Cretaceous. Similarly, the earliest sediments in the Paparoa Tectonic Zone – the Pororari Group – are non-marine, coarse clastics of mid-Cretaceous age that "are thought to have been laid down during a period of locally intense tectonic activity" (Suggate, 1978).

By the Late Cretaceous there had been a major reorganisation of regional tectonics as new spreading centres opened in the Tasman Sea and Southern Ocean, and activity in WARS and the Campbell Rift ceased (Fig. 2B). A possible schematic reconstruction for this time is shown in Fig. 3. Further major reorganisation occurred during the mid-Cenozoic (Fig. 2C) with an increased rate of spreading between Australia and east Antarctica, a change in spreading direction in the Southern Ocean to the south of New Zealand, and the northward propagation of new ocean crust in the Emerald and Solander Basins into the New Zealand region (Norris and Turnbull, 1993; Sutherland, 1995; Lamarche et al., 1997; Sutherland, 2000; Sutherland and Melhuish, 2000; Lebrun et al., 2003). The opening of the Emerald Basin was preceeded by a modest reactivation of WARS in the Adare Basin (Cande et al., 2000) and in the Ross Sea and Byrd Subglacial Basin (Sutherland, 2008).

## 4. Biological evidence

### *4.1. West Gondwanan endemism*

McLoughlin (2001) provided a summary of fossil Gondwanan floras, which included those of the critical mid-Cretaceous when angiosperms were diversifying. According to McLoughlin (2001) there was little differentiation shown by Jurassic or Lower Cretaceous floras beyond a possible latitudinal gradient separating those of the equatorial and mid-latitudes from those of high latitude (Sanmartin and Ronquist, 2004). By the mid-Cretaceous east and west Gondwanan floras had become distinct. McLoughlin (2001) recognised four floral provinces: two in east Gondwana (southeast Australia + New Zealand and India + northern Australia) and two in west Gondwana (southern South America + Antarctic Peninsula and Africa). The two east Gondwanan floras represented high latitude and equatorial assemblages and their distinctiveness can be attributed to latitudinal control, in part at least. Both west Gondwanan floras evolved in the temperate high latitudes. The southern South American + Antarctic Peninsula flora was the more diverse of the two and contained a number of endemic genera. The evolution of the two west Gondwanan floras can be understood in terms of vicariance and relative lengths of isolation. Southern South America and the Antarctic Peninsula were separated from Africa by sea-floor spreading in the south Atlantic, and from east Antarctica by the opening of the Weddell Sea and rifting in eastern WARS at 130 Ma. At this time only shallow seas separated Africa from east Antarctica, and complete isolation may only have been achieved at 118 Ma when the Mozambique Ridge (Fig. 1: MR) finally rifted from Dronning Maud Land, east Antarctica, and a deep water connection between the Weddell Sea and southwest Indian Ocean was formed (König and Jokat, 2006).

Further evidence of west Gondwanan endemism is provided by the biota of the New Zealand Subantarctic Islands, emergent parts of the Campbell Plateau. Michaux and Leschen (2005) have reviewed the biota of these islands and concluded that it is composite. One element is old and relictual, and shows closest relationships with southern South American taxa, while the second element is younger and composed of species that are either found on the mainland of New Zealand or are closely related to New Zealand species. They concluded

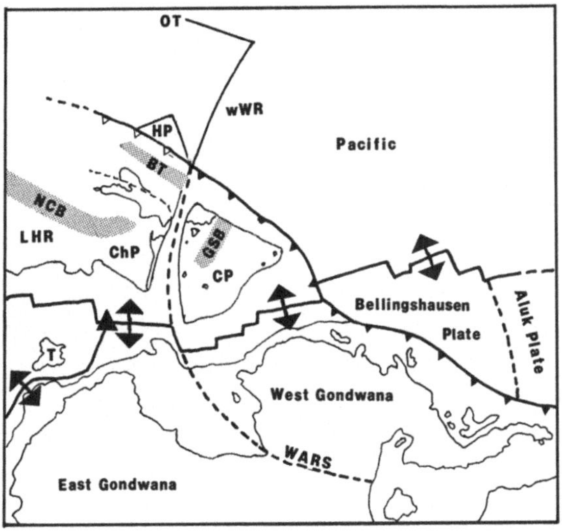


**Fig. 3.** Schematic Upper Cretaceous (70 Ma) reconstruction of polar Gondwana. BT = Bounty Trough, ChP = Challenger Plateau, CP = Campbell Plateau, GSB = Great South Basin, HP = Hikurangi Plateau, LHR = Lord Howe Rise, NCB = New Caledonia Basin, OT = Osbourne Trough, T = Tasmania, WARS = West Antarctic Rift System, wWR = west Wishbone Ridge. Dashed line = inactive WARS-Campbell Rift.

that the older fauna is west Gondwanan and dates from circa 88 Ma (Eagles et al., 2004) to 83 Ma (McAdoo and Laxon, 1997) when the Campbell Plateau separated from Marie Byrd Land, west Gondwana.

### 4.2. Polar Gondwanan biogeographic patterns

The nature of the relationship between southern South America, Australia and New Zealand is a classic biogeographic problem that remains unresolved to this day. Sanmartin and Ronquist (2004) have reviewed and summarised the phylogenies of 54 animal and 19 plant taxa whose distributions include southern South America, Australia and New Zealand. Based on these phylogenies, they identified three biogeographic patterns; the southern Gondwana pattern (SGP) in which southern south America is more closely related to Australia than to New Zealand – (NZ(sSA,Aus)) – the plant southern pattern (PSP) – (sSA(NZ,Aus)) – and the inverted southern pattern (ISP) – (Aus(NZ,sSA). This study, while exhaustive and thorough, suffers from a fundamental problem which precludes any solution to this classic biogeographic problem. This problem has recently been discussed by Giribit and Edgecombe (2006) and concerns the importance of analysing areas at the appropriate scale. Giribit and Edgecombe (2006) pointed out that treating 'Australia' as a biogeographic unit obscured finer biogeographic detail, and that artefactual results are a likely consequence of such an analysis (Nelson and Ladiges, 1996, 2001). Treating areas with differing histories as single entities can also lead to false paradoxes. For example, how can New Zealand be most closely related to Australia and southern South America simultaneously (PSP and ISP)? If New Zealand is biologically and geologically composite, as is argued here, then the apparent paradox vanishes if each element is treated separately.Wegener's (1929) observation that southern South America's relationship with New Zealand was via west Antarctica (ISP of Sanmartin and Ronquist, 2004) can now be refined as a relationship between southern South America and the Campbell Plateau via west Antarctica, which is a result of the original configuration of these areas into west Gondwana between circa 105 Ma and 88 Ma.

Using modern geography to describe endemic areas (the terminal nodes of the cladogram) rather than Cretaceous geography also leads to artefactual results. If the nodes of the three patterns described by Sanmartin and Ronquist (2004) are defined in terms of Cretaceous geography, then it becomes clear that only two of the patterns (ISP and PSP) are real. The ISP links two west Gondwanan areas (internal node date of 88 Ma) with east Gondwana (Australia). The basal node joining east and west Gondwana can be dated to 105 Ma. In the PSP, two east Gondwanan areas are linked (internal node dated to the opening of the Tasman Sea between 85 and 55 Ma — see following section), and the basal node representing the east-west Gondwans split at 105 Ma. In the SGP the internal node represents the east-west Gondwana split at 105 Ma rendering the basal node paralagous (Nelson and Ladiges, 1996). In other words, the SGP is a pattern based on widespread species and is equivalent to a polyphyletic (or non-natural) taxonomic group.

### 4.3. Biological dating of geological events

Phylogenies based on molecular data have an advantage over morphological-based phylogenies because it is possible to calibrate rates of change and date internal nodes, assuming that substitution/deletion rates remain constant for all lineages. While calibration is not straightforward and is a matter of some contention (Heads, 2005), there are a number of studies that are relevant to the timing of events related to the breakup of polar Gondwana. McDaniel and Shaw (2003) used the moss *Pyrrhobryum mnioides*, found in New Zealand, southeast Australia and Tasmania, southern South America, and the neotropics of South America, to estimate when east and west Gondwana separated. Their analysis, based on three chloroplast loci, produced a phylogeny in which Australasian and South American populations of this species grouped into two clades. Resolution in the Australasian clade was poor (perhaps the assumption of constant rates did not hold), but the South American taxa resolved into two distinct groups, a neotropical clade and a southern clade. McDaniel and Shaw (2003) calibrated their phylogeny using a 14 Ma age for the origination of the Atacama Desert, which these authors suggested represented the vicariant event responsible for splitting the southern and neotropical South American populations. Using this date yields an 80 Ma estimate for the split between east and west Gondwana. While this estimate is too imprecise to test the geological estimate of 105 Ma, it is within the right order of magnitude and shows the utility of their calibration method.

Renner et al. (2000) produced a molecular phylogeny for the southern sassafrases (Atherospermataceae) found in Australia, New Zealand, New Caledonia, New Guinea, and southern South America. Sassafras species exhibit a strong ecological association with the southern beeches (*Nothofagus*), and other old Gondwanan plant groups such as the podocarps and araucarians. The sassafrase phylogeny [(Aus(NG(Aust,NC)(sSA(sSA,NZ))))] grouped the southern South American and New Zealand species into a west Gondwanan clade that is a sister group to an east Gondwanan clade (Australia and New Caledonia species). Renner et al. (2000) used a variety of fossil and geological dates to calibrate the phylogeny. For example, they dated the node linking the New Zealand and Chilean species at 80 Ma based on the age of fossil *Laureliopsis*-like wood. This date is consistent with geological estimates of the age when the Campbell Plateau split from west Gondwana, and suggested a 109 Ma date for the node linking the east and west Gondwanan clades.

While modern Antarctica is a frozen wasteland there is a restricted terrestrial fauna present. Allegrucci et al. (2006) used a molecular-based study of Antarctic Chironomidae to show that these midges had originated on the continent and had remained *in situ* during climatic deterioration. This is a very important result because it means that age estimates of cladogram nodes can be obtained directly from endemic Antarctic species, rather than indirectly from species of surrounding areas. Allegrucci et al. (2006) used assumed substitution rates to obtain divergence ages that are consistent with geological estimates. For example, they estimated that the Antarctic Peninsula midges evolved at approximately 49 Ma as the Drake Passage opened and the Antarctic Peninsula became isolated from southern South America.

## 5. Late-Cretaceous disruption of east Gondwana

### 5.1. Melanesian Rift

Fig. 4 shows the main components of what I have here termed the Melanesian Rift. The Melanesian Rift is an elongate continental fragment that detached from east Gondwana during the Cretaceous and Palaeogene. Knowledge of this largely submerged continental fragment has increased rapidly in recent years as both New Zealand (Wright et al., 2002) and Australia (Stagg et al., 1999) carried out extensive marine surveys to delimit their respective continental shelves following ratification of the United Nations Convention on the Law of the Sea (UNCLOS) in 1996, and oil companies developed the technology to allow them to prospect in deep water, offshore basins.

Gaina et al. (1998) modelled the separation of the Lord Howe Rise (LHR) from the eastern margin of Australia. Their animation showed an asynchronous detachment of the Lord Howe Rise from the eastern margin of Australia, starting at approximately 85 Ma in the south (although extension probably started much earlier (Crawford et al., 2003)) and finishing as late as 52 Ma in the north when sea-floor spreading in the Coral Sea ceased. The southern section of the Lord Howe Rise, the Challenger Plateau (ChP), became isolated when the Bellona Trough (BT) opened between 84 and 73 Ma. The Challenger Plateau is composed of continental crust (Tulloch et al., 1991) only moderately thinned (Wood and Woodward, 2002). The thinness of sediment cover (Wood, 1993) indicates substantial periods of post-rift

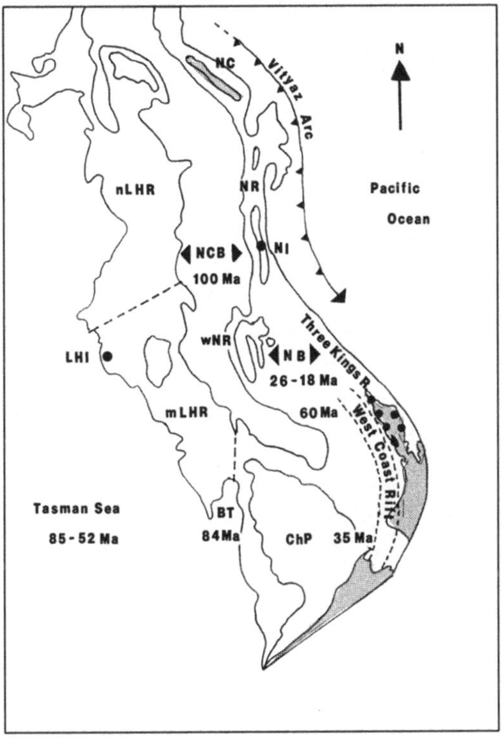


**Fig. 4.** Major units and structures of the Melanesian Rift. Outer contour = 2000 m depth, inner contour = 500 m depth. BT = Bellona Trough, ChP = Challenger Plateau, LHI = Lord Howe Island, mLHR = middle Lord Howe Rise, nLHR = northern Lord Howe Rise, NB = Norfolk Basin, NC = New Caledonia, NCB = New Caledonia Basin, NI = Norfolk Island, NR = Norfolk Ridge, wNR = west Norfolk Ridge. Closed circles = approximate extent of the Northland Allochthon. Dates indicate opening times.

emergence. According to Stagg et al. (1999) the mid- and northern sections of the Lord Howe Rise (mLHR, nLHR) show a complex structure of north–south trending platforms, ridges and rifts. The Lord Howe platform bordering the New Caledonian Basin (NCB) may have been emergent subsequent to detachment, particularly in the central parts where there was less crustal extension than in the north. A string of seamounts, dating from the mid-Cenozoic, run along the western edge of the central sector of the Lord Howe Rise (Lord Howe Island (LHI) is the youngest) and may also have been emergent (Quilty, 1993).

The New Caledonia Basin (NCB) runs approximately north–south for about 1000 km. Four kilometres of Cretaceous coals and non-marine clastics, and deep water Cenozoic sediments are recorded by Willcox et al. (2001) in the north, while a similar sequence and thickness of sediments has been reported from the southern New Caledonia Basin by Stagpoole et al. (2000). The New Caledonia Basin has probably been a deep waterway since at least the start of the Cenozoic, and is presently submerged to depths of up to 5 km. Age estimates vary from early-Cretaceous (Wood, 1993), to mid-Cretaceous (Stagpoole et al., 2000), or even, in parts, late-Cretaceous-Palaeocene (Lafoy et al., 2005). Willcox et al. (2001) reported a 3-fold structure within the New Caledonia Basin, and recognised northern, central, and southern sectors. Grobys et al. (2008b) reported that the crustal thickness beneath the New Caledonia Basin averaged 14 km, with thinner (11 km) crust between latitudes 25° S and 27° S. These thicknesses indicate highly thinned continental crust in the northern and southern sectors, with probable incipient sea-floor spreading within the central sector. Lafoy et al. (2005) also suggested that the central region of the New Caledonia Basin was underlain by oceanic crust that they argued was emplaced between 62 and 56 Ma. Klingelhoefer et al. (2007) suggested that the central section of the New Caledonian Basin was underlain by normal oceanic crust, but that the northern section was composed of serpentinised upper mantle material indicative of non-magmatic emplacement during ultra-slow spreading. It seems clear that the

New Caledonia Basin has acted to isolate New Caledonia (NC) and the Norfolk Ridge (NR) from the Lord Howe Rise for a considerable period.

The Norfolk Ridge is a complex structure that became disrupted during the mid-Cenozoic. Grobys et al. (2008b) reported average crustal thicknesses of 24 km for the West Norfolk Ridge (wNR) and 21 km for the Norfolk Ridge, while Klingelhoefer et al. (2007) reported a thickness of 17 km for a region south of New Caledonia. These results indicate that both sections of the Norfolk Ridge are composed of thinned continental crust. Granites, metamorphics, and mid-Cretaceous coal measures have been dredged from the West Norfolk Ridge, and Upper Cretaceous marine shales from the Norfolk Ridge south of New Caledonia (Herzer et al., 1999). The Norfolk Ridge was disrupted in the mid-Cenozoic by the opening of the Norfolk Basin (NB), forming the West Norfolk Ridge and translating the Three Kings Rise eastwards. Early models of the Norfolk Basin treated it as a back-arc basin resulting from westward dipping subduction along the eastern margin of the Three Kings Rise during the opening of the South Fiji Basin (e.g. Ballance et al., 1982). However, subduction-related, Miocene andesite volcanics are found on the western margin of the Three Kings Rise indicating eastwards subduction.

Recent studies of the Norfolk Basin suggest that its history is more complex than a simple back-arc origin (Sdrolias et al., 2004). Sdrolias et al., 2004 described a complex structure of plateaux and abyssal plains within the Norfolk Basin. They interpreted the plateaux as disrupted Cretaceous oceanic crust trapped within a matrix of later, mid-Cenozoic oceanic crust. They suggested that this older crust was formed in a back-arc basin during the mid-Cretaceous as Pacific crust was being subducted beneath east Gondwana. The opening of this back-arc basin separated a Loyalty-Three Kings Rise arc from the Pacific margin of east Gondwana. This marginal basin was subsequently closed as crust was subducted beneath the western margin of the Loyalty-Three Kings Rise arc in response to the northeastwards translation of the Melanesian Rift as the Tasman Sea opened. The closing of this marginal basin resulted in the collision of the Loyalty-Three Kings Rise arc with the Melanesian Rift causing, according to Sdrolias et al. (2004), the obduction of the New Caledonian ophiolites at 45 Ma and the trapping of Cretaceous oceanic crust between the Norfolk Ridge and Three Kings Rise. There was regional uplift and subaerial erosion on the Lord Howe Rise during this period of tectonic compression during the Upper Eocene (38 Ma) to Middle Oligocene (30 Ma) coincident with these events (Klingelhoefer et al., 2007). Following collision of the Loyalty-Three Kings Rise arc with the Melanesian Rift, subduction polarity switched to the west and migrated southwards (Vityaz Arc, Fig. 4), opening the South Fiji Basin and causing back-arc extension in the Norfolk basin. The presence of emergent islands along the Three Kings Rise and Norfolk Ridge between 38 and 31 Ma reported in Meffre et al. (2006) indicates uplift in this region at this time. Schellart et al. (2006) suggested that collision of the d'Entrecasteaux-Loyalty-Three Kings Rise-Northland Plateau arc was also responsible for the emplacement of the Northland Allochthon. In their model the collision is asynchronous, with earlier contact in the north (emplacement of the New Caledonian ophiolites between 38 and 34 Ma) and later collision in the south (emplacement of the Northland Allochthon between 25 and 22 Ma). This latter suggestion is in contrast to other models of the formation and emplacement of the Northland Allochthon (e.g. Rait, 2000; Bradshaw, 2004; Whattam et al., 2005; Mortimer et al., 2007).

Running south from the Norfolk Basin are a series of basins – Reinga (Herzer et al., 1997; Stagpoole et al., 2000), Northland (Isaac et al., 1994; Uruski et al., 2004) and north and south Taranaki (Kamp et al., 2004; Stagpoole et al., 2004) – that Kamp (1986) collectively termed the North Island Rift System (Fig. 4: West Coast Rift). This rift system was a major feature some 1200 km long and 100–200 km wide during the Upper Cretaceous (60 Ma) to mid-Cenozoic (35 Ma). It was surrounded by coastal plains that were terrestrial, in part at least, until the end of the Eocene. Subsidence and flooding of the rift proceeded from the north from about 60 Ma, finally inundating the southern portion at about 35 Ma, effectively isolating the North Island, New Zealand from the Challenger Plateau.

### *5.2. Biological evidence*

While the number of molecular phylogenies of Melanesian Rift taxa is small there are sufficient to illustrate the potential importance of such studies as aids in model reconstruction. The two most extensive studies of Melanesian Rift taxa are those of Ponder et al. (2003) for the flax snail *Placostylus*, and Boon et al. (2001) for the parakeet *Cyanoramphus*. *Placostylus* species are large (5–9 cm) pulmonate gastropods found in northern New Zealand, Three Kings Islands, New Caledonia, Loyalty Islands, Lord Howe Island, Vanuatu, Solomon Islands, and Fiji. While the study of Ponder et al. (2003) is weakened – biogeographically – by the inclusion of only one of the many New Caledonian taxa, and no species from the Loyalty Islands, Vanuatu, Solomons or Fiji, the resulting phylogeny [(TKI(NC(NZ, LHI)))] is fascinating, and if repeated for other groups could settle the controversy over the Cretaceous-Eocene boundary between the Pacific plate and the Melanesian Rift discussed by Schellart et al. (2006). The important (and unexpected) result was the basal placement of the Three Kings Islands species, making this clade the earliest lineage of all the *Placostylus* species. This result clearly supports the model of Sdrolias et al. (2003), which posited a Cretaceous origin for the Loyalty-Three Kings Rise arc, and thus by implication supports an eastwards dipping subduction zone between the Melanesian Rift and Pacific plate from the mid-Cretaceous to Eocene.

Fig. 5 is a simplified version of the *Cyanoramphus* phylogeny published in Boon et al. (2001). The basal position of *C. saisetti* shows that New Caledonia (NC) was the first area to be isolated from the rest of the Melanesian Rift. New Caledonia (plus New Zealand) is also basal in the phylogeny of owlet-nightjars (Dumbacher et al., 2003), reinforcing the view that much of New Caledonia's flora and fauna is old and relictual (Ladiges and Cantrill, 2007). The northern Norfolk

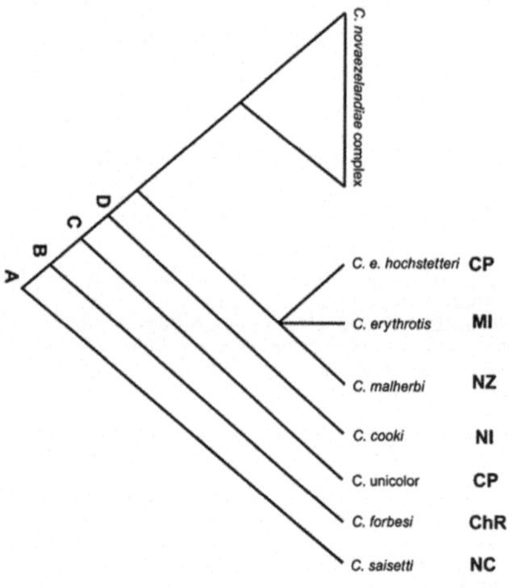


**Fig. 5.** Simplified phylogeny of *Cyanoramphus* parakeets after Boon et al. (2001). ChR = Chatam Rise, CP = Campbell Plateau, MI = Macquarie Island, NC = New Caledonia, NI = Norfolk Island, NZ = New Zealand.

Ridge, of which the present Island of New Caledonia is a part, was isolated from the Lord Howe Rise with the mid-Cretaceous opening of the New Caledonia Basin, and perhaps from the southern parts of the Norfolk Ridge by Late Cretaceous, as indicated by dredged marine shales of this age from this region. It is possible to calibrate this phylogeny using a 30–25 Ma date for the collision of the Campbell Plateau (CP) with the Melanesian Rift (Michaux and Leschen, 2005). Fixing node C in Fig. 4 at 30–25 Ma yields a date-range for node A, the date of isolation of New Caledonia, of 60–49 Ma, i.e. Upper Cretaceous or Lower Palaeocene. The phylogeny also shows the progressive isolation of first the Chatham Rise (ChR) (node B) and later Norfolk Island (NI) (node D). Age estimates for these nodes are 46–38 Ma for the Chatham Rise and 21–17 Ma for Norfolk Island. The age of node D is approximately the age of the opening of the Norfolk Basin, which was presumably responsible for the isolation of Norfolk Island from The Three Kings Rise and Northland, New Zealand. The early isolation of the Chatham Islands is an interesting result, which will be discussed in the next section.

Finally, Cooper et al. (2001) estimated that the divergence of the New Zealand kiwi from the Australian emu-cassowary clade occurred between 72 and 65 Ma. The monophyly of the kiwi-cassowary-emu clade has recently been confirmed by Harshman et al. (2008), even though this study has shown that ratites are a paraphyletic group unless South American tinamous are included. Cooper et al. (2001) considered that the divergence date of 72–65 Ma necessitated kiwi dispersal into New Zealand, repeating the common mistake of assuming the entire Melanesian Rift separated from Australia at 84 Ma, rather than a final separation date of 54 Ma. A 65–72 Ma divergence age range between kiwi (NZ) and the emu-cassowary clade (AUS) would indicate that the Melanesian Rift (or parts thereof) and cratonic Australia were still contiguous until approximately halfway through the separation process.

### 5.3. *Position of Chatham Rise in reconstruction models*

> "A knowledge of the New Zealand subcontinent's former shape and position within eastern Gondwana is a pre-requisite to any model of the plate tectonic evolution of the south west Pacific." (Kamp, 1986)

Kamp (1986) discussed seven different reconstruction models that had been proposed since 1975, and more variants have been published and continue to be published to the present day (e.g. Mortimer, 2008). Apart from a single model, discussed below, the Chatham Rise is always shown attached to the Campbell Plateau and west Antarctica. Because the pre-rift positions of the Campbell Plateau and the Lord Howe Rise are both well constrained, the arrangement of the internal blocks of the 'New Zealand subcontinent' has to be distorted to accommodate this fit. This distortion leads to a rather bizarre situation of postulating a lacuna between the Challenger Plateau and the rest of the South Island east of the Alpine fault. While various explanations have been proposed to account for how this internal distortion resolved itself into the present arrangement of blocks within the New Zealand subcontinent (e.g. Bradshaw et al., 1996; Sutherland, 1999b; Hall et al., 2004, Wood and Stagpoole, 2007), the alternative explanation that the arrangement is wrong does not seem to have been considered.

Skelley and Leschen (2006) have reviewed published phylogenies containing Chatham Island species. These phylogenies, based on morphology and/or molecular data, reveal that of the thirteen taxa studied six species were most closely related to widespread New Zealand species (i.e. occurring in both the North and South Islands), six were most closely related to South island taxa, and one had its sister taxon in the North Island. The majority of the Chatham Island species were derived, although some were more basal within their clades. Their suggestion that the flora and fauna are composite in terms of age and source area supports an earlier analysis of the Chatham Island biota by Craw (1988). The flora and fauna of Chatham Island group appear to be related to the east Gondwanan biota of the Melanesian Rift rather than to the west Gondwanan biota of the Campbell Plateau, as might be expected given the proximity of the Chatham Rise to the Campbell Plateau and Marie Byrd Land in virtually all reconstruction models. When the Campbell Plateau and Chatham islands share (east Gondwanan) taxa, available phylogenies either group the Campbell Plateau with mainland New Zealand taxa, and place the Chatham Island as either sister group (e.g. *Cyanoramphus novaezelandiae* subspecies, Boon et al. (2001)) or place the Chathams with other New Zealand species as sister group (e.g. nacellid limpets, Goldstein et al. (2006)). In neither case is there any direct linkage of the Chatham Islands with any Campbell Plateau area. The biological evidence suggests that the Chatham Islands should be part of the east Gondwanan Melanesian Rift, as previously suggested by Austin (1975), who placed the Chatham Rise adjacent to the east coast of the North Island, New Zealand.

The Chatham Rise is a predominantly shallow (<300 m) submarine platform some 1100 km long. According to Wood et al. (1989) the basement rocks are Upper Palaeozoic to Mesozoic greywackes and schists similar to those found in Canterbury and Otago respectively. The schists only occur east of longitude 180°, about halfway along the Rise. According to Adams et al. (2008) the basement rocks of the Chatham Islands and Forty Fours are composed of Early to Middle Triassic sediments and volcanics metamorphosed in the Lower Jurassic ($197 \pm 13$ Ma). These authors regarded the Matarakau greywackes as equivalent to the Torlesse of the South Island and lower North Island, New Zealand, and the metamorphosed volcanics of the Kaingaroa Formation and the greywackes of the Forty Fours as equivalent of the Caples terrane of the South Island, New Zealand. Mortimer et al. (2006) also suggested that the schists of the Stuttgart seamount, a fragment of Chatham Rise basement isolated within the Bounty Trough, as equivalent to the Caples terrane and Haast schists.

The basement rocks of the Chatham Rise have been extensively faulted parallel with the long axis in the mid-Cretaceous (Wood et al., 1989; Laird, 1993; Wood and Herzer, 1993). The stratigraphy of the Chatham Rise, based mainly on seismic profiles, was summarised by Wood and Herzer (1993). The lowest sedimentary unit in the marginal basins along the flanks of the Chatham Rise is a 5–7 km thick, Lower Cretaceous sequence that lies unconformably on basement rocks. Above this basal unit are some 2–5 km of mid-Cretaceous (100–80 Ma) terrestrial and shallow marine clastics with associated rift-related, alkali basalts (dated to 85–82 Ma), similar to the Tapuaenuku igneous complex of Marlborough (Panter et al., 2006). There are up to 1500 m of Upper Cretaceous transgressive silts and muds on the flanks and within topographic lows on the rise crest, which shows evidence of subaerial relief at this time. An analysis of detrital zircons from these Upper Cretaceous cover sediments shows that they were not derived from the Bounty Island granites, but rather from reworked local eastern province rocks (Adams et al., 2008). The early-Cenozoic history of the rise is one of progressive marine transgression and decrease in clastic sedimentation until the Upper Eocene, when there was extensive subaerial eruption of alkali basalts along the eastern part of the Chatham Rise. Panter et al. (2006) dated these volcanic to 40–35 Ma. A widespread unconformity existed between the Upper Eocene and Oligocene with numerous erosion surfaces evident from the seismic profiles. The later Cenozoic is characterised by stable shelf conditions, reduced sediment supply, absence of sediments along the crest of the rise, and by at least four major unconformities; middle Late Miocene, Miocene/Pliocene boundary, mid-Pliocene, and latest Pliocene. A further period of volcanic activity occurred at about 5 Ma.

The Chatham Rise is flanked by the Bounty Trough to the south and the Hikurangi Plateau to the north. The extensive, east-west striking faulting on the Chatham Rise parallels the strike of the Bounty Trough, which is a deep water basin running in an east-west direction. No equivalent parallel faulting has been reported from the Campbell Plateau. Carter et al. (1994) described a three-fold structure within the Bounty Trough consisting of three sub-basins separated by N–S striking faults. The age and origin of this basin are yet to be resolved. Kamp (1986) suggested that it was a mid-Cretaceous rift ($105 \pm 5$ Ma), while Eagles et al. (2004) argued that it started as a continental strike-slip zone, transforming into a rift structure at 83 Ma that became incorporated into the Pacific-Antarctica plate boundary. Larter et al. (2002) also favoured an 83 Ma opening of the Bounty Trough, and likened it to the modern day Gulf of California as a seaway developed between the Chatham Rise and west Antarctica. Gohl et al. (2007) linked the opening of the Bounty Trough to rifting in the Amundsen Sea, west Antarctica, at or before 90 Ma. Grobys et al. (2007) produced a seismic profile across the middle sub-basin showing the crust to be only 9 km thick. These authors suggested that the basement of this sub-basin was thinned continental crust intruded by mafic bodies, which they interpreted as indicating rifting rather than back-arc spreading. Grobys et al. (2008b) also suggested a two phase origin for the Bounty Trough consisting of mid-Cretaceous (115 Ma) back-arc spreading in response to subduction along the northern margin of the Chatham Rise, followed by a later (90–87 Ma) rifting phase producing oceanic crust in the outer sub-basin and the thinning and intrusion of mafic magmas into the middle sub-basin.

The Hikurangi Plateau is a LIP (Large Igneous Province) with crust much thicker than normal oceanic crust (Hoernle et al., 2004). Davy and Wood (1994) described the complex structure of the Hikurangi Plateau including a basin-high-basin unit filled with Mesozoic sediments which runs subparallel to the Chatham Rise, a northern margin of shallow basement, and a southeast basement complex overlain by several kilometres of sediments which are deformed by volcanic intrusives. They recognised two episodes of volcanism, an older event (99–89 Ma according to Davy et al. (2008)) which produced large, geographically isolated centres now exposed as seamounts or buried by sediments, and younger Neogene activity resulting in small centres and intrusives. Wood and Davy (1994) showed that the crustal thickness varied from 10 km along the northern margin to 15 km adjacent to the Chatham Rise. They placed the eastern margin of the Hikurangi Plateau at 176°W as they were unsure whether the crust to the east of this was Hikurangi Plateau or normal oceanic crust thickened by sediment. Davy et al. (2008) united the Hikurangi Plateau with the Manihiki Plateau at 125–120 Ma when spreading in the Osbourne Trough (Mortimer et al., 2006) separated them and translated the Hikurangi Plateau southwards to collide with the Chatham Rise at circa 100 Ma (Luyendyk, 1995). Davy et al. (2008) reported that Hikurangi Plateau crust could be traced southwards beneath the Chatham Rise for 50–100 km. According to them the asymmetric collision of the Hikurangi Plateau with the northern margin of the Chatham Rise choked the subduction zone and resulted in the welding of the Hikurangi Plateau to the Chatham Rise.

However the Chatham Rise is to be placed in relation to the rest of New Zealand, it seems clear that it should be treated as part of a unit which includes the Hikurangi Plateau and the Bounty Trough. Any reconstruction also has to retain the approximately 60° anticlockwise rotation that the Chatham Rise has undergone since the Cretaceous (Grindley et al., 1977). A possible schematic reconstruction for the Chatham Rise has been indicated in Fig. 3. In this reconstruction the Chatham Rise (plus Bounty Trough and Hikurangi Plateau) has been placed outboard of the Melanesian Rift to reconcile the east Gondwanan fauna and flora of the Chatham Islands with its geological position. While additional, *ad hoc* hypotheses, such as complete submergence and reinvasion from an east Gondwanan source area might also explain such an absence, this hypothesis can be rejected initially with reference to the principle of Ockham's Razor. Palaeomagnetism preserved in the mid-Cenozoic volcanics of the Chatham Rise indicate that it was indeed north of its present position during the mid-Cenozoic (Grindley et al., 1977). Attachment to the Melanesian Rift also explains the similarity of the Cretaceous volcanics on the Chatham Islands to other east Gondwanan layered igneous complexes (Panter et al., 2006), and their absence from the Campbell Plateau and other west Gondwanan areas.

Fig. 6, based on King (2000), shows a series of schematic reconstructions for the Chatham Rise at 40 Ma, 30 Ma, 20 Ma and 10 Ma. In Fig. 6A (40 Ma) the Chatham Rise (CR) and Bounty Trough (BT) are aligned parallel with the east coast of the South Island, New Zealand (eSI). This arrangement aligns the boundaries of the Torlesse and Caples terranes with their equivalents on the Chatham Rise, and the schists of central Otago with those of the eastern Chatham Rise, obviating the need for any oroclinal bending within the New Zealand subcontinent. At 40 Ma the Campbell Plateau (CP) is shown isolated from the southern portion of the Melanesian Rift by the Campbell Rift proposed earlier. The geographical separation of the Campbell Plateau from the Chatham Rise is certainly one possible explanation for the lack of detrital zircons derived from the Bounty Island granites in the Cretaceous cover sediments of the Chatham Rise. While activity along the Vityaz Arc is well to the north of New Zealand at 40 Ma, a connection between this active plate boundary and the Campbell Rift is indicated in Fig. 6A by a series of strike-slip faults marking the boundaries of the east coast North Island block (eNI) and the east coast South Island block (eSI). Active subduction of old Cretaceous crust beneath the western margins of the Three Kings Rise (TKR) and Northland Plateau is also indicated in Fig. 6A.

Fig. 6B (30 Ma) represents a time of major tectonic reorganisation as active subduction from the north and a spreading system from the south migrated into the New Zealand region. In Fig. 6B it is suggested that these changes resulted in both the collision of the Campbell Plateau with the Challenger Plateau and the detachment of the east coast of the South Island and the Chatham Rise from the Melanesian Rift along strike-slip boundary faults. In Fig. 6C (20 Ma) the east coast of the South Island and the Bounty Trough have amalgamated with the Campbell Plateau. The boundary of this collision zone is the Livingstone Fault (and offshore equivalents), a major feature that extends to the base of the crust (Mortimer et al., 2002). The occurrence of a widespread Upper Eocene to Early Miocene (40–23 Ma) unconformity along the Chatham Rise and the isolation and speciation recorded in *Cyanoramphus* parakeets (46–38 Ma) may be related to these events. To the south of New Zealand sea-floor spreading had become increasingly oblique and an active transform fault was propagated through New Zealand to connect with a subduction zone along the east coast of the North Island, New Zealand. Active spreading in the Norfolk Basin (NB) translated the Three Kings Rise (TKR) along the Vening-Meinesz Fracture Zone (VMZ). By 10 Ma (Fig. 6D) the South Fiji Basin (sFB) had opened and the New Zealand region had attained a recognisably modern configuration.

## 6. Conclusions

While geology and biology remain separate and distinct disciplines, with their own methodologies and modes of analyses, I hope I have demonstrated that there is (potentially) much to be gained from a 'reciprocal illumination' (Hennig, 1966) between geological approaches (tectonic models) and biological approaches (constructing phylogenies) to the reconstruction of history. This is particularly important, in my view, for areas such as the New Zealand region where sharp and well defined, simple plate boundaries have not, in general, existed. Because of the prolonged period of high heat flow and complex interactions, plate boundaries in the New Zealand region may be better described as 'sticky zones'. In such areas of high geological complexity, it may be possible, given phylogenies of the appropriate sensitivity and resolution, to both test and refine reconstruction models. A number of examples have been used in this paper to illustrate this approach, which was advocated by

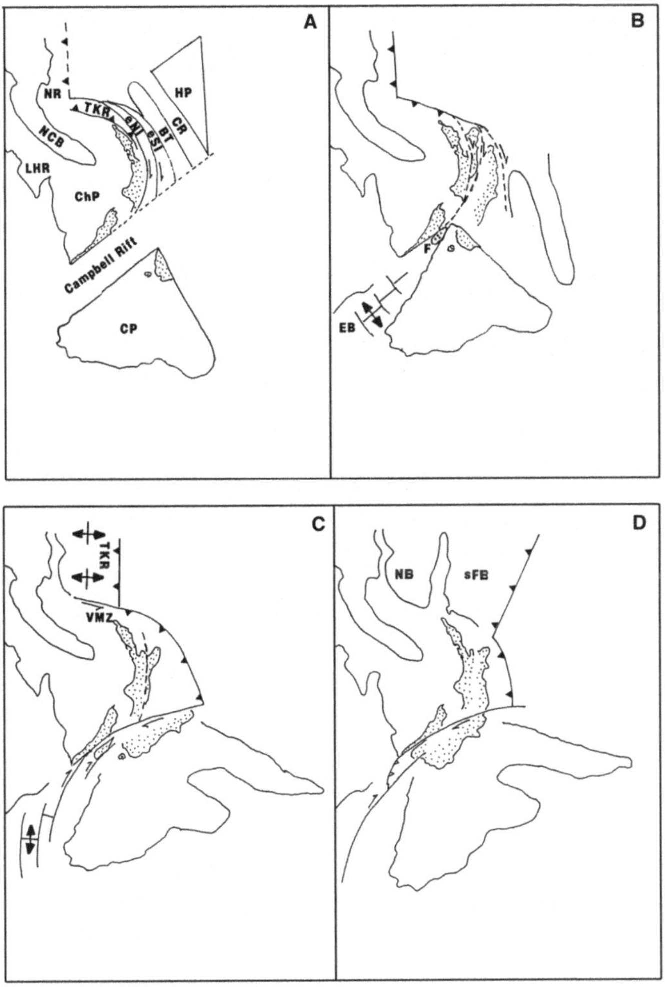


**Fig. 6.** Schematic reconstructions of the Chatham Rise at four time intervals. A = 40 Ma, B = 30 Ma, C = 20 Ma, D = 10 Ma. BT = Bounty Trough, CP = Campbell Plateau, CR = Chatham Rise, ChP = Challenger Plateau, EB = Emerald Basin, eNI = east coast North Island, eSI = east coast South Island, F = Fiordland, HP = Hikurangi Plateau, LHR = Lord Howe Rise, NB = Norfolk Basin, NCB = New Caledonia Basin, NR = Norfolk Ridge, TKR = Three Kings Rise, sFB = South Fiji Basin, VMZ = Vening-Meinesz Fracture Zone.

Wegener at the very start of the subject, but which seems to have been lost as scientists became increasingly specialised.

Biologists ignore or sideline the results of the geophysicist at their peril. While no *a priori* role is, in my view, a legitimate use of reconstruction models, any phylogeny that makes no sense in the light of well established geological facts is most probably erroneous, and points to the need for additional data (characters) to be sought and new analyses to be undertaken. Perhaps the most important use of such models is, however, in the calibration of molecular phylogenies. Date estimates from well constrained tectonic events can be used to estimate ages of other nodes within a phylogeny, and consistency of dates obtained from different phylogenies could establish the veracity of such estimates. Examples are also provided to illustrate some of these points.

## Acknowledgements

My thanks to two anonymous referees, and to Jan Grobys and M. Santosh for constructive and useful comments on an earlier draft of this paper.

## References

Adams, C.J., 2007. Geochronology of Palaeozoic terranes at the Pacific Ocean margin of Zealandia. Gondwana Research 13, 250–258.

Adams, C.J., Campbell, H.J., Graham, I.J., Mortimer, N., 1998. Torlesse, Waipapa and Caples suspect terranes of New Zealand: integrated studies of their geological history in relation to neighbouring terranes. Episodes 21, 235–240.

Adams, C.J., Pankhurst, R.J., Maas, R., Millar, I.L., 2005. Nd and Sr isotopic signatures of metasedimentary rocks around the South Pacific margin and implications for their provenance. Geological Society, London, Special Publications 246, 113–141.

Adams, C.J., Campbell, H.J., Griffin, W.J., 2008. Age and provenance of basement rocks of the Chatham Islands: an outpost of Zealandia. New Zealand Journal of Geology and Geophysics 51, 245–259.

Allegrucci, G., Carchini, G., Todisco, V., Convey, P., Sbordoni, V., 2006. A molecular phylogeny of Antarctic chironomidae and its implications for biogeographical history. Polar Biology 29, 320–326.

Austin, P.M., 1975. Paleogeographic and paleotectonic models for the New Zealand geosyncline in eastern Gondwanaland. Geological Society of America Bulletin 86, 1230–1234.

Ballance, P.F., Pettinga, J.R., Webb, C., 1982. A model of the Cenozoic evolution of northern New Zealand and adjacent areas of the southwest Pacific. Tectonophysics 87, 37–48.

Beggs, J.M., 1993. Depositional and tectonic history of the great south basin. In: Ballance, P.F. (Ed.), South Pacific Sedimentary Basins. In: Sedimentary Basins of the World, vol. 2. Elsevier, Amsterdam, pp. 365–373.

Behrendt, J.C., LeMasurier, W.E., Cooper, A.K., et al., 1991. Geophysical studies of the West Antarctic Rift System. Tectonics 10, 1257–1273.

Benedetto, J.L., 2004. The allochthony of the Argentine Precordillera ten years later (1993–2003): a new paleobiogeographic test of the microcontinental model. Gondwana Research 7, 1027–1039.

Bialas, R.W., Studinger, M., Fitzgerald, P.G., 2007. Plateau collapse model for the Transantarctic-West Antarctic Rift System: insights from numerical experiments. Geology 35, 687–690.

Biram, J. (Translator), 1966. The Origin of Continents and Oceans (English edition of 4th revised edition). Dover Publications, New York.

Bishop, D.J., 1992. Extensional tectonism and magmatism during the middle Cretaceous to Paleocene, North Westland, New Zealand. The New Zealand Journal of Geology and Geophysics 35, 81–91.

Boon, W.M., Kearvell, J.C., Daugherty, C.H., Chambers, G.K., 2001. Molecular Systematics and Conservation of *kakariki* (*Cyanoramphus* spp.). In: Science for Conservation, vol. 176. Department of Conservation, Wellington.

Bradshaw, J.D., 2004. Northland Allochthon: an alternative hypothesis of origin. New Zealand Journal of Geology and Geophysics 47, 375–382.

Bradshaw, J.D., 2008. The Ross Orogen and Lachlan Fold Belt in Marie Byrd Land, Northern Victoria Land and New Zealand: implications for the tectonic setting of the Lachlan Fold Belt in Antarctica. In: Cooper, A.K., Raymond, C.R., et al. (Eds.), Antarctica: a Keystone in a Changing World. doi:10.3133/of2007-1047.srp059.

Bradshaw, J.D., Weaver, S.D., Muir, R.J., 1996. Mid-Cretaceous oroclinal bending of New Zealand terranes. New Zealand Journal of Geology and Geophysics 39, 461–468.

Bradshaw, J.D., Pankhurst, R.J., Storey, B.C., et al., 1997. New Zealand superterranes recognised in Marie Byrd Land and Thurston Island. In: Ricci, C.A. (Ed.), The Antarctic Region: Geological Evolution and Processes. Siena, Terra Antarctica, pp. 429–436.

Bremer, K., Gustafsson, M.H.G., 1997. East Gondwanan ancestry of the sunflower alliance of families. Proceedings of the National Academy of Sciences of the United States America 94, 9188–9190.

Brundin, L., 1966. Transantarctic relationships and their significance, as evidenced by chironomid midges, with a monograph of the subfamily Podonominae and Aphroteniinae and the austral Heptagyniae. Kongliga Svenska Ventenskaps Academiens Handlinger 11, 1–472.

Cande, S.C., Stock, J.M., Müller, R.D., Ishihara, T., 2000. Cenozoic motion between East and West Antarctica. Nature 404, 145–150.

Carter, R.M., Carter, L., Davy, B., 1994. Seismic stratigraphy of the Bounty Trough, southwest Pacific Ocean. Marine and Petroleum Geology 11, 79–93.

Cook, R.A., Sutherland, R., Zhu, H., 1999. Cretaceous-Cenozoic geology and petroleum systems of the Great South Basin, New Zealand. Institute of Geological & Nuclear Sciences Monograph 20 190 pp.

Coombs, D.S., Landis, C.A., Norris, R.J., et al., 1976. The Dun Mountain ophiolite belt, New Zealand, its tectonic setting, constitution, and origin, with special reference to the southern portion. American Journal of Science 276, 561–603.

Cooper, A., Lalueza-Fox, C., Anderson, S., et al., 2001. Complete mitochondrial genome sequence of two extinct moas clarify ratite evolution. Nature 409, 704–707.

Craw, R.C., 1988. Continuing the synthesis between panbiogeography, phylogenetic systematic and geology as illustrated by empirical studies on the biogeography of New Zealand and the Chatham Islands. Systematic Zoology 37, 291–310.

Crawford, A.J., Meffre, S., Symonds, P., 2003. 120 to 0 Ma tectonic evolution of the southwest Pacific and analogous geological evolution of the 600 to 220 Ma Tasman fold belt system. Geological Society of Australia Special Publication 22, 379–397.

Davy, B., Wood, R., 1994. Gravity and magnetic modelling of the Hikurangi Plateau. Marine Geology 118, 139–151.

Davy, B., Hoernle, K., Werner, R., 2008. Hikurangi Plateau: crustal structure, rifted formation, and Gondwana subduction history. Geochemistry, Geophysics, Geosystems 9. doi:10.1029/2007GC001855.

Dettman, M.E., Pocknall, D.J., 1990. *Nothofagidites* Erdtman ex Potonié, 1960; a catalogue of species with notes on the paleogeographic distribution of *Nothofagus* Bl. (Southern Beech). New Zealand Geological Survey Paleontological Bulletin 60 (Wellington).

Dumbacher, J.P., Pratt, T.K., Fleischer, R.C., 2003. Phylogeny of the owlet-nightjars (Aves: Aegothelidae) based on mitochondrial DNA sequence. Molecular Phylogenetics and Evolution 29, 540–549.

Eagles, G., Gohl, K., Larter, R.D., 2004. High resolution animated tectonic reconstruction of the South Pacific and West Antarctic margin. Geochemistry, Geophysics, Geosystems 5. doi:10.1029/2003GC000657.

Elliot, D.H., Fleming, T.H., 2000. Weddell triple junction: the principal focus of Ferrar and Karoo magmatism during initial breakup of Gondwana. Geology 28, 539–542.

Fitzgerald, P., 2002. Tectonics and landscape evolution of the Antarctic plate since the breakup of Gondwana, with emphasis on the West Antarctic Rift System and the Transantarctic Mountains. Royal Society of New Zealand Bulletin 35, 453–469.

Gaina, C., Roest, W.R., Müller, R.D., Symonds, P., 1998. The opening of the Tasman Sea: a gravity anomaly animation. Earth Interactions 2-004 (Electronic Journal).

Giribit, G., Edgecombe, G.D., 2006. The importance of looking at small-scale patterns when inferring Gondwanan biogeography: a case study of the centipede *Paralamyctes* (Chilopoda, Lithobiomorpha, Henicopidae). Biological Journal of the Linnean Society 89, 65–78.

Gohl, K., Teterin, D., Eagles, G., et al., 2007. Geophysical survey reveals tectonic structures in the Amundsen Sea embayment, West Antarctica. In: Cooper, A.K., et al. (Ed.), Antarctica's Continent-Ocean Transitions: Consequences for Tectonic Reconstructions: Antarctica: a Keystone in a Changing World. The United States Geological Society Open-File Report. doi:10.3133/of2007-1047-srp047.

Goldstein, S.J., Gemmell, N.J., Schiel, D.R., 2006. Molecular phylogenetics and biogeography of the nacellid limpets of New Zealand (Mollusca: Patellogastropoda). Molecular Phylogenetics and Evolution 38, 261–265.

Grindley, G.W., Adams, C.J., Lumb, J.T., Watters, W.A., 1977. Palaeomagnetism, K-Ar dating, and tectonic interpretation of Upper Cretaceous and Cenozoic volcanic rocks of the Chatham Islands, New Zealand. New Zealand Journal of Geology and Geophysics 20, 425–467.

Grobys, J.W.G., Gohl, K., Barker, D., et al., 2007. Is the Bounty Trough off eastern New Zealand an aborted rift? Journal of Geophysical Research 112. doi:10.1029/2005JB004229.2007.

Grobys, J.W.G., Gohl, K., Uenzelmann-Neben, G., et al., 2008a. Extensional and magmatic nature of the Campbell Plateau and Great South basin from deep crustal studies. Tectonophysics. 472, 213–225 doi:10.1016/j.tecto.2008.05.003.

Grobys, J.W.G., Gohl, K., Eagles, G., 2008b. Quantitative tectonic reconstructions of Zealandia based on crustal thickness estimates. Geochemistry, Geophysics, Geosystems 9. doi:10.1029/2007gc001691.

Hall, L.S., Lamb, S.H., Niocaill, C.M., 2004. Cenozoic distributed rotational deformation, South Island, New Zealand. Tectonics 23. doi:10.1029/2002TC001421.

Harshman, J., Braun, E.L., Braun, M.J., et al., 2008. Phylogenetic evidence for multiple losses of flight in ratite birds. Proceedings of the National Academy of Sciences 105, 13462–13467.

Heads, M., 2005. Dating nodes on molecular phylogenies: a critique of molecular biogeography. Cladistics 21, 62–78.

Hedges, S.B., Parker, P.H., Sibley, C.G., Kumar, S., 1996. Continental breakup and the ordinal diversification of birds and mammals. Nature 381, 226–229.

Hennig, W., 1966. Phylogenetic Systematics, (Davis, D.D., Zangeri, R., Translators). Urbana, University of Illinois Press.

Herzer, R.H., Chaproniere, G.C.H., Edwards, A.R., et al., 1997. Seismic stratigraphy and structural history of the Reinga basin and its margin, southern Norfolk Ridge system. New Zealand Journal of Geology and Geophysics 40, 425–451.

Herzer, R.H., Sykes, R., Killops, S.D., et al., 1999. Cretaceous carbonaceous rocks from the Norfolk Ridge System, Southwest Pacific: implications for regional petroleum potential. New Zealand Journal of Geology and Geophysics 42, 57–73.

Hoernle, K., Hauff, F., Werner, R., Mortimer, N., 2004. New Insights into the Origin and Evolution of the Hikurangi Oceanic Plateau (Southwest Pacific) from Multi-beam Mapping and Sampling EOS: Transactions of the American Geophysical Union, vol. 85, pp. 401–408.

Isaac, M.J., Herzer, R.H., Brook, F.J., Hayward, B.W., 1994. Cretaceous and Cenozoic sedimentary basins of Northland, New Zealand. Institute of Geological and Nuclear Sciences Monograph, vol. 8. Institute of Geological and Nuclear Sciences, Lower Hutt.

Isozaki, Y., 2009. Illawarra Reversal: the fingerprint of a superplume that triggered Pangean breakup and the end-Guadalupian (Permian) mass extinction. Gondwana Research 15, 421–432.

Kamp, P.J., 1986. Late Cretaceous-Cenozoic tectonic development of the southwest Pacific region. Tectonophysics 121, 225–251.

Kamp, P.J., Vonk, A.J., Nelson, C.S., et al., 2004. Constraints on the evolution of the Taranaki Fault from thermochronology and basin analysis. Implications for the Taranaki Fault play. 2004 New Zealand Petroleum Conference Proceedings, 7–10 March, 2004.

King, P.R., 2000. Tectonic reconstructions of New Zealand: 40 Ma to the Present. New Zealand Journal of Geology and Geophysics 43, 611–638.

Klingelhoefer, F., Lafoy, Y., Collot, J., et al., 2007. Crustal structure of the basin and ridge system west of New Caledonia (southwest Pacific) from wide-angle and reflection seismic data. Journal of Geophysical Research 112. doi:10.1029/2007JB005093.

König, M., Jokat, W., 2006. The Mesozoic breakup of the Weddell sea. Journal of Geophysical Research 111. doi:10.1029/2005JB004035.

Ladiges, P.Y., Cantrill, D., 2007. New Caledonia–Australian connections: biogeographic patterns and geology. Australian Systematic Botany 20, 383–389.

Lafoy, Y., Brodien, I., Vially, R., Exon, N.F., 2005. Structure of the basin and ridge system west of New Caledonia (Southwest Pacific): a synthesis. Marine Geophysical Researches 26, 37–50.

Laird, M.G., 1993. Cretaceous Continental Rifts: New Zealand Region. In: Ballance, P.F. (Ed.), South Pacific Sedimentary Basins. In: Sedimentary Basins of the World, vol. 2. Elsevier, Amsterdam, pp. 37–49.

Lamarche, G., Collot, J.-Y., Wood, R.A., et al., 1997. The Oligocene-Miocene Pacific-Australia plate boundary, south of New Zealand: evolution from oceanic spreading to strike-slip faulting. Earth and Planetary Science Letters 148, 129–139.

Larter, R.D., Cunningham, A.P., Barker, P.F., et al., 2002. Tectonic evolution of the Pacific margin of Antarctica 1. Late Cretaceous tectonic reconstructions. Journal of Geophysical Research 107. doi:10.1029/2000JB000052.

Lebrun, J.-F., Lamarche, G., Collot, J.-Y., 2003. Subduction initiation at a strike-slip plate boundary: the Cenozoic Pacific-Australian plate boundary south of New Zealand. Journal of Geophysical Research 108, 1–18.

Luyendyk, B.P., 1995. Hypothesis for Cretaceous rifting of East Gondwana caused by subducted slab capture. Geology 23, 373–376.

Luyendyk, B.P., Wilson, D.S., Siddoway, C.S., 2003. Eastern margin of the Ross Sea Rift in western Marie Byrd Land, Antarctica: crustal structure and tectonic development. Geochemistry, Geophysics, Geosystems 4. doi:10.1029/2002GC000462.

McAdoo, D., Laxon, S., 1997. Antarctic tectonics: constraints from ERS-1 satellite marine gravity field. Science 276, 556–560.

McDaniel, S.F., Shaw, A.J., 2003. Phylogeographic structure and cryptic speciation in the trans-Antarctic moss *Pyrrhobryum mnioides*. Evolution 57, 205–215.

McLoughlin, S., 2001. The breakup history of Gondwana and its impact on pre-Cenozoic floristic provincialism. Australian Journal of Botany 49, 271–300.

Meffre, S., Crawford, A.J., Quilty, P.G., 2006. Arc-continent collision forming a large island between New Caledonia and New Zealand in the Oligocene. AESC 2006, Melbourne, Australia.

Michaelson, W., 1922. Die Verbreitung der Oligochätea im Lichte der Wegenerschen Theorie der Kontinentenverschiebung und andere Verbreitung dieser Tiergruppe. Verhandlungen des Naturwissenschaftlichen Veriens zu Hamburg im 1922. Hamburg. 37 pp.

Michaux, B., 2008. Alfred Russel Wallace – biogeographer. In: Smith, C., Beccaloni, G. (Eds.), Natural Selection and Beyond: Celebrating the Intellectual Legacy of Alfred Russel Wallace. Oxford University Press, UK, pp. 166–185.

Michaux, B., Leschen, R.A.B., 2005. East meets west: biogeology of the Campbell Plateau. Biological Journal of the Linnean Society 2005, 95–115.

Mildenhall, D.C., 1980. New Zealand Late Cretaceous and Cenozoic Plant Biogeography: a contribution. Palaeogeography, Palaeoclimatology, Palaeogeography 31, 197–233.

Mortimer, N., 2008. Zealandia. Arizona Geological Society Digest 22, 227–233.

Mortimer, N., Tulloch, A.J., Spark, R.N., et al., 1999. Overview of the Median Batholith, New Zealand: a new interpretation of the geology of the Median Tectonic Zone and adjacent rocks. Journal of African Earth Sciences 29, 257–268.

Mortimer, N., Davey, F.J., Melhuish, A., et al., 2002. Geological interpretation of a deep seismic reflection profile across the Eastern Province and Median Batholith: crustal architecture of an extended Phanerozoic convergent orogen. New Zealand Journal of Geology and Geophysics 45, 349–363.

Mortimer, N., Hoernle, K., Hauff, F., et al., 2006. New constraints on the age and evolution of the Wishbone Ridge, southwest Pacific Cretaceous microplates, and Zealandia–West Antarctica breakup. Geological Society of America 34, 185–188.

Mortimer, N., Herzer, R.H., Gans, P.B., et al., 2007. Oligocene–Miocene tectonic evolution of the South Fiji Basin and Northland Plateau, SW Pacific Ocean: evidence from petrology and dating of dredged rocks. Marine Geology 237, 1–24.

Nelson, G., Ladiges, P.Y., 1996. Paralogy in cladistic biogeography and analysis of paralogy-free subtrees. American Museum Novitates 3167 58 pp.

Nelson, G., Ladiges, P.Y., 2001. Gondwana, vicariance biogeography and the New York School revisited. Australian Journal of Botany 49, 389–409.

Norris, R.J., Turnbull, I.M., 1993. Cenozoic basins adjacent to an evolving transform plate boundary, southwest New Zealand. In: Ballance, P.F. (Ed.), South Pacific Sedimentary Basins. Sedimentary Basins of the World, vol. 2. Elsevier, Amsterdam, pp. 251–270.

Norvick, M.S., Smith, M.A., 2001. Mapping the plate tectonic reconstruction of southern and southeastern Australia and implications for petroleum systems. Australian Petroleum Production and Exploration Association Journal 15–35.

Palmowski, D., Hill, K.C., Hoffman, N., Bernecker, T., 2001. Otway Basin rifting to sea-floor spreading – hydrocarbon implications. PESA Eastern Australasian Basins Symposium, Melbourne 25–28 November, pp. 499–505.

Panter, K.S., Blusztajn, J., Hart, S.R., et al., 2006. The origin of HIMU in the SW Pacific: evidence from intraplate volcanism in southern New Zealand and subantarctic Islands. Journal of Petrology 17, 1673–1704.

Ponder, W.F., Colgan, D.J., Gleeson, D.M., Sherley, G.H., 2003. Relationships of *Placostylus* from Lord Howe Island: an investigation using the mitochondrial cytochrome *c* oxidase 1 gene. Molluscan Research 23, 159–178.

Powell, C.McA., Roots, S.R., Veevers, J.J., 1988. Pre-breakup continental extension in East Gondwanalanda and the early opening of the eastern Indian Ocean. Tectonophysics 155, 261–283.

Quilty, P.G., 1993. Tasmantid and Lord Howe seamounts: biostratigraphy and palaeoceanographic significance. Alcheringa 17, 28–53.

Rait, G.J., 2000. Thrust transport directions in the Northland Allochthon, New Zealand. New Zealand Journal of Geology and Geophysics 43, 271–288.

Randall, D.E., Niocaill, C.M., 2004. Cambrian paleomagnetic data confirm a Natal Embayment location for the Ellsworth-Whitmore Mountains, Antarctica, in Gondwana reconstructions. Geophysical Journal International 157, 105–116.

Renner, S.S., Foreman, D.B., Murray, D., 2000. Timing transantarctic disjunctions in the Atherospermataceae (Laurales): evidence from coding and noncoding chloroplast sequences. Systematic Biology 49, 579–591.

Riley, T.R., Knight, K.B., 2001. Age of pre-break-up Gondwana magmatism. Antarctic Sciences 13, 99–110.

Sanmartin, I., Ronquist, F., 2004. Southern hemisphere biogeography inferred by event-based models: plant versus animal patterns. Systematic Biology 53, 216–243.

Schellart, W.P., Lister, G.S., Toy, V.G., 2006. A Late Cretaceous and Cenozoic reconstruction of the southwest Pacific region: tectonics controlled by subduction and slab rollback processes. Earth-Science Reviews 76, 191–233.

Sdrolias, M., Müller, R.D., Gaina, C., 2003. Tectonic evolution of the Southwest Pacific using constraints from back arc basins. In: Hills, R.R., Müller, R.D. (Eds.), Evolution and Dynamics of the Australian Plate: Geological Society of Australia Special Publication, vol. 22, pp. 343–359.

Sdrolias, M., Müller, R.D., Mauffret, A., Bernardel, G., 2004. Enigmatic formation of the Norfolk Basin, SW Pacific: a plume influence on back-arc extension. Geochemistry, Geophysics, Geosystems 5. doi:10.1029/2003GC000643.

Siddoway, C.S., 2008. Tectonics of the West Antarctic Rift System: new light on the history and dynamics of distributed intracontinental extension. In: Cooper, A.K., et al. (Ed.), Antarctica: a Keystone in a Changing World. The National Academies Press, Washington, DC, pp. 91–110.

Skelley, P.E., Leschen, R.A.B., 2006. Erotylinae (Insecta: Coleoptera: Cucujoidea: Erotylidae): taxonomy and biogeography. Fauna of New Zealand 59 58 pp.

Spalletti, L.A., Artabe, A.E., Morel, E.M., 2003. Geological factors and evolution of southwestern Gondwana Triassic plants. Gondwana Research 6, 119–134.

Stagg, H.M.J., Borissova, I., Alcock, M., Moore, A.M.G., 1999. Tectonic provinces of the Lord Howe Rise 'Law of the Sea' study has implications for frontier hydrocarbons. AGSO Research Newsletter 31 (http://www.agso.gov.au/information/publications/renews/).

Stagpoole, V., Herzer, R., Funnell, R., Uruski, C., 2000. Deep water frontier basins of northwestern NZ – burial history and hydrocarbon maturation. 2000 NZ Petroleum Conference Proceedings, 19–22 March, 2000.

Stagpoole, V., Dravitzki, S., Ingham, M., Bennie, S., Bibby, H., 2004. Structure of the Taranaki Fault from the interpretation of magneto-telluric observations. 2004 New Zealand Petroleum Conference Proceedings, 7–10 March, 2004.

Suggate, R.P. (Ed.), 1978. The Geology of New Zealand, 2 Vols. Government Printer, Wellington.

Sutherland, R., 1995. The Australian boundary and Cenozoic plate motions in the SW Pacific: some constraints from Geosat data. Tectonics 14, 819–831.

Sutherland, R., 1999a. Basement geology and tectonic development of the greater New Zealand region: an interpretation from regional magnetic data. Tectonophysics 308, 341–362.

Sutherland, R., 1999b. Cenozoic bending of the New Zealand basement terranes and Alpine Fault displacement: a brief review. New Zealand Journal of Geology and Geophysics 42, 295–301.

Sutherland, R., 2000. Recent progress in understanding the structure and petroleum potential of southern South Island's frontier basins. 2000 New Zealand Petroleum Conference Proceedings, 19–22 March, 2000, pp. 153–162.

Sutherland, R., 2008. The significance of Antarctica for studies of global geodynamics. In: Cooper, A.K., et al. (Ed.), Antarctica: a Keystone in a Changing World. The National Academies Press, Washington, DC, pp. 115–124.

Sutherland, R., Melhuish, A., 2000. Formation of the Solander Basin, southwestern South Island, New Zealand, controlled by a major fault in continental crust and upper mantle. Tectonophysics 19, 44–61.

Torsvik, T.H., Gaina, C., Redfield, T.F., 2008. Antarctica and global paleogeography: from Rodinia, through Gondwanaland and Pangea, to the birth of the Southern Ocean and the opening of gateways. In: Cooper, A.K., et al. (Ed.), Antarctica: a Keystone in a Changing World. The National Academies Press, Washington, DC, pp. 125–140.

Tulloch, A.J., Kimbrough, D.L., Wood, R.A., 1991. Carboniferous granite basement dredged from a site on the southwestern margin of the Challenger Plateau, Tasman Sea. New Zealand Journal of Geology and Geophysics 34, 121–126.

Uruski, C.I., Cook, R.A., Herzer, R.H., Isaac, M.J., 2004. Petroleum geology of the Northland sector of the greater Taranaki Basin. New Zealand Petroleum Conference Proceedings, pp. 7–10. March, 2004.

Vaughan, A.P.M., Pankhurst, R.J., 2008. Tectonic overview of the West Gondwana margin. Gondwana Research 13, 150–162.

Vaughan, A.P.M., Storey, B.C., 2000. The eastern Palmer Land shear zone: a new terrane accretion model for the Mesozoic development of the Antarctic Peninsula. Journal of the Geological Society (London) 157, 1243–1256.

Vaughan, A.P.M., Leat, P.T., Pankhurst, R.J., 2005. Terrane processes at the margins of Gondwana: introduction. Geological Society, London, Special Publications 246, 1–21.

Veevers, J.J., 2006. Updated Gondwana (Permian-Cretaceous) earth history of Australia. Gondwana Research 9, 231–260.

Wallace, A.R., 1858. On the tendency of varieties to depart indefinitely from the original type. Proceedings of the Linnean Society 3, 53–62.

Wandres, A.M., Bradshaw, J.D., 2005. New Zealand tectonostratigraphy and implications from conglomeratic rocks for the configuration of the SW Pacific margin of Gondwana. Geological Society, London, Special Publications 246, 179–216.

Wegener, A., 1929. Die Entstehung der Kontinente und Ozeane, 4th revised edition. Friederich Vieweg und Sohn, Braunschweig.

Whattam, S.A., Malpas, J., Ali, J.R., et al., 2005. Formation and emplacement of the Northland ophiolite, northern New Zealand: SW Pacific tectonic implications. Journal of the Geological Society of London 162, 225–241.

Willcox, J.B., Sayers, J., Stagg, H.M.J., van den Beuque, S., 2001. Geological framework of the Lord Howe Rise and adjacent ocean basins. Eastern Australasian Basins Symposium, pp. 211–227.
Winberry, J.P., Anandakishnan, S., 2004. Crustal structure of the West Antarctic rift system and Marie Byrd Land hotspot. Geology 32, 977–980.
Wood, R., 1993. The Challenger Plateau. In: Ballance, P.F. (Ed.), South Pacific Sedimentary Basins – Sedimentary Basins of the World. Elsevier, Amsterdam, pp. 351–364.
Wood, R., Davy, B., 1994. The Hikurangi Plateau. Marine Geology 118, 153–173.
Wood, R., Herzer, R.H., 1993. The Chatham Rise, New Zealand. In: Ballance, P.F. (Ed.), South Pacific Sedimentary Basins – Sedimentary Basins of the World. Elsevier, Amsterdam, pp. 329–349.
Wood, R.A., Stagpoole, V.M., 2007. Validation of tectonic reconstructions by crustal volume balance: New Zealand through the Cenozoic. Geological Society of America Bulletin 119, 933–943.
Wood, R., Woodward, D., 2002. Sediment thickness and crustal structure of offshore western New Zealand from 3D gravity modelling. New Zealand Journal of Geology and Geophysics 45, 243–255.
Wood, R., Andrews, P.B., Herzer, R.H., et al., 1989. Cretaceous and Cenozoic geology of the Chatham Rise region, South Island, New Zealand: New Zealand Geological Survey Basin Studies, vol. 3. 76pp.
Wood, R., Lamarche, G., Herzer, R., Delteil, J., Davy, B., 1996. Paleogene seafloor spreading in the south east Tasman Sea. Tectonics 15, 966–975.
Wright, I., Sheppard, J., Turner, R., et al., 2002. New Zealand Continental Shelf Project – a status report of the survey programme. NZ Petroleum Conference proceedings, 24–27 February, 2002.

# Attribution

**Michaux, B. 2009. Reciprocality between biology and geology: Reconstructing polar Gondwana. *Gondwana Research*, 16:655–668.**

First Published in 2009 in *Gondwana Research* 16(3):655–668, doi.org/10.1016/j.gr.2009.06.002. Reprinted with permission from Elsevier.

# *chapter seven*

# *The Great South Land*

Many Europeans, from early Greek philosophers to Renaissance cartographers to eighteenth-century capitalists, believed in the existence of a Great Southern Land (*Terra Australis Incognita*) that counterbalanced the northern hemisphere landmasses, thereby ensuring the earth rotated smoothly like some perfectly tuned flywheel. So entrenched in European thought was this idea that the British government was prepared to finance an expedition, ostensibly to observe the transit of Venus from Tahiti, to try and find it. This was not driven by some lofty motive of discovery for discovery's sake, but by the lure of trade and profit (Armstrong and Martin 2015). Alexander Dalrymple, an employee of the East India Company before joining the Admiralty, estimated that this unknown landmass had a population of some 50 million and convinced the British government that trading opportunities and potential profit were worth a little investment. In response, the British government instructed the Admiralty to send Cook and various scientists, including the botanists Banks and Solander, to observe the transit of Venus from Tahiti. But the secret purpose of the expedition was to search for an unknown land to the west of Tahiti as far as 40° 22′ S and claim it for Britain, should it be found (Armstrong and Martin 2015).

Although Cook was unsuccessful in the quest for *Terra Australis Incognita* and eventually had to turn north and sail westwards, he did claim *Terra Nullius* – Australia – for the British crown. While he extensively mapped the east coast of Australia and put Botany Bay on the map, this was rather imperialistic given that Australia was already populated and had been continuously so by aboriginal tribes for about 50,000 years (Rasmussen et al. 2011). Also, the western and northern regions of Australia were well known to Malay, Chinese, Portuguese and Dutch traders. A second Pacific expedition, also led by Cook, finally put paid to the idea of a populated and prosperous unknown Great South Land when Cook and the crew of the *Resolution* reached 71° 10′ S without seeing any sign of it. In Cook's words, 'this Southern Continent (supposing there is one) must lay within the Polar Circle where the sea is so pestered with ice that the land is thereby inaccessible'.

How Cook might marvel at the accessibility and activity of Antarctica today. In 2016, almost 4500 scientists, technicians and service personnel worked on the frozen continent during the summer season (CIA

World Factbook: https://www.cia.gov/library/publications/the-world-factbook/geos/ay.html) and increasing numbers of tourists are visiting (International Association of Antarctic Tour Operators: https://iaato.org/tourism-statistics). While Antarctica may not have become an important global trading centre, it has proved to be centrally important to many scientific research programmes, including geophysical research.

## *Plate motion circuits*

Plate motion circuits connect plates that share boundaries, and well-constrained plate motions from one part of a circuit can be used to infer plate motions of other plates on the circuit whose motions are either unknown or uncertain. Plate motions about spreading centres are well constrained and the directions and rates of movement can be accurately calculated from fracture zones and magnetic anomalies. However, plates that share subduction, diffuse or transform boundaries are poorly constrained because the information needed to calculate their motions directly has either been destroyed (subduction boundaries) or is uncertain and often ambiguous (diffuse or transform boundaries). Motions between plates that share such boundaries can be inferred by vector summation around appropriate parts of the plate motion circuit. A circuit is said to be closed when all plate motions between connected plates have been calculated. The most up-to-date global plate motion model (MORVEL) is based on mid-ocean ridge velocities (DeMets et al. 2010). Antarctica is central to this model in terms of the number of connections it has within the global circuit, replacing earlier models (NUVEL and variants) that were centred on Africa (DeMets et al. 1990, 1994). The Antarctic plate thus plays a pivotal role in this global network.

The reconstruction of past plate configurations has long fascinated me. The sheer audacity of attempting to recreate the deep past has to rank as one of the great intellectual achievements of modern geology. Of course, these models have their practical uses, for example, in understanding the genesis and distribution of natural resources at various times in the past, but to see global maps of deep time or tectonic animations is truly remarkable. But how accurate are they? Certainly, scale is one factor in accuracy – a global reconstruction of the Carboniferous world gives a generalised view that becomes less accurate and more vaguely defined as one zooms in to look at a finer and finer resolution. Time is another factor – the more modern the reconstruction, the more accurate it is likely to be because fewer data have been destroyed by the recycling of oceanic crust and modification of continental crust. But the greatest factor affecting accuracy is the nature of the data themselves. Reconstruction of past plate configurations becomes more accurate the better constrained the models are, and just as in the case of modern plate motion circuits, the best data come from that part of the circuit where plate boundaries are spreading centres that leave well-defined magnetic anomalies and fracture zones. Reconstruction of past plate configurations are also beset by additional problems such as the presence of undetected tectonic boundaries; uncertainty about the location and characteristics of poorly defined boundaries; internal deformation of plates; or the degree of stretching of continental crust leading to uncertainty in the position of the original continental edge. Such difficulties can result in anomalous fits producing overlaps or gaps between adjacent continental blocks.

## Polar Gondwana at 100 Ma

Matthews and her colleagues (Matthews et al. 2012) argued that 100 Ma was a time of major global plate rearrangement and a critical time in the tectonic development of East Gondwana. They attributed these near global tectonic changes to the cessation of subduction along the Australian/New Zealand margin of East Gondwana, resulting in a change from compressional to extensional tectonics in east Australia, New Zealand and West Antarctica. Onshore geology correlated with this event included uplift and erosion in east Australia, the generation of A-type magmas in New Zealand and Marie Byrd Land, compression in the Antarctic Peninsula and uplift in Alexander Island. Further north, the Andean trench became active with deformation recorded in the northern Andes. Although the collision of the Hikurangi Plateau with the Chatham Rise also dates from this time, Matthews et al. (2012) attributed the cessation of subduction and initiation of a strike-slip boundary to the oblique collision between a paired ocean ridge system – similar perhaps to the modern Lau-Colville and Kermadec Ridges – with the offshore subduction zone.

Extensional tectonics initiated at circa 100 Ma resulted in increasing fragmentation of the Australasian sector of East Gondwana and completed the decoupling of West from East Antarctica. Early phases of extension included: rifting between Australia and East Antarctica that was characterised by slow and sporadic spreading throughout the Cretaceous (Powell et al. 1988; Norvick and Smith 2001), and the development of a strike-slip boundary between Tasmania and East Antarctica by circa 85 Ma (Lamb et al. 2016); rifting between East and West Antarctica along the West Antarctic Rift System (WARS) between 105 and 85 Ma (Fitzgerald 2002) with the main phase of rifting between 100 and 90 Ma (Siddoway 2008); and the opening of the Bounty Trough as a result of the Phoenix Plate spreading centre propagating westwards at 90 Ma (Gohl et al. 2013; Kipf et al. 2014). Continued extension eventually resulted in the opening of ocean basins and geodispersal of continental fragments. The Tasman Sea opened in the south at 85 Ma and continued spreading until 63 Ma (Sdrolias et al. 2001) or 55 Ma

(Schellart et al. 2006) in the north, resulting in the Lord Howe Rise completely separating from Australia. Further to the south, the Campbell Plateau separated from Marie Byrd Land (West Antarctica) along the Antipodes Rift between 84 and 80 Ma as the growing Bellingshausen Sea Plate propagated southwestwards from the Amundsen Sea, West Antarctica, (Wobbe et al. 2012; Gohl et al. 2013), forming the Southern Ocean between West Antarctica and the Campbell Plateau.

Many models have been proposed to describe the tectonic development of the southwest Pacific from 100 Ma, but as Matthews et al. (2015) have discussed, there is little consensus between different models due to a paucity of data and its often-ambiguous nature, and a poorly constrained and therefore non-robust southwest Pacific plate circuit for this time period. Although three spreading centres – Southeast Indian Ocean, Tasman Sea and Amundsen Sea – were active during the Late Cretaceous, the former is connected to the Pacific via the diffuse Australian-East Antarctic spreading centre that lacks ocean crust in the central and eastern portions from which magnetic lineations might be recovered, and the other two are only useful after 80 Ma. There are also a number of other possible plate boundaries present in the southwest Pacific of the time that could account for some portion of relative plate motions. Matthews et al. (2015) identified two such boundaries in the southwest Pacific region – WARS and a Lord Howe Rise-Pacific junction. In Michaux (2009), I suggested that a major tectonic boundary – the Campbell Fault – also existed in this part of East Gondwana.

## *Age of the Alpine Fault*

As Mortimer (2018) noted, the idea of a substantial pre-Eocene movement along a precursor to the modern Alpine Fault is not a new idea and has been around since the 1960s (Suggate 1963). Neither is it unknown for older faults to be reactivated at a later date (Sutherland et al. 2000). However, the idea that the Alpine Fault predates the Eocene has been somewhat controversial but one that has never really gone away because there is a conundrum regarding the amount of displacement along the Alpine Fault. Using onshore geology and looking at the displacement of markers either side of the fault leads to an estimation of about 450 km of dextral displacement. Analysis of the offshore Pacific-Australian-Antarctica plate circuit leads to an estimation of 850 km of relative displacement between the Australian and Pacific plates.

Traditionally, this discrepancy has been accounted for by deformation across a broad zone leading to the bending of basement rocks in the South Island, New Zealand, into an S-shaped orocline. Under this model, basement terranes were straight and contiguous at the end of the Cretaceous and bending and displacement occurred in the Cenozoic. Lamb et al. (2016)

have proposed an alternative model in which terrane deformation is pre-Cenozoic and all subsequent plate motion has been accommodated along the Alpine Fault. The model requires that there was a large sinistral movement (<300 km) along a proto-Alpine Fault between 105 and 70 Ma. Their proposed proto-Alpine Fault acted as a tectonic boundary that separated Northwest Zealandia from Southeast Zealandia, and formed one arm of a triple junction with WARS and the Australian-East Antarctic Rift. Mortimer (2018) detailed the evidence for a pre-Eocene proto-Alpine Fault and also suggested that this divided Zealandia into northern and southern portions in the Late Cretaceous. The Campbell Rift of Michaux (2009: Figure 2) is the equivalent of the proto-Alpine Fault of these authors.

## *Extent and position of continental fragments*

Figure 7.1 is a schematic representation of polar Gondwana at about 100 Ma. The position and possible movement along tectonic boundaries are indicated in Figure 7.1, but the question arises as to the extent and positioning of the various continental fragments within the New Zealand sector. In Figure 7.1, the Eastern Province terranes and Median Batholith (Mortimer et al. 2014) are confined to East Gondwanan New Zealand and are absent from the West Gondwanan Campbell Plateau (Figure 7.1: CP) and Marie Byrd Land (Figure 7.1: MBL).

As discussed in the previous chapter, there is really no evidence to suggest the presence of Eastern Province terranes anywhere on the Campbell Plateau; in fact, sedimentary samples recovered from around the Bounty Islands or from bore holes in the Great South Basin appear to be Western Province rocks similar to the Greenland Group of Westland and Northwest Nelson or Palaeozoic sedimentary rocks of the Ford Ranges, Marie Byrd Land (Adams et al. 1995; Adams et al. 2008). Neither is there convincing evidence for the presence of other New Zealand basement markers extending onto the Campbell Plateau such as the Stokes Magnetic Anomaly (Beggs et al. 1990) or the Median Batholith. An analysis of zircon age profiles by Adams et al. (2017) established that detrital zircons found in the Great South Basin are much younger that those recovered from west coast basins of the South Island. The older zircons (130–110 Ma) of the west coast basins were derived from the Median Batholith, but the younger Great South Basin zircons (106–100 Ma) must have been derived from A-type granites associated with the change to extension tectonics discussed earlier. They concluded that the Median Batholith was absent from the source area of the Great South Basin sediments. Adams et al. (2017) also noted that the overall zircon profile from the Great South Basin differs from the west coast basins by its lack of Devonian and Carboniferous zircons and presence of Proterozoic and Palaeozoic zircons, also indicating a different source area for the zircons and suggesting a different history for the Great South Basin.

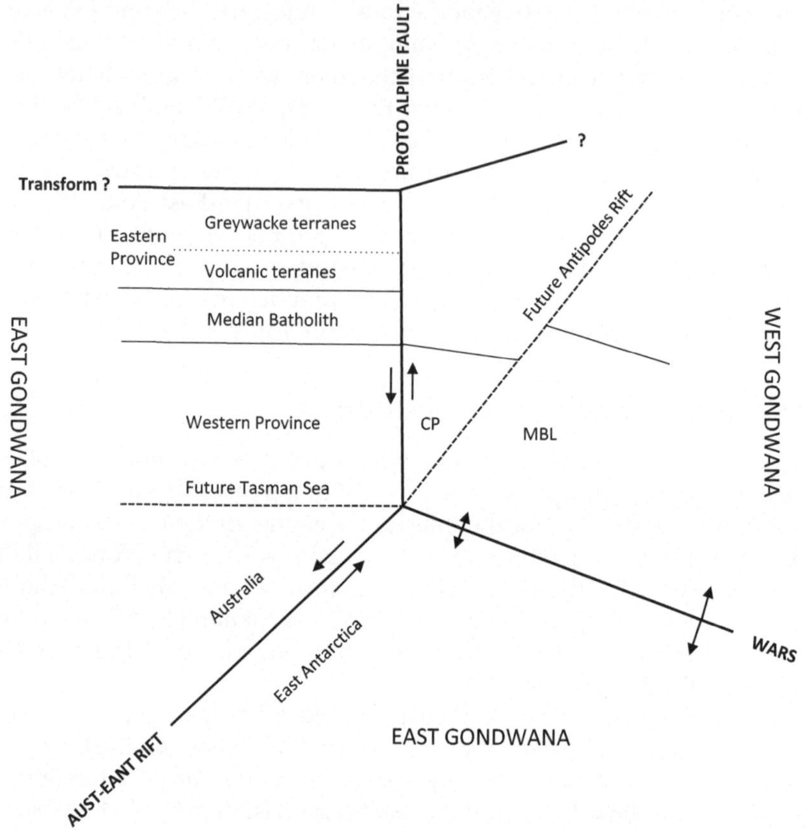


*Figure 7.1* Polar Gondwana at 100 Ma. Solid line = active tectonic boundaries with possible movements indicated. Dashed line = Late Cretaceous tectonic boundary; ? = unknown boundary type. AUST-EANT = Australia-East Antarctica; CP = Campbell Plateau; MBL = Marie Byrd Land; WARS = West Antarctic Rift System.

Strogen et al. (2017) identified two distinct phases of faulting within the Taranaki Basin. The older faults (105–83 Ma) trend northwest or west northwest while the younger faults (80–55 Ma) trend northeast. These ages and trends correlate with those of the Bounty Trough (after taking into account an anticlockwise rotation of the Bounty Trough at 105 Ma) (Davy 2014: Figure 3) and the Great South Basin respectively. It has always seemed structurally odd to me that the Bounty Trough and Great South Basin are more or less orthogonal to each other, but this conundrum is resolved if the Bounty Trough and Chatham Rise were not originally juxtaposed with the Campbell Plateau. Given its age and original orientation, the Bounty Trough would appear to be related to Lord Howe Rise structures such as the Taranaki and New Caledonian basins. The younger northeast strike correlates with the trend of the proto-Alpine Fault and may represent a period of extension along this boundary on both the Challenger Plateau (East Gondwana) and the Campbell Plateau (West Gondwana).

The geology of the Chatham Rise, which is situated along the northern margin of the Bounty Trough, clearly links it to the Eastern Province of the North and South Islands of New Zealand. Adams et al. (2008) suggested the greywackes and schists of the Chatham Islands were equivalent to the Rakaia terrane, and Andrews et al. (1978) correlated the schists of the Forty Fours, a group of small islets to the east of the Chathams, with those of the Caples terrane of Otago. In light of the foregoing discussion and the distinctly East Gondwanan biota of the Chatham Islands discussed in the previous chapter, I would suggest that at 100 Ma both the Chatham Rise and the Bounty Trough were part of the Australian plate, that is, part of East Gondwana. Removing the Chatham Rise and Bounty Trough from the Campbell Plateau, which is unequivocally part of West Gondwana, removes the need for the rather strained reconstructions that such an arrangement necessitates (e.g. Sutherland 1999: Figure 3; King 2000: Figure 1; Crampton et al. 2003: Figure 5; Wood and Stagpoole 2007: Figure 3).

## *What's in a name?*

Zealandia is a term that was first introduced by Bruce Luyendyk (1995) to describe an extensive area of largely submerged continental crust off the east coast of Australia that some have claimed should be treated as a continent in its own right (Mortimer et al. 2017). Although many New Zealand scientists have enthusiastically adopted the name, I have always regarded it with some suspicion. While I now wish I'd paid more attention in my Latin classes at school, I think the name should be spelt Zelandia. Usage suggests that 'Zealandia', 'Zelandia' and 'Zelanda' are all acceptable spellings, and after all isn't this just the way languages 'evolve', the fact is that Latin is a dead language, and therefore, its grammar is fixed. Grammatical quibbles aside, is Zealandia a name that is just a simple signifier, a label that has no more significance than the name 'Asia' or 'Europe' or should a name be something else? I would argue that Zealandia, in a biogeological sense, is a polyphyletic entity because it is a composite of two continental blocks – the East Gondwanan Melanesian Rift and the West Gondwanan Campbell Plateau – each of which have had separate histories before their combination. Ebach and Michaux (2017) discussed the role of monophyly in the discovery of natural taxonomic groupings and by analogy extended this idea to discovering natural biogeographic areas. These authors suggested that the advances made in systematics when poly- and paraphyletic taxa were identified and removed could also accrue to biogeography if poly- and paraphyletic areas were also identified and eliminated. In my view, the name Zealandia serves no useful purpose and simply supports the continued use of polyphyly in biogeography. A more detailed discussion of the nature of natural biogeological areas is addressed in the following chapter.

The LINNEAN SOCIETY of London BIOLOGICAL Journal of the Linnean Society

*Biological Journal of the Linnean Society*, 2010, **101**, 193–212. With 5 figures

# Biogeology of Wallacea: geotectonic models, areas of endemism, and natural biogeographical units

BERNARD MICHAUX*

*Private Bag, Kaukapakapa, New Zealand*

*Received 10 December 2009; revised 11 March 2010; accepted for publication 12 March 2010*

The concepts of biogeographical regions and areas of endemism are briefly reviewed prior to a discussion of what constitutes a natural biogeographical unit. It is concluded that a natural biogeographical unit comprises a group of endemic species that share a geological history. These natural biogeographical units are termed Wallacean biogeographical units in honour of the biogeographer A.R. Wallace. Models of the geological development of Indonesia and the Philippines are outlined. Areas of endemism within Wallacea are identified by distributional data, and their relationship to each other and to the adjacent continental regions are evaluated using molecular phylogenies from the literature. The boundaries of these areas of endemism are in broad agreement with earlier works, but it is argued that the Tanimbar Islands are biologically part of south Maluku, rather than the Lesser Sundas, and that Timor (plus Savu, Roti, Wetar, Damar, and Babar) and the western Lesser Sundas form areas of endemism in their own right. Wallacean biogeographical units within Wallacea are identified by congruence between areas of endemism and geological history. It is concluded that although Wallacea as a whole is not a natural biogeographical region, neither is it completely artificial as it is formed from a complex of predominantly Australasian exotic fragments linked by geological processes within a complex collision zone. The Philippines are argued to be an integral part of Wallacea, as originally intended. © 2010 The Linnean Society of London, *Biological Journal of the Linnean Society*, 2010, **101**, 193–212.

ADDITIONAL KEYWORDS: molecular phylogenies – Philippines – Wallacean biogeographical units.

## INTRODUCTION

Wallacea (Fig. 1) is a biogeographical region separating Asia from Australasia. It and the adjacent areas of Sundaland (Borneo, Java, Sumatra, and Peninsula Malaya) and the Philippines are recognized as internationally important biodiversity hotspots by Conservation International (http://www.biodiversityhotspots.org). Hotspots are defined on the basis of biodiversity, degree of endemicity, and level of threat (ICBP, 1992; Cardillo *et al.*, 2006). Although the faunas of many Wallacean islands are highly endemic, hierarchical patterns of relationship are evident both within some islands, and between islands and the adjacent continental areas. For example, Sulawesi is recognized as an important centre of endemism within Wallacea, but a more detailed analysis reveals it to be composed of a number of endemic subregions, which are themselves related on a broader scale to other areas within and beyond Wallacea. Indeed, the primary purpose of this paper is to describe this hierarchical pattern, and to examine its congruence with rift-drift-docking events in this complex modern orogen. Understanding the historical development of present-day biodiversity can provide a broader perspective for those deciding conservation priorities.

### NATURAL BIOGEOGRAPHICAL UNITS

In order to describe any pattern it is first necessary to define the units that make up that pattern. Because biogeographers are dealing with natural processes such as the evolution of organisms and the results of tectonic change, those units should be natural, i.e. represent the product of these processes, rather than being chosen for other reasons such as utility (Michaux, 2008). Traditionally, the fundamental units

*E-mail: michaux@xtra.co.nz

 

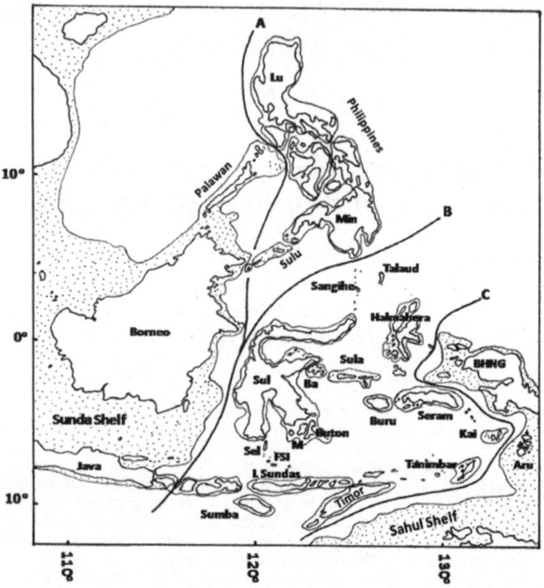


**Figure 1.** Locality Map of Wallacea. Wallacea was originally defined as the area between Huxley's Line (A) and Lydekker's Line (C), i.e. the region bounded by the Sunda and Sahel shelves. Subsequently, Wallacea was redefined as the area between Wallace's Line (B) and the Sahul Shelf. Ba, Banggai Island; BHNG, Bird's Head New Guinea; FSI, Flores Sea Islands; L Sundas, Lesser Sundas; Lu, Luzon; Min, Mindanao; M, Muna; Sel, Selayer; Sul, Sulawesi.

of biogeography have been regions (Nelson, 1978). Morrone (2002) discussed how many of these traditionally recognized regions have been criticized by panbiogeographers and vicariance biogeographers as non natural because their analyses suggested that the regions were biologically composite. Any conclusions based on a biogeographical analysis of such areas run the risk of being spurious. Panbiogeographers have gone even further, and have argued that natural biogeographical units should not be based on present-day land areas at all, but instead should be based on ocean basins (e.g. Craw, Grehan & Heads, 1999).

Cox (2001) has also reviewed the concept of biogeographical regions, and discussed Indonesia in some detail. He argued that Wallace's Line, the boundary between the Oriental and Australasian regions, is artificial, and a result of Wallace's over-reliance on mammal distributions. Cox (2001) suggested that if Wallace had used flowering plant distributions instead he would not have recognized any boundary (although see van Welzen, Slik & Alahuhta, 2005). Cox (2001) favoured a more realistic (i.e. natural) solution by defining the boundaries of the Oriental and Australasian regions in terms of the edges of the Sunda and Sahul shelves, respectively, and recognizing the intervening islands as a distinct region (Wallacea). Interestingly, this analysis led Cox (2001) to include the Philippines as part of Wallacea.

Following the advent of vicariance biogeography in particular, the search for natural biogeographical units has focused on areas of endemism, which at their simplest can be defined as the 'congruent distributional limits of two or more species' (Platnick, 1991). Identifying areas of endemism by parsimony analysis of a species/area matrix (Rosen, 1988) continues to be a technique widely used in biogeographical analysis (e.g. Löwenberg-Neto & de Carvalho, 2009). Parsimony analysis of endemism (PAE) has been criticized by Harold & Mooi (1994) because the technique lacks phylogenetic input, and therefore would not necessarily yield 'natural' areas. According to these authors 'areas of endemism are fundamentally historical entities, not distributional ones, and their definition should take history into account'. The historical input, in their view, is the cladogram – a summation of a taxon's biological history. Clearly the distribution and overlap of endemic species is a necessary part of any definition of what constitutes a natural biogeographical unit, but endemism alone is not sufficient in my view, as I explain below. What seems to be missing from contemporary treatments of endemism is any sense of the temporal and spatial continuity of the area itself. As A.R. Wallace suggested many years ago, natural biogeographical units will have '. . . formed a single more or less continuous area while separated either by geographical, climatical or biological barriers from the adjacent areas' (Wallace, 1894: 612).

For Wallace, the two processes of isolation in space and biological differentiation through time were inseparable, because one (isolation) led to the other (speciation). Wallace's view of what constituted natural – the dual criteria of biological and geological uniqueness – has some important implications for how natural biogeographical units are identified. Because Wallace was the first to suggest a geological/historical component to the identification of natural biogeographical areas. I propose to name such entities Wallacean biogeographical units.

Figure 2 is a graphical representation of biological–geological space. The geological axis is discrete: the upper half represents areas that have a shared geological history; the lower half represents areas that do not. The biological axis is also divided by the presence or absence of endemic species, but the presence of endemics is a continuum, ranging from zero to complete endemism. Areas that share a geological history but do not contain endemic species are found in the top-left quadrant. Such a situation could arise if all local endemics were extirpated, for example if the land was completely submerged (in the case of terrestrial endemics), or no endemics had yet evolved. The space occupied by Wallacean biogeographical units is represented by the top-right quadrant. All areas of endemism in this space have a shared geological history, but the practical identification of Wallacean biogeographical units becomes more difficult as the number of endemic species decrease. This space is identified as 'not recognized' in Figure 2. An important conclusion is that Wallacean biogeographical units may be defined by the presence of only a few endemic species. The long-standing dispute about whether the Philippines are part of Wallacea (Dickerson *et al.*, 1923; White & Bruce, 1986; Coates & Bishop, 1997), which dates back to the mid-19$^{th}$ century, and is immortalized in the positioning of Huxley's and Wallace's lines (Fig. 1A, B), has essentially been an argument about the relative number of endemics shared by the Philippines and Sundaland/Asia or Wallacea, respectively.

The bottom-right quadrant represents areas of endemism that do not share a geological history, for example they may be geologically composite.

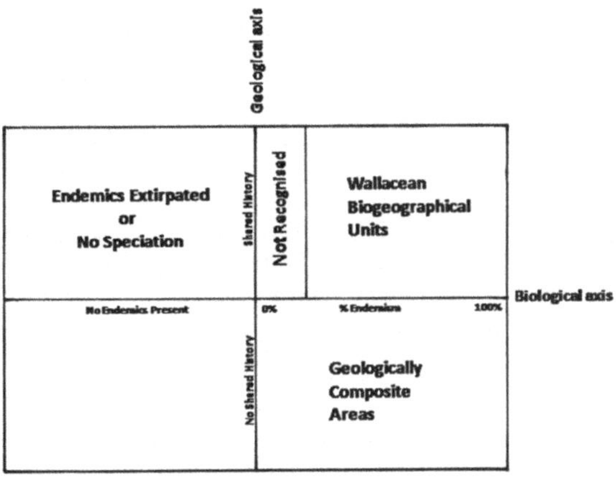


**Figure 2.** Analysis of biological–geological space. Areas of endemism are natural or Wallacean biogeographical units if they share a common biological and geological history. Areas of endemism that are geologically composite are not Wallacean biogeographical units. See text for discussion.

Although these areas of endemism are not Wallacean biogeographical units, they may be composed of more geographically restricted areas that are. These could be identified as such if, for example, sufficient numbers of endemics were still restricted to their original range. Even if endemic species were no longer confined to their original range, any biogeographical signal should still be retained in a fully resolved cladogram. This might be an example of what Henderson (1991) referred to as biogeography without. The analytical task here is to tease out individual biogeographical signals from the mixture (Harold & Mooi, 1994; Michaux & Leschen, 2005). Wallacea, with its complex geological history, is a prime candidate for such an analysis. An important conclusion to be drawn from this discussion is that the presence of endemic species is not sufficient to identify Wallacean biogeographical units.

## GEOLOGICAL DEVELOPMENT OF THE INDONESIAN OROGEN

Modern Indonesia is a region of great geological complexity resulting from the collision of the Australian Plate margin with Eurasia (Sundaland) and the Philippine Sea Plate. It is the site of a developing modern orogen or mountain building event. Describing such a complex situation in a way that is both relevant and useful to biogeographers is a task fraught with potential dangers, not only because many of the details are not yet clear, but also because there is an inherent risk of oversimplification, to the point of being misleading. However, it is also important to bear in mind that many of the details, which may be critical to geological understanding, may not be so critical to the biogeographer because of a lack of one-to-one correspondence between geological models and biogeographical analysis. Different geological models may be equally consistent with a given biogeographical interpretation, and vica versa.

The theoretical perspective that underlies the biogeographical analyses in this paper is that speciation is linked to the formation of barriers to dispersal. Only when genetic interchange between populations of a species is disrupted can speciation occur. The origins of barriers to dispersal are regarded as predominantly geological: rifting, the opening of basins, mountain building, or other vicariant events relatedor to geological change. Models of the geological development of the Indonesian orogen are clearly relevant to understanding speciation patterns and the development of endemism in the region. Biogeographical analysis of endemic species sometimes reveals that a fauna and flora is composite. Composite faunas and floras can result when independent geological entities such as continental fragments or island arcs are translated, dispersed, and amalgamated. The dynamic nature of the Indonesian orogen is known to include the dispersal of Australian crust westwards. Although it may never be possible to know if this translated crust remained above water, the congruence between biological and geological patterns and the presence of palaeoendemics are seen as compelling evidence that it did so.

### SULAWESI

*Cretaceous tectonics of central Indonesia*

Hamilton (1988) was the first to suggest that the island of Sumba (Fig. 1) was part of Sundaland. This suggestion was subsequently confirmed by both palaeomagnetic (Wensink, 1994) and isotope (Vroon, van Bergen & Forde, 1996) data. Rangin *et al.* (1990a) linked Sumba with western Sulawesi and east Java in what they termed the Sumba Block, which they suggested had amalgamated with Sundaland by 'the early Tertiary'. More recent work by Smyth *et al.* (2007) showed that east Java's arc volcanics and ophiolites are underlain by continental basement containing Archaean zircons. Smyth *et al.* (2007) suggested that the continental crust underlying east Java was originally adjacent to west Australia, because this region has source rocks from which zircons with the appropriate age structure could have been derived. Proterozoic and Palaeozoic zircons obtained from volcanic rocks (Bergman *et al.*, 1996), and Archaean and Proterozoic zircons and isotopic signatures of metamorphic rocks (van Leeuwen *et al.*, 2007) from western Sulawesi, indicate that it too is underlain by Australian-derived continental crust. Hall, Clements & Smyth (2009a) combined east Java, western Sulawesi, and the crust underlying the southern Makassar Straits and east Java Sea as the Argo fragment (or fragments), a terrane derived from north-western Australia in the mid-Cretaceous and sutured to south-west Borneo between 92 and 80 Ma. According to Hall *et al.* (2009a), Sundaland became an elevated and mostly emergent continental region following this collision.

*Cenozoic tectonics of western Indonesia*

The reconstruction of Hall *et al.* (2009a) predicted north-west subduction beneath Sumba and west Sulawesi between 63 and 50 Ma, where appropriately aged volcanics are found. The relationship between this renewed tectonic activity along the eastern margin of Sundaland and Late Cretaceous volcanic activity along the western Pacific margin is unknown, but is too early to be a consequence of Australia's northward drift. The earliest phase of the orogen proper dates from *c.* 45 Ma, when rifting opened the

Makassar Strait separating western Sulawesi from the east coast of Kalimantan, and isolating it from Sundaland (Hall *et al.*, 2009b).

In most reconstruction models, Australasian-derived microcontinental blocks became caught up in the collision zone and translated westwards to collide with this Eurasian fragment between *c.* 20 and 5 Ma (e.g. Ali *et al.*, 1996; Charlton, 1996; Hall, 1996, 1998; Hall & Wilson, 2000). This process accounted for the Australasian continental crust found in Banggai-Sula and Buton/Tukang Besi. However, present fieldwork in Sulawesi is casting increasing doubt on the reality of these microcontinental fragments (R. Hall, pers. comm.). Indeed, rather than modelling these Australasian fragments as isolated microcontinents that became amalgamated with Sulawesi, Hall and his colleagues favour a single structure (the Sula Spur), which subsequently became fragmented after collision with Sundaland. Although details of the Sula Spur have yet to be published, Hall *et al.* (2009a) used it in their latest reconstruction model, where it appears to be an extension of the Bird's Head region of New Guinea, and consists of the south-east peninsula of Sulawesi/Buton, the north-east peninsula of Sulawesi, Banggai/Sula, and south Maluku (Buru, Seram, and possibly Obi). For the biogeographer, the important point is that Sulawesi is geologically composite, and the way this was achieved is not as important as it is to the geologist. This is a good example of the lack of one-to-one correspondence discussed earlier. However, the probability of tectonically transporting terrestrial Australasian taxa on a structurally coherent entity such as the Sula Spur would appear to be greater than on isolated microcontinental fragments.

The third and final geological component of modern Sulawesi is the Northern Arm, an island arc terrane formed above the subduction zone that accommodated Australia's northward drift, and is related to other arcs produced over this subduction system, and now found in the Philippines and Halmahera (Hall, 1998). It probably amalgamated with the Eurasian continental fragment at much the same time as the arrival of the Australasian continental Sula Spur.

### MALUKU

Maluku is geologically composite, with south Maluku (Buru and Seram) composed of Australasian continental basement, and north Maluku (Halmahera) composed of island arc volcanics. In most reconstructions, south Maluku is composed of a series of microcontinents grouped with the Bird's Head microcontinent (e.g. Vroon *et al.*, 1996; Hall, 1998). In the light of the foregoing discussion of the Sula Spur, the continental basement found on Buru and Seram might best be interpreted as part of this continental extension from western New Guinea. According to Linthout *et al.* (1996), ophiolites in western Seram were emplaced following a collision with the Banda Arc between 8 and 5.4 Ma, although other workers dispute this interpretation (R. Hall. pers. comm.).

North Maluku is compositionally dominated by oceanic and island arc terranes (Hakim & Hall, 1991). An extensive Late Cretaceous to Eocene (*c.* 94–45 Ma) arc system built on older ophiolite basement is exposed on East Halmahera (Hall *et al.*, 1991). Hall *et al.* (1991) interpreted east Halmahera as part of the Philippine Sea Plate margin, and suggested that similarly aged arcs built on ophiolites could be traced north into the Philippines (Mindanao) and east into northern and central New Guinea. West Halmahera is an active arc built on this older arc system from *c.* 11 Ma. Although much of north Maluku is arc-derived, continental basement rocks are known from Bacan and Obi (Ali & Hall, 1995). In Bacan, these continental rocks and the ophiolites of the Philippine Sea Plate have been juxtaposed since about 22 Ma. Whether this juxtaposition was brought about by the collision of the Australian margin with the Halmahera arc or whether the continental fragments of Bacan and Obi were transported westwards to their present position via the Sorong Fault is uncertain (Malaihollo & Hall, 1996).

### TIMOR AND THE LESSER SUNDAS

Timor is composed of an Australian continental rift fragment tectonically overlain by Asian material obducted during collision with the Banda Arc at 2.4 Ma (Richardson & Blundell, 1996). The Lesser Sunda Islands are usually regarded as relatively recent products of Neogene volcanism, related to subduction along the Banda Arc (e.g. Hamilton, 1988). However, the intriguing possibility that the western Lesser Sundas are built on Australian continental rocks as adjacent areas such as west Sulawesi and east Java are, leaves questions about their geological origin open.

### PHILIPPINES

Figure 3 shows the major tectonic features of the Philippines. The area between the eastward dipping subduction zones to the west and the westwards dipping subduction zones to the east is known as the Philippine Mobile Belt. This feature, a series of island arcs and ophiolite suites, is divided along its entire length by the Philippine Fault. The Philippine fault has been interpreted as a suture zone between lateral extensions of the Sangihe and Halmahera arcs (Moore & Silver, 1982), but as the fault is considerably younger than these two arcs (Rangin, Jolivet &

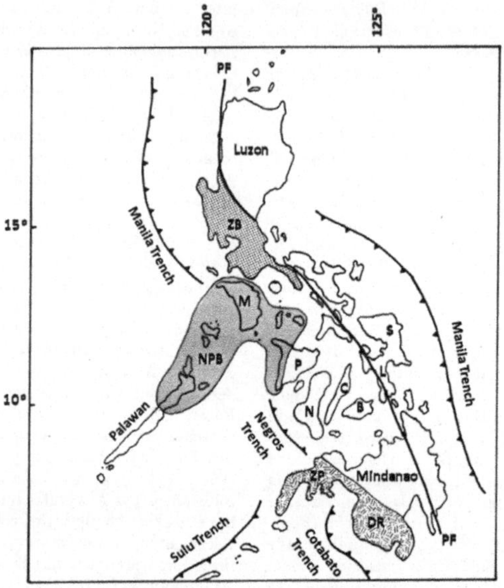


**Figure 3.** Simplified tectonic map of the Philippines. B, Bohol; C, Cebu; DR, Daguma Ranges; L, Leyte; M, Mindoro; N, Negros; NPB, North Palawan Block; P, Panay; PF, Philippine Fault; S, Samar; ZB, Zambales Block; ZP, Zamboanga Peninsula. NPB (fine stipple) and ZP + DR (irregular hatching) are continental terranes of Asian and possibly Sundaic in origin, respectively; ZB (coarse stipple) is an island arc complex of possible Sundaic origin. The region between the western and eastern arc systems is known as the Philippine Mobile Belt.

Pubellier, 1990b; Yumul *et al.*, 2008) this interpretation is not generally accepted (Yumul *et al.*, 2004). For example, Pubellier *et al.* (1996) interpreted the fault as a boundary between fragmentation (west) and dispersion (east) zones within the Philippines, with the former responding to stresses by fragmentation through basin formation, and the latter responding to stresses by translation northwards.

The ophiolite complexes of the Philippine Mobile Belt consist of a Cretaceous suite (> 65 Ma) in the central and eastern part and a younger Eocene–Oligocene suite (*c.* 50–30 Ma) in the west (Yumul *et al.*, 2008). The central and eastern Cretaceous ophiolites have been translated from the south-east since at least 45 Ma (Rangin *et al.*, 1990b), as part of the Philippine Sea plate (Hall, 1998), and, as suggested by Hall *et al.* (1991), may be equivalent to the similarly aged arc/ophiolite suite of eastern Halmahera. Yumul *et al.* (2008) divided the younger ophiolite suites into subduction related and collision related. In their model, the collision-related ophiolites were obducted when the leading edge of the Philippine Sea Plate came into contact with the Eurasian/Sundaic fragments (Dimalanta & Yumul, 2006), shown in Figure 3 as the Zambales Block (ZB), the North Palawan Block (NPB), and Zamboanga Peninsula (ZP)/Daguma Ranges (DR).

The Zambales Ophiolite Complex consists of a back arc basin/island arc sequence (Defant, de Boer & Oles, 1988). It is not clear if the Zambales Block is exotic or if it was once a marginal basin/island arc complex adjacent to Sundaland at 60–45 Ma. Subduction of this age is predicted to have occurred along the south-eastern margin of Sundaland by Hall *et al.* (2009a). The other two fragments are continental. The North Palawan Block has a metamorphic basement that extends as far south as the Ulugan Bay Fault (Hutchison, 1989). South Palawan consists of obducted ophiolite sheets formed during the collision of Asian exotic blocks with the west Bornean margin during the opening of the South China Sea. The continental crust of the North Palawan Block can be traced northwards into western Mindoro, Tablas, Sibuyan, and western Panay (Rangin *et al.*, 1990b; Yumul *et al.*, 2008). It may also be part of an extensive continental basement that underlies the Dangerous Grounds of the South China Sea, to the west of Borneo (R. Hall, pers. comm.). These continental basements were originally adjacent to the Asian mainland and translated southwards during the Neogene opening of the South China Sea (Dimalanta & Yumul, 2006), and emplaced between *c*. 15–10 Ma (Packham, 1996). The Zamboanga Peninsula/Daguma Range Block is also continental (Yumul *et al.*, 2004), and is in tectonic contact with the adjacent rocks of central Mindanao. This contact dates from about 25 Ma. Although the Zamboanga Peninsula/Daguma Range Block has been interpreted as a Sundaic fragment by authors such as Rangin *et al.* (1990a) and Clements *et al.* (2009), Yumul *et al.* (2004) suggested that it might have been part of the North Palawan Block, and thus an Asian exotic terrane. The possibility that it may have been part of the Argo fragment should also be considered.

## AREAS OF ENDEMISM AND WALLACEAN BIOGEOGRAPHICAL UNITS IN WALLACEA

In the following section, areas of endemism within Wallacea are identified by describing the distribution and overlap of (primarily) endemic avian and mammalian species. Distributional data for other major groups, such as flowering plants (van Welzen *et al.*, 2005), appear to identify a broadly similar set of endemic areas within Wallacea. Because this approach is both descriptive and based on a limited taxonomic sampling, the endemic areas identified by this method are regarded as provisional hypotheses only. Relationships between areas of endemism are described using the data from published molecular phylogenies, and biologically composite faunas are identified. Figure 4 shows five areas of endemism within Indonesia (areas 1–4) and the Philippines (area 5) that will be discussed below, and their status as Wallacean biogeographical units will also be evaluated.

### SULAWESI

Sulawesi and its satellite islands of Buton and Muna, off the south-east peninsula, the Banggai and Sula groups found to the east, and the Sangihe Archipelago and Talaud group off the northern peninsula (Fig. 1) form an area of endemism because the islands share many endemic species with mainland Sulawesi, or have their own endemic subspecies of Sulawesian endemics (Coates & Bishop, 1997; Indrawan, Masala & Pesik, 1997; Riley, 1997; Stones *et al.*, 1997; Koch

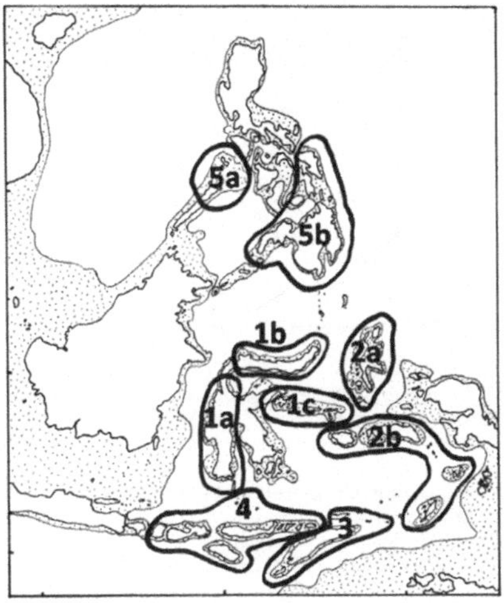


**Figure 4.** Some areas of endemism in Indonesia and the Philippines. 1, Sulawesi: 1a = south-west peninsula; 1b, North Arm; 1C, east Sulawesi/Banggai/Sula; 2, Maluku: 2a, north Maluku; 2b, south Maluku; 3, Timor, Roti, Weta, and Babar; 4, western Lesser Sundas; 5, Philippines: 5A, north Palawan; 5B, Mindanao, Bohol, Leyte, and Samar. See text for discussion.

*et al.*, 2007). Banggai, Sula, and Sangihe are also phylogenetically linked to Sulawesi as sister areas in the cladogram of the cicada genus *Dilobopyga* (Duffels, 1990). The southern boundary of the Sulawesian area of endemism passes between Selayar and the Flores Sea Islands (Duffels, 1990; Dutson, 1995; Auliya *et al.*, 2002).

A summary of molecular phylogenetic data for 33 Sulawesian taxa is presented in Figure 5 (See Appendix S1 for details). The impression that Sulawesi's fauna has an ancient component (White & Bruce, 1986; Musser, 1987; Groves, 2001; Keast, 2001) is borne out by the proportion of primitive species. A primitive species is defined here as one that is either basal on the cladogram or is basal in its clade. Many of Sulawesi's endemic birds and mammals, as well as other groups listed in Appendix S1, have been shown to be primitive using modern molecular methods. For example, the emblematic maleo (*Macrocephalon maleo*) is basal to an Australian–New Guinean clade, and is the earliest evolved of all extant megapodes (Birks & Edwards, 2002). Likewise, the babirusa – an enigmatic pig – is basal to all other Suidae, and is

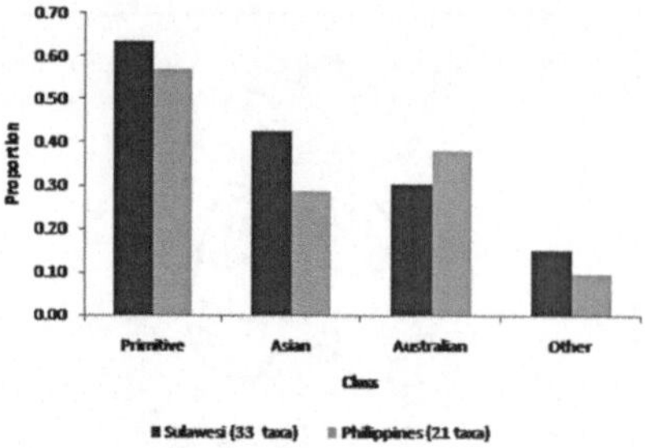


**Figure 5.** Sulawesi and Philippine faunal categories based on a sample of molecular phylogenies (listed in Appendices S1 and S2). Primitive species are defined as basal on a cladogram or basal to a clade. Asian or Australasian species are defined by membership of a clade in which the constituent species are exclusively distributed in that geographical category, or are basal to such a clade. Other: Africa/Sulawesi/Philippines.

therefore a very early product of suid evolution (Price, Bininda-Emonds & Gittleman, 2005; Funk *et al.*, 2007; Marcot, 2007).

The geographically mixed nature of the Sulawesian fauna (Musser, 1987; Cranbrook, 1991; Coates & Bishop, 1997; Groves, 2001; Keast, 2001) and flora (Coates & Bishop, 1997; van Welzen *et al.*, 2005), which so impressed Wallace when he first visited the island (Michaux, 2008), is also well supported by the molecular data. Sulawesian endemics were classified into geographical types on the basis of their inclusion within geographically uniform clades, or if they were basal to such a clade. For example, species classified as Australian are either embedded in clades composed of species that are of exclusively Australasian distribution or are basal to such clades. Figure 5 shows that Sulawesi is biologically composite, and composed of a mix of predominantly Asian and Australian species.

The distribution of endemics within Sulawesi is non-uniform and shows evidence of regionality (Musser, 1987; Groves, 2001; Riley *et al.*, 2003; Evans *et al.*, 2003a; Merker *et al.*, 2009). Although the endemic areas recognized by these different authors are not identical, and the boundaries of endemic areas within Sulawesi are still in need of clarification, it seems clear from these studies that distinct centres of endemism are associated with the south-western peninsula (Fig. 4: 1A), the Northern Arm (Fig. 4: 1B), and eastern Sulawesi/Banngai/Sula (Fig. 4: 1C).

Sulawesi is clearly not a Wallacean biogeographical unit because it is both biologically and geologically composite. However, the south-west peninsular, North Arm, and eastern Sulawesi/Banggai/Sula are more localized areas of endemism with distinct geological histories, and therefore qualify as Wallacean biogeographical units. Table 1 lists the endemics hypothesized to have originally been associated with each area of endemism shown in Figure 4, the geological processes responsible for the origin of these areas of endemism, and an age estimation over which these processes operated. The Asian endemics of Sulawesi are inferred to have been part of the original fauna of the south-west peninsula. It is proposed that the more primitive Asian taxa colonized the south-west peninsula following the amalgamation of the Argo fragment with the margin of the Sunda Shelf, at some time after *c.* 80 Ma, and that the more derived endemics evolved in response to their isolation on the south-west peninsula following the opening of the Makassar Strait at *c.* 45 Ma. Likewise, it is proposed that the Australasian endemics were tectonically dispersed into Sulawesi by one or other (or both) Australasian fragments (North Arm of Sulawesi and eastern Sulawesi/Banggai/Sula) when they were amalgamated with the south-west peninsula at *c.* 25 Ma.

## MALUKU

North and south Maluku (Fig. 4: 2A and 2B) have been recognized as separate areas of endemism since Wallace visited them in the 1850s. Their distinctiveness is shown by the lack of shared marsupial species (Cranbrook, 1991; Flannery, 1995), and differing avifaunas (White & Bruce, 1986; Bowler & Taylor, 1989; Lambert & Young, 1989; Jepson, 1993; Lambert, 1994; Linsley, 1995; Coates & Bishop, 1997). For example, Coates & Bishop (1997) listed 64 endemic bird species for the whole of Maluku, but of these only five are shared by north and south Maluku. Of the 18 endemics found on Obi, 13 are shared with north Maluku, and only one with south Maluku, linking Obi with north Maluku. Twenty seven endemic bird species and four endemic genera are present in north Maluku, including the monotypic birds-of-paradise genera *Lycocorax* and *Semioptera*, which are restricted to Halmahera.

South Maluku is also a centre of avian endemism, with 33 endemic species and two endemic genera (Coates & Bishop, 1997). Although Coates & Bishop (1997) included the Kai islands as part of south Maluku, they placed the Tanimbar group with the Lesser Sundas. Why they should do so is unclear, because of the 13 endemic species listed for Tanimbar, only two are shared with the western Lesser Sundas. White & Bruce (1986) described the Tanimbar avifauna as '. . . mixed, partly generalized, and strongly

**Table 1.** Summary of Wallacean biogeology listing proposed areas of endemism, the endemics inferred to be associated with those areas, the geological processes underlying their formation, and an estimation of the timescales over which those processes operated, and an evaluation of their status as Wallacean biogeographical units (WBUs)

| | Area of endemism | Associated endemics | Geological process | Docking | Origin | WBU |
|---|---|---|---|---|---|---|
| Sulawesi | South-west peninsula | Asian endemics | Opening of Makassar strait | | 45 Ma | Y |
| | | Paleoendemics | Mobilism of Sunda taxa onto Sumba Block | | 90–80 Ma | |
| | North Arm | Australasian endemic | Collision of island arc | *c.* 25 Ma | 45 Ma Island Arc | Y |
| | Sula Banggai | Australasian endemics | Collision of Sula Spur | *c.* 25–10 Ma | Continental extension | Y |
| Philippines | North Palawan | Mainland Asian endemics | Southwards translation of North Palawan Block | *c.* 15–10 Ma | 35 Ma s China fragment | Y |
| | South Palawan | Sundaic endemics | Mobilism in west Borneo collision zone | | *c.* 20 Ma | N |
| | South Philippines | Palaeoendemics | Mobilism of Sunda taxa onto Zamboanga terrane | | 90–80 Ma | N |
| | | Australasian endemics | Philippine Mobile Belt | *c.* 25 Ma–recent | > 45 Ma Island arcs | |
| Maluku | Maluku | Maluku endemics | Continent–arc collision, mobilism | *c.* 23 Ma | Australian fragment | N |
| | North Maluku | North Maluku endemics | Formation of East Halmahera Arc | | 45 Ma | N |
| | South Maluku | South Maluku endemics | Fragmentation of Sula Spur | | < 25 Ma | Y |
| Lesser Sundas | Western Lesser Sundas | Asian endemics | Mobilism of Sunda taxa onto Sumba Block | | 60 Ma | Y |
| | | Australian endemics | Mobilism following Timor–Banda Collision | | 2.4 Ma | |
| Timor | Timor/Wetar/Babar | Australian endemics | Rifting of Australian continental margin | | ?100 Ma NW Australia | N |
| | | Asian endemics | Mobilism following Timor–Banda collision | | 2.4 Ma | |

See text for discussion.

Mollucan. . .', and suggested that Tanimbar is biogeographically part of south Maluku. The Tanimbar checklist of Bishop & Brickle (1999) supports this claim. In Figure 4, the boundaries of the two subregions have been drawn so that Obi is included in north Maluku, and Tanimbar is included in south Maluku.

The fauna of Maluku is distinctly Australasian in character, as indicated by the presence of marsupials (nine species) and placentals related to Australasian species (14 of 47 bats and six of 11 murids; Flannery, 1995). Phylogenetic data confirms this Australasian relationship, although there are differences in specific patterns. For Maluku as a whole, there is a sister-group relationship between north and south Maluku, with northern Australia and New Guinea as an out-area (Duffels & de Boer, 1990; de Jong, 1998; Brown & Toft, 1999; Birks & Edwards, 2002; Duffels & Turner, 2002; Jønsson *et al.*, 2008). This relationship confirms not only the phylogenetic link between Maluku and Australasia, but also that Maluku as a whole is an area of endemism. North Maluku shows a sister-area relationship to New Guinea/Melanesia (Holloway, 1984; de Boer & Duffels, 1996; Irestedt *et al.*, 2009), whereas south Maluku shows a sister-area relationship with both Australia and the continental regions of New Guinea (Holloway, 1991; Wright *et al.*, 2008; Nyári *et al.*, 2009), and Sulawesi (Duffels, 1990; Butlin *et al.*, 1998; Holloway, 1998; Evans *et al.*, 2003b; Outlaw & Voelker, 2006; Lohman *et al.*, 2008).

Maluku as a whole is not a Wallacean biogeographical unit because, like Sulawesi, it is biologically and geologically composite. North Maluku is also an area of endemism that is geologically composite, with Australasian continental crust present in Bacan and Obi. In Table 1 it is hypothesized that north Maluku is also biologically composite because of the juxtaposition of arc terrane endemics, which show phylogenetic linkage to New Guinean and/or northern Melanesian species, with endemics tectonically dispersed into north Maluku as the Australian continental margin collided with this arc system. For example, *Lycocorax* and *Semioptera* are restricted to Halmahera, and are found in separate clades in the birds-of-paradise phylogeny of Irestedt *et al.* (2009). *Lycocorax* is the more primitive genus, and is found in a clade containing otherwise exclusively New Guinean taxa. *Semioptera* is more derived and found in a clade that includes Australian as well as New Guinean species. This congruence between biological and geological patterns leads to the interpretation that *Lycocorax* was part of the original island arc fauna, whereas *Semioptera* was tectonically transported into north Maluku. Although north Maluku is not a Wallacean biogeographical unit, south Maluku is because its endemics all appear to share a common geological history as part of a Sula Spur fauna (Table 1). The sister-area relationship between south Maluku and Sulawesi described above can also be interpreted in terms of this same shared geological history.

## TIMOR

Although Timor is often included as part of the Lesser Sundas, the distribution of endemic birds suggests it is not. Of the endemic species listed in Coates & Bishop (1997), only four are shared with the western Lesser Sundas, whereas 13 are shared with Wetar (and sometimes Damar). Rather than linking Timor with the western Lesser Sundas, the avifauna indicates that Timor, Weta, and possibly Damar, form a distinct area of endemism. A molecular study of *Liasis* pythons also linked Timor with Savu, Wetar, Semau, and Barbar (Rawlings, Barker & Donnellan, 2004). This grouping of islands is shown as an area of endemism in Figure 4.

The fauna of Timor shows a mix of Australasian and Asian affinities. The bats are predominantly Australasian, with only one – *Pipistrellus javanicus* – linking Timor with Asia (Cranbrook, 1991). The avifauna has approximately equal proportions of Asian and Australasian resident, non-endemic birds (Trainor, Coates & Bishop, 2007). Timor and Australia are sister areas in the molecular phylogenies reported in Rawlings *et al.* (2004) and Norman *et al.* (2007). Molecular phylogenies linking Timorese taxa with Asian/Sundaic species are found in Watts & Baverstock (1994), McGuire & Heang (2001), Creer *et al.* (2006), Outlaw & Voelker (2006), Yago *et al.* (2008), and Nyári *et al.* (2009). A number of fossil mammals are known from Timor including an Eocene artiodactylid (*Anthracotherium*) (von Koenigswald, 1967) and a Pleistocene pygmy stegodont, which are both Asian-related, and at least six subfossil murid genera of Australasian affinity (Musser, 1981).

The Timor area of endemism (Fig. 4: 3) is not a Wallacean biogeographical unit because it is geologically and biologically composite. In Table 1, the Asian element in the Timorese fauna is inferred to have dispersed into the region following the collision with the Banda Arc at *c*. 2.4 Ma. Biological confirmation that a phase of faunal and floral mobilism followed this collision is provided by the molecular study of Payn, Dvorak & Myburg (2007). According to Payn *et al.* (2007), *Eucalyptus urophylla* haplotypes are most diverse, and therefore have had the longest history of residence, on Timor and Wetar. Islands to the west have populations that show a progressive decrease in haplotype variability, until a single haplotype is fixed in populations on Flores and Lomblen. These authors interpreted this pattern of decreasing haploid variability to westward dispersal of the species from Timor and Wetar. Although the presence of *Stegodon* on Timor may also reflect a period of mobilism, Ducrocq (1996) and Lihoreau & Ducrocq (2007) have argued that the Eocene rocks containing *Anthracotherium* are allochthonous and tectonically transported in the Asian thrust sheets that occupy the upper stratigraphic levels on Timor. The presence of Australian taxa on an island purported to be only a few million years old is more problematic, despite its close proximity to Australia. The phenomenon of species older than the age of the island they occur on (Beverley & Wilson, 1985) shows that a broader perspective is required for the interpretation of the flora and fauna on young islands. Beverley & Wilson (1985) reconciled the occurrence of the ancient (40 Ma) picture-winged *Drosophila* on modern Hawaiian islands by proposing dispersal from older, eroding islands of the chain onto newly erupting centres. Thus, the persistence of an ancient lineage was maintained in a dynamic setting. In the case of the Timor area of endemism there is a large Australian component to its flora and fauna, much of which shows a degree of endemism sufficient to suggest long isolation. The biological evidence implies that some part of the rifted structure underlying the Timor area of endemism had remained above water. The island of Timor itself may be recent, but the larger structure is not. A modern example from New Zealand is the presence of an *Oligosoma* skink on the Chatham Islands that Chapple, Ritchie & Daugherty (2009) suggested is ten million years old, considerably older than the island itself. Chatham Island is part of a much larger structure – the Chatham Rise – which could have provided ample areas for the skink to inhabit prior to the emergence of the modern island at *c.* 3 Ma.

## THE LESSER SUNDAS

The western Lesser Sundas (Fig. 4: 4) is defined as an area of endemism mainly by its avifauna. It is a major centre of avian biodiversity, with 29 species (and 17 subspecies) endemic to the western Lesser Sunda Islands (Coates & Bishop, 1997; Linsley, Jones & Marsden, 1999; Verhoeye & Holmes, 1999; Olsen *et al.*, 2002; Trainor, 2003a, b; Trainor *et al.*, 2006). In general, the fauna as a whole can be described as depauperate and unbalanced when compared with that of the adjacent Sunda Shelf. These characteristics are demonstrated by the indigenous mammalian fauna, which lacks marsupials, and is restricted to shrews (two species), murid rodents (two species), and bats (19 species) (Musser, 1981; Cranbrook, 1991; Kitchener, How & Maharadatunkamsi, 1991; Helgen & Wilson, 2002). Fossil remains of up to three stegodont species (van den Bergh, 1999) and five species of murid rodent (Musser, 1981) indicate that a more diverse mammalian fauna existed in the past.

The mammalian fauna is Asian, with little evidence of Australasian affinity. The avifauna is also predominantly Asian, although an Australasian influence is more pronounced. After discounting species that are either endemic or widely distributed (i.e. Indo-Australian), 72% of Lesser Sunda birds are Asian and 29% Australasian (distributional data from Coates & Bishop, 1997). The few phylogenies available for western Lesser Sunda species support the conclusion that although the fauna is geographically mixed, it is predominantly Asian. *Eucalyptus* (Payn *et al.*, 2007), megapodes (Birks & Edwards, 2002), cockatoos (Brown & Toft, 1999), and *Ninox* owls (Wink *et al.*, 2008) are Australasian elements of the flora and fauna, as is the most iconic species of the western Lesser Sundas, the Komodo dragon (*Varanus komodoensis*; Schulte, Melville & Larson, 2003). Interestingly, the other Lesser Sunda relative (*Varanus salvator*) is a member of an Asian clade. Other Asian/Sundaic taxa include gastropods (Hausdorf, 1995), butterflies (Yago *et al.*, 2008), fish (Takehana, Naruse & Sakaizumi, 2005), amphibians (Evans *et al.*, 2003b), reptiles (McGuire & Heang, 2001; Auliya *et al.*, 2002), and birds (Outlaw & Voelker, 2006; Pasquet *et al.*, 2007; Illera *et al.*, 2008; Jønsson *et al.*, 2008; Nyári *et al.*, 2009).

The western Lesser Sunda area of endemism (Fig. 4: 4) is classified in Table 1 as a Wallacean biogeographical unit. Although it is biologically composite, being composed of both Asian and Australasian species, it is not geologically composite. The Asian element of the fauna is interpreted as autochthonous to the western Lesser Sundas, and therefore these species share a common geological history. The Australasian species are interpreted as the product of mobilism following the collision of the Banda Arc with the Australian margin.

## THE PHILIPPINES

The Philippines, which includes the islands of the Sulu Archipelago and Palawan, form an area of endemism that is characterized by a diverse and highly endemic biota. Heaney *et al.* (2002) listed 172 terrestrial mammals native to the Philippines, of which 111 (64%) are endemic. The number of endemics is probably the highest of any mammalian fauna (adjusted for area), and includes 22 endemic genera. There has also been a remarkable radiation of murid rodents, with 55 species (53 endemic) in 21 genera (15 endemic) presently recognized. Between 1992 and 2002 some 22 new mammal species were discovered by biologists, indicating that the rich mammalian fauna is still only partially known, and more species await discovery. The more diverse Philippine avifauna (572 species) also shows significant endemism, with 30% of the species being endemic (Kennedy *et al.*, 2000). Van Welzen *et al.* (2005) reported higher than expected levels of endemism among flowering plants, based on a regression analysis of percentage endemism versus land area in Malesia.

An analysis of mammalian and avian endemic distributions shows evidence of regionality within the Philippines. Heaney *et al.* (2002) recognized six major areas of endemism on the basis of mammal distributions (Luzon, Samar + Leyte + Mindanao, Panay + Negros + Cebu, Sulu, Mindoro, and Palawan), which they interpreted in terms of Pleistocene island-configurations. These mammalian-defined areas of endemism are supported by lepidopteran distributional patterns (Vane-Wright, 1990). Avian distributions show strong support for the biological separation of Palawan from the Philippines (of the 16 endemics only one is shared with Mindoro) and some support (13 species) for a southern grouping of Mindanao + Leyte + Samar ± Bohol (distributional data from Kennedy *et al.* 2000). These two areas of endemism are shown in Figure 4 (5a and 5b).

Palawan's endemic mammals are quite distinct from those of the rest of the Philippines, being more closely related to Sunda Shelf or mainland Asian species (Heaney *et al.*, 2002; Esselstyn, Widmann & Heaney, 2004). For example, five placental families widely distributed in Asia and the Sunda Shelf are restricted to Palawan (porcupines, cats, mustelids, mongooses, and mouse deer); in addition the tamarau, a native bovid, is restricted to Mindoro. Palawan and Mindoro share a sister-group relationship in the cladogram of the amphibian *Limnonectes* (Evans *et al.*, 2003b). A further nine Asian/Sundaic genera are also restricted to Palawan. Molecular phlogenies that show Palawan species to be more closely related to Asian/Sundaic rather than other Philippine species are found in Gamauf & Haring (2004), Malhotra & Thorpe (2004), Outlaw & Voelker (2006), Campbell *et al.* (2007), Jønsson *et al.* (2008), Esselstyn & Brown (2009), and Nyári *et al.* (2009).

A summary of molecular phylogenic data for 21 Philippine taxa is shown in Figure 5 (see Appendix S2 for details). A comparison of the Philippine and Sulawesian faunas show striking parallels in general characteristics, such as the proportion of primitive taxa and their geographically composite nature. The murid rodent phylogeny of Jansa, Barker & Heaney (2006) exemplifies these characteristics particularly clearly. There are three broad groupings of murids: a group of derived genera spread over three clades, two of which are Sundaic/Asian (*Bullimus* and *Crunomys*) and one clade of possible Australasian origin (*Limnomys* + *Tarsomys* + *Rattus*); a group of genera that are the sister group to an Old World clade (Steppan *et al.*, 2005), which together are sister to the derived group; a group of genera that are basal to all other Murinae and of no obvious geographical affinity. Thus, the Philippine murids appear to have been derived from different areas at different times: the palaeoendemic murids having evolved long enough ago to be without obvious relatives, apart from a generalized sister-group relationship with other Murinae.

Clearly the Philippines as a whole is not a Wallacean biogeographical unit, because it is both biologically and geologically composite. Neither of the areas of endemism discussed above is recognized as Wallacean biogeographical units in Table 1, but dividing Palawan into its constituent geological parts does result in the recognition of North Palawan as a Wallacean biogeographic unit. The Palawan fauna is a mixture of older, tectonically transported mainland Asian species, and younger Bornean/Sundaic species that dispersed into south Palawan during a period of mobilism following the creation of new land during the Neogene. The Australasian element in the Philippine fauna is interpreted to have been tectonically dispersed on the island arcs of the Philippine Mobile Belt. The palaeoendemics are hypothesized to have been associated with the Daguma Ranges/Zamboanga Peninsula continental fragment, and/or the Zambales Block island arc. The Zamboanga Peninsula and Zambales Block are interpreted as original parts of the Sundaland margin (or in close proximity to it) and might be associated, respectively, with the Argo fragment and Cretaceous–Palaeocene subduction system described previously.

## DISCUSSION

### Delimiting Wallacea

It is obvious that if component regions such as Sulawesi and Maluku are not Wallacean biogeographical units, then Wallacea as a whole cannot be one either. And yet neither is it completely artificial, because all the areas currently recognized as parts of Wallacea are tectonically derived, predominantly Australasian fragments caught up in a complex collision zone. Although it may not meet the strict criteria for 'naturalness', Wallacea is composed of a number of Wallacean biogeographical units that are linked by geological processes resulting from the interplay between the Eurasian and Australasian continents and the Philippine Sea Plate. Regarded in this light, the Philippines are also part of Wallacea because the arcs of the Philippine Mobile Belt, the Daguma Ranges–Zamboanga Peninsula continental fragment, and the Zambales Block are parts of this collision zone. The North Palawan Block is a terrane derived from the Asian mainland that was translated southwards when the South China Sea opened: it is distinct with respect to both source area and tectonic genesis from other Philippine/Wallacean terranes, and is not part of Wallacea. Huxley's Line comes closest to delimiting Wallacea from Sundaland, but in reality no boundary can be drawn, as none exists in modern geography.

### IMPORTANCE OF IDENTIFYING WALLACEAN BIOGEOGRAPHICAL UNITS

Although biogeographers may not universally agree on the best way of identifying areas of endemism, there is agreement that these areas form the basis from which biogeographical analysis should proceed. In my view there is a further step to take, and that is to confirm that areas of endemism are also Wallacean biogeographical units. These are the fundamental units for further biogeographical analysis because they are natural entities, not human constructs. For example, 'Sulawesi' is an area of endemism, but not a Wallacean biogeographical unit: it is a human geopolitical construct that has no biogeographical reality. Any attempt to use the area 'Sulawesi' in biogeographic analysis is doomed to failure.

If an area of endemism is found to be both biologically and geologically composite, then it may be possible to identify more geographically restricted areas that are Wallacean biogeographical units, and to use these in any analysis. In the example of Sulawesi, the south-west peninsula and its inferred Asian/Sundaic-related endemics can be used as one input into the analysis, and Banggai/Sula and/or the North Arm with their associated Australasian derived endemics as another. This avoids paradoxical outcomes, such as implied multiple sister-area relationships for an area of endemism, by resolving these multiple relationships into their constituent parts.

## ACKNOWLEDGEMENTS

I would like thank Robert Hall and three anonymous reviewers for their thoughtful comments on an earlier version of this paper.

## REFERENCES

**Ali JR, Hall R. 1995.** Evolution of the boundary between the Philippine Sea Plate and Australia: palaeomagnetic evidence from eastern Indonesia. *Tectonophysics* **251:** 251–275.

**Ali JR, Milsom J, Finch EM, Mubroto B. 1996.** SE Sundaland accretion: palaeomagnetic evidence of large Plio-Pleistocene thin-skin rotations in Buton. *Geological Society Special Publication* **106:** 431–444.

**Auliya M, Mausfeld P, Schmitz A, Böhme W. 2002.** Review of the reticulated python (*Python reticulatus* Schneider, 1801) with the description of new subspecies from Indonesia. *Naturwissenschaften* **89:** 201–213.

**van den Bergh GD. 1999.** The late Neogene elephantoid-bearing faunas of Indonesia. *Scripta Geologica* **117:** 1–419.

**Bergman SC, Coffield DQ, Talbot JP, Garrard RA. 1996.** Tertiary tectonic and magmatic evolution of western Sulawesi and the Makassar Strait, Indonesia: evidence for a Miocene continent-continent collision. *Geological Society Special Publication* **106:** 391–429.

**Beverley SM, Wilson AC. 1985.** Ancient origin for Hawaiian Drosophilinae inferred from protein comparisons. *Proceedings of the National Academy of Sciences* **82:** 4753–4757.

**Bininda-Emonds OR, Gittleman JL, Purvis A. 1999.** Building large trees by combining phylogenetic information: a complete phylogeny of the extant Carnivora (Mammalia). *Biological Reviews* **74:** 143–175.

**Birks SM, Edwards SV. 2002.** A phylogeny of the megapodes (Aves: Megapodiidae) based on nuclear and mitochondrial DNA sequences. *Molecular Phylogenetics and Evolution* **23:** 408–421.

**Bishop D, Brickle NW. 1999.** An annotated checklist of the birds of the Tanimbar Islands. *Kukila* **10:** 115–150.

**de Boer AJ, Duffels JP. 1996.** Historical biogeography of the cicadas of Wallacea, New Guinea and the West pacific: a geotectonic explanation. *Palaeogeography, Palaeoclimatology, Palaeoecology* **124:** 153–177.

**Bowler J, Taylor J. 1989.** An annotated checklist of the birds of Manusela National Park, Seram. (Birds recorded on the Operation Raleigh Expedition). *Kukila* **4:** 3–29.

**Brown DM, Toft CA. 1999.** Molecular systematics and biogeography of the cockatoos (Psittaciformes: Cacatuidae). *The Auk* **116:** 141–157.

**de Bruyn M, Wilson JA, Mather PB. 2004.** Huxley's line demarcates extensive genetic divergence between eastern and western forms of the giant freshwater prawn Macrobrachium rosenbergii. *Molecular Phylogenetics and Evolution* **30:** 251–257.

**Butlin RK, Walton C, Monk KA, Bridle JR. 1998.** Biogeography of Sulawesi grasshoppers, genus *Chitura*, using DNA sequence data. In: Hall R, Holloway JD, eds. *Biogeography and geological evolution of SE Asia*. Leiden: Backhuys, 355–359.

**Campbell P, Putnam AS, Bonney C, Bilgin R, Morales JC, Kunz TH, Ruedas LA. 2007.** Contrasting patterns of genetic differentiation between endemic and widespread species of fruit bats (Chiroptera: Pteropodidae) in Sulawesi, Indonesia. *Molecular Phylogenetics and Evolution* **44:** 474–482.

**Cardillo M, Mace GM, Gittleman JL, Purvis A. 2006.** Latent extinction risk and the future battlegrounds of mammal conservation. *Proceedings of the National Academy of Sciences* **103:** 4157–4161.

**Chapple DG, Ritchie PA, Daugherty CH. 2009.** Origin,

diversification, and systematic of the New Zealand skink fauna (Reptilia: Scincida). *Molecular Phylogenetics and Evolution* **52:** 470–487.

**Charlton TR. 1996.** Correlation of the Salawati and Tomori Basins, eastern Indonesia: a constraint on left-lateral displacements of the Sorong fault zone. *Geological Society Special Publications* **106:** 465–482.

**Clements B, Hall R, Smyth HR, Cottam MA. 2009.** Thrusting of a volcanic arc: a new structural model for Java. *Petroleum Geoscience* **15:** 159–174.

**Clements KD, Alfaro ME, Fessler JL, Westneat MW. 2004.** Relationships of the temperate Australasian labrid fish tribe Odacini (Perciformes; Teleostei). *Molecular Phylogenetics and Evolution* **32:** 575–587.

**Coates BJ, Bishop KD. 1997.** *Birds of Wallacea.* Queensland, Australia: Dover Publications, 535 pp.

**Cox CB. 2001.** The biogeographic regions reconsidered. *Journal of Biogeography* **28:** 511–523.

**Cranbrook E. 1991.** *Mammals of South-East Asia*, 2nd edn. Singapore: Oxford University Press, 96 pp.

**Craw RC, Grehan JR, Heads MJ. 1999.** *Panbiogeography: tracking the history of life.* Oxford Biogeography Series 11.New York: Oxford University Press, 231 pp.

**Creer S, Pook CE, Malhotra A, Thorpe RS. 2006.** Optimal intron analyses in the *Trimeresurus* radiation of Asian Pitvipers. *Systematic Biology* **55:** 57–72.

**Defant MJ, de Boer JZ, Oles D. 1988.** The western Central Luzon volcanic arc, the Philippines: two arcs divided by rifting. *Tectonophysics* **145:** 305–317.

**Dickerson RE, Merrill ED, McGregor RC, Schultze W, Taylor EH, Herre AW. 1923.** Distribution of life in the Philippines. *Manila Bureau of Science Monograph* **21:** 1–322.

**Dimalanta CB, Yumul GP. 2006.** Magmatic and amagmatic contributions to crustal growth in the Philippine island arc system: comparison of the cretaceous and post-cretaceous periods. *Geosciences Journal* **10:** 321–329.

**Drovetski SV, Zink RM, Fadeev IV, Nesterov EV, Koblik EA, Red'kin YA, Rohwer S. 2004.** Mitochondrial phylogeny of *Locustella* and related genera. *Journal of Avian Biology* **35:** 105–110.

**Ducrocq S. 1996.** The Eocene terrestrial mammal from Timor, Indonesia. *Geological Magazine* **133:** 763–766.

**Duffels JP. 1990.** Biogeography of Sulawesi cicadas (Homoptera: Cicadoidea). In: Knight WJ, Holloway JD, eds. *Insects and the rain forests of South East Asia (Wallacea).* London: Royal Entomological Society, 63–72.

**Duffels JP, de Boer AJ. 1990.** Areas of endemism and composite areas in East Malesia. In: Baas P *et al.*, ed. *The plant diversity of Malesia.* Netherlands: Kluwer, 249–272.

**Duffels JP, Turner H. 2002.** Cladistic analysis and biogeography of the cicadas of the Indo-Pacific subtribe Cosmopsaltriina (Hemiptera: Cicadoidea: Cicadidae). *Systematic Entomology* **27:** 235–261.

**Dutson G. 1995.** The birds of Salayer and the Flores Sea Islands. *Kukila* **7:** 129–141.

**Esselstyn JA, Brown RM. 2009.** The role of repeated sea-level fluctuations in the generation of shrew (Soricidae: *Crocidura*) diversity in the Philippine Archipelago. *Molecular Phylogenetics and Evolution* **53:** 171–181.

**Esselstyn JA, Widmann P, Heaney LR. 2004.** The mammals of Palawan Island, Philippines. *Proceedings of the Biological Society of Washington* **117:** 271–302.

**Evans BJ, Brown RM, McGuire JA, Supriatna J, Andayani N, Diesmos A, Iskandar D, Melnick DJ, Cannatella DC. 2003b.** Phylogenetics of fanged frogs: biogeographicalal hypotheses at the interface of the Asian and Australian faunal zones. *Systematic Biology* **52:** 794–819.

**Evans BJ, Supriatana J, Andayani N, Setiadi MI, Cannatella DC, Melnick DJ. 2003a.** Monkeys and toads define areas of Endemism on Sulawesi. *Evolution* **57:** 1436–1443.

**Flannery T. 1995.** *Mammals of the South-West Pacific and Moluccan Islands.* Chatswood, NSW, Australia: Reed Books, 464 pp.

**Funk SM, Verma SK, Larson G, Prasad K, Singh L, Narayan G, Fa JE. 2007.** The pygmy hog is a unique genus: 19th century taxonomists got it right first time round. *Molecular Phylogenetics and Evolution* **45:** 427–436.

**Gamauf A, Haring E. 2004.** Molecular phylogeny and biogeography of Honey-buzzards (genera *Pernis* and *Henicopernis*). *Journal of Zoological Systematics and Evolutionary Research* **42:** 145–153.

**Groves C. 2001.** Mammals in Sulawesi: where did they come from and when, and what happened to them when they got there. In: Metcalfe I, Smith JMB, Morwood M, Davidson I, eds. *Faunal and floral migrations and evolution in SE Asia-Australasia.* Lisse: AA Balkema, 333–342.

**Hakim AS, Hall R. 1991.** Tertiary volcanic rocks from the Halmahera Arc, Eastern Indonesia. *Journal of Southeast Asian Earth Sciences* **6:** 271–287.

**Hall R. 1996.** Reconstructing Cenozoic SE Asia. *Geological Society Special Publications* **106:** 153–184.

**Hall R. 1998.** The plate tectonics of Cenozoic SE Asia and the distribution of land and sea. In: Hall R, Holloway JD, eds. *Biogeography and geological evolution of SE Asia.* Leiden: Backhuys, 99–131.

**Hall R, Clements B, Smyth HR. 2009a.** Sundaland: basement character, structure and plate tectonic development. Proceedings, Indonesian Petroleum Association, May 2009. Available at http://searg.rhul.ac.uk/pubs/

**Hall R, Cloke IR, Nur'aini S, Puspita SD, Calvert SJ, Elders CF. 2009b.** The North Makassar Straits: what lies beneath? *Petroleum Geoscience* **15:** 147–158.

**Hall R, Nichols G, Ballantyne P, Charlton T, Ali J. 1991.** The character and significance of basement rocks of the southern Molucca sea region. *Journal of Southeast Asian Earth Sciences* **6:** 249–258.

**Hall R, Wilson MEJ. 2000.** Neogene sutures in eastern Indonesia. *Journal of Asian Earth Sciences* **18:** 781–808.

**Hamilton WB. 1988.** Plate tectonics and island arcs. *Geological Society of America Bulletin* **100:** 1503–1527.

**Harold AS, Mooi RD. 1994.** Areas of endemism: definition and recognition criteria. *Systematic Biology* **43:** 261–266.

**Hausdorf B. 1995.** A preliminary phylogenetic and biogeographical analysis of the Dyakiidae (Gastropoda:

Stylommatophora) and a biogeographical analysis of other Sundaland taxa. *Cladistics* **11:** 359–376.

**Heaney LR, Balete DS, Dolar ML, Acala AC, Dans ATL, Gonzales PC, Ingle N, Lepiten M, Oliver W, Ong PS, Rickart EA, Tabaranza BR, Utzurrum RCB. 2002.** A synopsis of the mammalian fauna of the Philippine Islands. *Fieldiana, Zoology New Series* **88:** Online edition. Available at http://www.fieldmuseum.org/Philippine_Mammals/

**Helgen KM, Wilson DE. 2002.** The bats of Flores, with remarks on Asian *Tadarida*. *Breviora* **511:** 1–12.

**Henderson IM. 1991.** Biogeography without area? *Australian Systematic Botany* **4:** 59–72.

**Holloway JD. 1984.** Lepidoptera and the Melanesian Arcs. *Bishop Museum Special Publications* **72:** 129–169.

**Holloway JD. 1991.** Patterns of moth speciation in the Indo-Australian Archipelago. In: Dudley EC, ed. *Unity of evolutionary biology*. Portland, OR: Dioscorides Press, 340–372.

**Holloway JD. 1998.** Geological signal and dispersal noise in two contrasting insect groups in the Indo-Australian tropics: R-mode analysis of pattern in Lepidoptera and cicadas. In: Hall R, Holloway JD, eds. *Biogeography and geological evolution of SE Asia*. Leiden: Backhuys, 291–314.

**Honda M, Ota H, Murphy RW, Hikida T. 2005.** Phylogeny and biogeography of water skinks of the genus *Tropidophorus* (Reptilia: Scincidae): a molecular approach. *Zoologica Scripta* **35:** 85–95.

**Hutchison CS. 1989.** Displaced terranes of the southwest Pacific. In: Ben-Avraham B, ed. *Oxford monographs in geology and geophysics 11*. New York: Oxford University Press, 161–175.

**ICBP. 1992.** *Putting diversity on the map: priority areas for global conservation*. Cambridge, UK: International Council for Bird Preservation.

**Illera JC, Richardson DS, Helm B, Atienza JC, Emerson BC. 2008.** Phylogenetic relationships, biogeography and speciation in the avian genus *Saxicola*. *Molecular Phylogenetics and Evolution* **48:** 1145–1154.

**Indrawan M, Masala Y, Pesik L. 1997.** Recent bird observations from the Banggai Islands. *Kukila* **9:** 61–70.

**Irestedt M, Jønsson KA, Fjeldså J, Christidis L, Ericson PGP. 2009.** An unexpectedly long history of sexual selection in birds-of-paradise. *BMC Evolutionary Biology* **9:** 235–245.

**Jansa SA, Barker FK, Heaney LR. 2006.** The pattern and timing of diversification of Philippine endemic rodents: evidence from mitochondrial and nuclear gene sequences. *Systematic Biology* **55:** 73–88.

**Jepson P. 1993.** Recent ornithological observations from Buru. *Kukila* **6:** 85–109.

**de Jong R. 1998.** Halmahera and seram: different histories, but similar butterfly faunas. In: Hall R, Holloway JD, eds. *Biogeography and geological evolution of SE Asia*. Leiden: Backhuys, 315–326.

**Jønsson KA, Irestedt M, Fuchs J, Ericson PGP, Christidis L, Bowie RCK, Norman JA, Pasquet E, Fjeldså J. 2008.** Explosive avian radiations and multi-directional dispersal across Wallacea: evidence from the Campephagidae and other Crown Corvida (Aves). *Molecular Phylogenetics and Evolution* **47:** 221–236.

**Keast A. 2001.** The vertebrate fauna of the Wallacean Island interchange zone: the basis of imbalance and impoverishment. In: Metcalfe I, Smith JMB, Morwood M, Davidson I, eds. *Faunal and floral migrations and evolution in SE Asia-Australasia*. Lisse: AA Balkema, 287–310.

**Kennedy RS, Gonzales PC, Dickinson EC, Miranda HC, Fisher TH. 2000.** *Birds of the Philippines*. New York: Oxford University Press, 369 pp.

**Kirsch JAW, Wolman MA. 2001.** Molecular relationships of the bear cuscus, *Ailurops ursinus* (Marsupialia: Phalangeridae). *Australian Mammalogy* **23:** 23–30.

**Kitchener DJ, How RA, Maharadatunkamsi. 1991.** *Paulamys* sp. Cf *P. naso* (Musser, 1981) (Rodentia: Muridae) from Flores Island, Nusa Tenggara, Indonesia – description from a modern specimen and a consideration of its phylogenetic affinities. *Records of the Western Australian Museum* **15:** 171–189.

**Koch A, Arida E, Riyanto A, Böhme W. 2007.** Islands between the realms: a revised checklist of the herpetofauna of the Talaud Archipelago, Indonesia, with a discussion about its biogeographical affinities. *Bonner zoologische Beiträ* **56:** 107–129.

**von Koenigswald GHR. 1967.** An upper Eocene mammal of the family Anthracotheriidae from the island of Timor, Indonesia. *Proceedings of the Koninklijk Nederlandse Akademie van Wetenschappen, Series B* **70:** 529–533.

**Köhler F, Glaubrecht M. 2007.** Out of Asia and into India: on the molecular phylogeny and biogeography of the endemic freshwater gastropod *Paracrostoma* Cossmann, 1990 (Caenogastropoda: Pachychilidae). *Biological Journal of the Linnean Society* **91:** 627–651.

**Köhler F, Schultze K-J, G?nther R, Plötner J. 2008.** On the genetic diversity in the mitochondrial 12S rRNA gene of Platymantis frogs from Western new Guinea (Anura: Ceratobatrachidae). *Journal of Zoological Systematics and Evolutionary Research* **46:** 177–185. doi: 10.1111/j.1439-0469.2007.00442.x.

**Lambert F. 1994.** Notes on the avifauna of Bacan, Kasiruta and Obi, North Moluccas. *Kukila* **7:** 1–9.

**Lambert F, Young D. 1989.** Some recent bird observations from Halmahera. *Kukila* **4:** 30–33.

**van Leeuwen TM, Allen CM, Kadarusman A, Elburg M, Michael PJ, Muhardjo, Suwijanto. 2007.** Petrologic, isotopic, and radiometric age constraints on the origin and tectonic history of the Malino Metamorphic Complex, NW Sulawesi, Indonesia. *Journal of Asian Earth Sciences* **29:** 751–777.

**Lihoreau F, Ducrocq S. 2007.** Family Anthracotheriidae. In: Prothero DR, Foss SE, eds. *Evolution of the Artiodactyls*. Baltimore, MD: Johns Hopkins University Press, 89–105.

**Linsley MD. 1995.** Some bird records from Obi, Maluku. *Kukila* **7:** 142–151.

**Linsley MD, Jones MJ, Marsden SS. 1999.** A review of the Sumba avifauna. *Kukila* **10:** 60–90.

**Linthout K, Helmers H, Wijbrans JR, van Wees JD. 1996.**

$^{40}Ar/^{39}Ar$ constraints on the obduction of the Seram ultramafic complex: consequences for the evolution of the southern Banda Sea. *Geological Society Special Publication* **106:** 455–464.

**Lohman DJ, Peggie D, Pierce NE, Meier R. 2008.** Phylogeography and genetic diversity of a widespread Old World butterfly, *Lampides boeticus* (Lepidoptera: Lycaenidae). *BMC Evolutionary Biology* **8:** 301–314.

**Lovette IJ, Rubenstein DR. 2007.** A comprehensive molecular phylogeny of the starlings (Aves: Sturnidae) and mockingbirds (Aves: Mimidae): congruent mtDNA and nuclear trees for a cosmopolitan avian radiation. *Molecular Phylogenetics and Evolution* **44:** 1031–1056.

**Löwenberg-Neto P, de Carvalho CJB. 2009.** Areas of endemism and spatial diversification of the Muscidae (Insecta: Diptera) in the Andean and Neotropical regions. *Journal of Biogeography* **36:** 1750–1759.

**McGuire JA, Heang KB. 2001.** Phylogenetic systematics of Southeast Asian flying lizards (Iguania: Agamidae: *Draco*) as inferred from mitochondrial DNA sequence data. *Biological Journal of the Linnean Society* **72:** 203–229.

**Malaihollo JF, Hall R. 1996.** The geology and tectonic evolution of the Bacan region, east Indonesia. *Geological Society Special Publication* **106:** 483–498.

**Malhotra A, Thorpe RS. 2004.** A phylogeny of four mitochondrial gene regions suggests a revised taxonomy for Asian pitvipers (*Trimeresurus* and *Ovophis*). *Molecular Phylogenetics and Evolution* **32:** 83–100.

**Marcot J. 2007.** Molecular phylogeny of terrestrial artiodactyls conflicts and resolution. In: Prothero DR, Foss SE, eds. *Evolution of the Artiodactyls*. Baltimore, MD: Johns Hopkins University Press, 4–18.

**Marks BD, Weckstein JD, Moyle RG. 2007.** Molecular phylogenetics of the bee-eaters (Aves: Meropidae) based on nuclear and mitochondrial DNA sequence data. *Molecular Phylogenetics and Evolution* **45:** 23–32.

**Mercer JM, Roth VL. 2003.** The effects of Cenozoic global change on squirrel phylogeny. *Science* **299:** 1568–1572.

**Merker S, Driller C, Perwitasari-Farajallah D, Pamungkas J, Zischler H. 2009.** Elucidating geological and biological processes underlying the diversification of Sulawesi tarsiers. *Proceedings of the National Academy of Sciences* **106:** 8459–8464.

**Michaux B. 2008.** Alfred Russel Wallace, biogeographer. In: Smith CH, Beccaloni G, eds. *Natural selection and beyond, the intellectual legacy of Alfred Russel Wallace*. New York: Oxford University Press, 166–185.

**Michaux B, Leschen RAB. 2005.** East meets west: biogeology of the Campbell Plateau. *Biological Journal of the Linnean Society* **86:** 95–115.

**Moore GF, Silver EA. 1982.** Collision processes in the north Molucca Sea. *Geophysical Monograph Series* **27:** 360–372.

**Morrone JJ. 2002.** Biogeographic regions under track and cladistic scrutiny. *Journal of Biogeography* **29:** 149–152.

**Moyle RG. 2006.** A molecular phylogeny of kingfishers (Alcedinidae) with insights into early biogeographical history. *The Auk* **123:** 487–499.

**Moyle RG, Fuchs J, Pasquet E, Marks BD. 2007.** Feeding behaviour, toe count, and the phylogenetic relationships among alcedinine kingfishers (Alcedininae). *Journal of Avian Biology* **38:** 317–326.

**Musser GG. 1981.** The giant rat of flores and its relatives east of Borneo and Bali. *Bulletin of the American Museum of Natural History* **169:** 71–175.

**Musser GG. 1987.** The mammals of Sulawesi. In: Whitmore TC, ed. *Biogeographicalal evolution of the Malay Archipelago*. Oxford: Clarendon Press, 73–93.

**Nelson G. 1978.** From Candolle to Croizat: comments on the history of biogeography. *Journal of Historical Biology* **11:** 269–305.

**Norman JA, Rheindt FE, Rowe DL, Christidis L. 2007.** Speciation dynamics in the Australo-Papuan *Meliphaga* honeyeaters. *Molecular Phylogenetics and Evolution* **42:** 80–91.

**Nyári AS, Peterson AT, Rice NH, Moyle RG. 2009.** Phylogenetic relationships of flowerpeckers (Aves: Dicaeidae): novel insights into the evolution of a tropical passerine clade. *Molecular Phylogenetics and Evolution* **53:** 613–619. doi:10.1016/j.ympev.2009.06.014.

**Olsen J, Wink M, Sauer-G?rth H, Trost S. 2002.** A new *Ninox* from Sumba, Indonesia. *Emu* **102:** 223–231.

**Outlaw DC, Voelker G. 2006.** Systematics of *Ficedula* flycatchers (Muscicapidae): a molecular reassessment of a taxonomic enigma. *Molecular Phylogenetics and Evolution* **41:** 118–126.

**Packham G. 1996.** Cenozoic SE Asia: reconstructing its aggregation and reorganisation. *Geological Society Special Publications* **106:** 123–152.

**Pasquet E, Pons J-M, Fuchs J, Cruaud C, Bretagnolle V. 2007.** Evolutionary history and biogeography of the drongos (Dicruridae), a tropical Old World clade of corvoid passerines. *Molecular Phylogenetics and Evolution* **45:** 158–167.

**Payn KG, Dvorak WS, Myburg AA. 2007.** Chloroplast DNA phylogeny reveals the island colonisation route of *Eucalyptus urophylla* (Myrtaceae). *Australian Journal of Botany* **55:** 673–683.

**Peña C, Wahlberg N, Weingartner E, Kodandaramaiah U, Nylin S, Freitas AVL, Brower AUZ. 2006.** Higher level phylogeny of Satyrinae butterflies (Lepidoptera: Nymphalidae) based on DNA sequence data. *Molecular Phylogenetics and Evolution* **40:** 29–49.

**Pitra C, F?rbass R, Seyfert H-M. 1997.** Molecular phlogeny of the tribe Bovini (Mammalia: Artiodactyla): alternative placement of the Anoa. *Journal of Evolutionary Biology* **10:** 589–600.

**Platnick NI. 1991.** On areas of endemism. *Australian Systematic Botany* **4:** Commentary.

**Price SA, Bininda-Emonds OR, Gittleman JL. 2005.** A complete phylogeny of the whales, dolphins and even-toed hoofed mammals (Cetartiodactyla). *Biological Reviews* **80:** 445–473.

**Pubellier M, Quebral R, Aurelio M, Rangin C. 1996.** Docking and post-docking tectonics in the southern Philippines. *Geological Society Special Publication* **106:** 511–524.

**Rangin C, Jolivet L, Pubellier M. 1990b.** A simple model

for the tectonic evolution of southeast Asia and Indonesia region for the past 43 m.y. *Bulletin of the Geological Society of France* **8:** 889–905.

**Rangin C, Pubellier M, Azéma J, Briais A, Chotin P, Fontaine H, Huchon P, Joliet L, Maury R, Muller C, Rampnoux J-P, Stephan J-F, Tournon J, Cottereau N, Dercourt J, Ricou L-E. 1990a.** The quest for Tethys in the western Pacific. 8 paleogeodynamic maps for Cenozoic time. *Bulletin of the Geological Society of France* **8:** 907–913.

**Rawlings LH, Barker D, Donnellan SC. 2004.** Phylogenetic relationship of the Australo-Papuan *Liasis* pythons (Reptilia: Macrostomata), based on mitochondrial DNA. *Australian Journal of Zoology* **52:** 215–227.

**Richardson AN, Blundell DJ. 1996.** Continental collision in the Banda arc. *Geological Society Special Publication* **106:** 47–60.

**Riley J. 1997.** The birds of Sangihe and Talaud, North Sulawesi. *Kukila* **9:** 3–36.

**Riley J, Hunowu Y, Mole J, Wangko MF. 2003.** Noteworthy records of birds from the Panau nature reserve. *Kukila* **12:** 17–26.

**von Rintelen T, Glaubrecht M. 2006.** Rapid evolution of sessility in an endemic species flock of the freshwater bivalve *Corbicula* from ancient lakes on Sulawesi, Indonesia. *Biological Letters* **2:** 73–77.

**Rosen BR. 1988.** From fossils to earth history: applied historical biogeography. In: Myers AA, Giller PS, eds. *Analytical biogeography: an integrated approach to the study of animal and plant distributions.* New York: Chapman & Hall, 437–481.

**Ruedi M, Auberson M, Savolainen V. 1998.** Biogeography of Sulawesian shrews: testing for their origin with a parametric bootstrap on molecular data. *Molecular Phylogenetics and Evolution* **9:** 567–571.

**Schaefer H, Renner SS. 2008.** A phylogeny of the oil bee tribe Ctenoplectrini (Hymenoptera: Anthophila) based on mitochondrial and nuclear data: evidence for early eocene divergence and repeated out-of-Africa dispersal. *Molecular Phylogenetics and Evolution* **47:** 799–811.

**Schreiber A, Seibold I, Nötzoid G, Wink H. 1999.** Cytochrome *b* gene haplotypes characterize chromosomal lineages of Anoa, the Sulawesi Dwarf Buffalo (Bovidae: *Bubalus* sp.). *The American Genetic Association* **90:** 165–176.

**Schulte JA, Melville J, Larson A. 2003.** Molecular phylogenetic evidence for ancient divergence of lizard taxa on either side of Wallace's Line. *Proceedings of the Royal Society of London B* **270:** 597–603.

**Smyth HR, Hamilton PJ, Hall R, Kinny PD. 2007.** The deep crust beneath island arcs: inherited zircons reveal a Gondwana continental fragment beneath East Java, Indonesia. *Earth and Planetary Science Letters* **258:** 269–282.

**Sparks JS, Smith WL. 2004.** Phylogeny and biogeography of the Malagasy and Australasian rainbowfishes (Teleostei: Melanotaenioidei): Gondwanan vicariance and evolution in freshwater. *Molecular Phylogenetics and Evolution* **33:** 719–734.

**Steppan SJ, Adkins RM, Spinks PQ, Hale C. 2005.** Multigene phylogeny of the Old World mice, Murinae, reveals distinct geographic lineages and the declining utility of mitochondrial genes compared to nuclear genes. *Molecular Phylogenetics and Evolution* **37:** 370–388.

**Stones AJ, Lucking RS, Davidson PJ, Raharjaningtrah W. 1997.** Checklist of the birds of the Sula Islands (1991), with particular references to Taliabu Island. *Kukila* **9:** 37–55.

**Su Y-C, Chang Y-H, Lee S-C, Tso I-M. 2007.** Phylogeny of the giant wood spider (*Nephila pilipes*, Araneae) from Asian-Australian regions. *Journal of Biogeography* **34:** 177–191.

**Su YC, Saunders RM. 2009.** Evolutionary divergence times in the Annonaceae: evidence of a late Miocene origin of *Pseuduvaria* in Sundaland with subsequent diversification in New Guinea. *BMC Evolutionary Biology* **9:** 153–171.

**Takehana Y, Naruse K, Sakaizumi M. 2005.** Molecular phylogeny of the medaka fishes genus *Oryzias* (Beloniformes: Adrianichthyidae) based on nuclear and mitochondrial DNA sequences. *Molecular Phylogenetics and Evolution* **36:** 417–428.

**Trainor CR. 2003a.** The birds of Lembata (Lomblen), Lesser Sundas. *Kukila* **12:** 39–53.

**Trainor CR. 2003b.** Recent bird observations from Kisar Island, Lesser Sundas. *Kukila* **12:** 54–58.

**Trainor CR, Benstead PJ, Martin K, Lesmana D, Agista D, Benstead MC, Drijvers R, Setiawan I. 2006.** New bird records for Nusa Tenggara Islands: Sumbawa, Mayo, Sumba, Flores, Pulau Besar and Timor. *Kukila* **13:** 6–22.

**Trainor CR, Coates BJ, Bishop KD. 2007.** *The birds of Timor-Leste.* Jakarta: Bird Life International Publication, No. 53, 113 pp.

**Vane-Wright RI. 1990.** The Philippines – key to the biogeography of Wallacea? In: Knight WJ, Holloway JD, eds. *Insects and the rain forests of South East Asia (Wallacea).* London: Royal Entomological Society, 19–34.

**Verhoeye J, Holmes DA. 1999.** The birds of the island of Flores – a review. *Kukila* **10:** 3–59.

**Vroon PZ, van Bergen MJ, Forde EJ. 1996.** Pb and Nd isotope constraints on the provenance of tectonically dispersed continental fragments in eastern Indonesia. *Geological Society Special Publication* **106:** 445–454.

**Wallace AR. 1894.** What are zoological regions? *Nature* **49:** 610–613.

**Wanntorp L, Kocyan A, Renner SS. 2006.** Wax plants disentangled: a phylogeny of Hoya (Marsdenieae, Apocynaceae) inferred from nuclear and chloroplast DNA sequences. *Molecular Phylogenetics and Evolution* **39:** 722–733.

**Watts CH, Baverstock PR. 1994.** Evolution of some South-East Asian murinae (Rodentia), as assessed by microcomplement fixation of albumen, and their relationship to Australian murines. *Australian Journal of Zoology* **42:** 711–722.

**van Welzen PC, Slik JWF, Alahuhta J. 2005.** Plant distribution patterns and plate tectonics in Malesia. *Biologiske Skrifter* **55:** 199–217.

**Wensink H. 1994.** Paleomagnetism of rocks from Sumba:

tectonic implications since the late Cretaceous. *Journal of Southeast Asian Earth Sciences* **9:** 51–65.

**White CMN, Bruce MD. 1986.** *The birds of Wallacea*. British Ornithologists' Union Check-list No. 7, 524 pp.

**Wink M, Heidrich P, Sauer-Gürth H, Elsayed A-A, Gonzalez J. 2008.** Molecular phylogeny and systematics of owls (Strigiformes). In: König C, Weick F, eds. *Owls of the world*. London: Christopher Helm, 42–63.

**Wright TF, Schirtzinger EE, Matsumoto T, Eberhard JR, Graves GR, Sanchez JJ, Capelli S, Müller H, Scharpegge J, Chambers GK, Fleischer RC. 2008.** A multilocus molecular phylogeny of Parrots (Psittaciformes): support for a gondwanan origin during the cretaceous. *Molecular Biology and Evolution* **25:** 2141–2156.

**Yago M, Hirai N, Kondo M, Tanikawa T, Ishii M, Wang M, Williams M, Ueshima R. 2008.** Molecular systematic and biogeography of the genus *Zizina* (Lepidoptera: Lycaenidae). *Zootaxa* **1746:** 15–38.

**Yumul GP, Dimalanta CB, Maglambayan VB, Marquez EJ. 2008.** Tectonic setting of a composite terrane: a review of the Philippine island arc setting. *Geosciences Journal* **12:** 7–17.

**Yumul GP, Dimalanta CB, Tamayo RA, Maury RC, Bellon H, Polve M, Maglambayan VB, Querubin CL, Cotten J. 2004.** Geology of the Zamboanga Peninsula, Mindanao, Philippines: an enigmatic South China continental fragment? *Geological Society Special Publication* **226:** 289–312.

**Ziegler T, Abegg C, Meijaard E, Perwitasari-Farajallah D, Walter L, Hodges JK, Roos C. 2007.** Molecular phylogeny and evolutionary history of Southeast Asian macaques forming the *M. silenus* group. *Molecular Phylogenetics and Evolution* **42:** 807–816.

**Zuccon D, Cibois A, Pasquet E, Ericson PGP. 2006.** Nuclear and mitochondrial sequence data reveal the major lineages of starlings, mynas and related taxa. *Molecular Phylogenetics and Evolution* **41:** 333–344.

## SUPPORTING INFORMATION

Additional Supporting Information may be found in the online version of this article:

**Appendix S1.** List of Sulawesian taxa, category membership (see Fig. 5, definitions as defined previously), and references.

**Appendix S2.** List of Philippine taxa, category membership (see Fig. 5, definitions as defined previously), and references.

Please note: Wiley-Blackwell are not responsible for the content or functionality of any supporting materials supplied by the authors. Any queries (other than missing material) should be directed to the corresponding author for the article.

## APPENDIX 1

List of Sulawesian taxa, category membership (see Fig. 5, definitions as defined previously), and references.

| | Primitive | Asian | Australian | African | Philippine | Reference |
|---|---|---|---|---|---|---|
| **Plants** | | | | | | |
| *Hoya* | | | X | | | Wanntorp Kocyan & Renner (2006) |
| Annonaceae | | | X | | | Su & Saunders (2009) |
| **Invertebrates** | | | | | | |
| Molluscs | | | | | | |
| *Paracrostoma* | X | | X | | | Köhler & Glaubrecht (2007) |
| *Corbicula* | X | X | | | | von Rintelen & Glaubrecht (2006) |
| Insects | | | | | | |
| Butterflies | | | | | | |
| Amathusiini | X | X | | | | Peña *et al.* (2006) |
| Satyrinae (clade 3) | | | | X | | Peña *et al.* (2006) |
| Hymenoptera | | | | | | |
| *Ctenoplectra* | X | | | | | Schaefer & Renner (2008) |
| **Vertebrates** | | | | | | |
| Fish | | | | | | |
| Rainbow fish | | | X | | | Sparks & Smith (2004) |
| *Oryzias* | X | X | | | | Takehana *et al.* (2005) |

APPENDIX 1 *Continued*

| | Primitive | Asian | Australian | African | Philippine | Reference |
|---|---|---|---|---|---|---|
| Amphibians | | | | | | |
| *Limnonectes* | | X | | | X | Evans *et al.* (2003b) |
| Reptiles | | | | | | |
| *Python reticulatus* | X | X | | | | Auliya *et al.* (2002) |
| *Tropidophorus* | | | | | X | Honda *et al.* (2005) |
| *Draco* | X | X | | | | McGuire & Heang (2001) |
| Birds | | | | | | |
| *Macrocephalon* | X | | X | | | Birks & Edwards (2002) |
| Cockatoos | | | X | | | Brown & Toft (1999) |
| *Ceyx fallax* | X | | | | | Moyle *et al.* (2007) |
| *Meropogon* | X | | | X | | Marks Weckstein & Moyle (2007) |
| *Cittura* | | | X | | | Moyle (2006) |
| *Coracina* | | | | X | | Jønsson *et al.* (2008) |
| *Lalage* | | | X | | | Jønsson *et al.* (2008) |
| *Locustella* | | X | | | | Drovetski *et al.* (2004) |
| *Basilornis* | X | | | | | Lovette & Rubenstein (2007) |
| *Scissirostrum* | X | | | | | Lovette & Rubenstein (2007) |
| *Enodes* | X | | | | | Lovette & Rubenstein (2007) |
| *Acridotheres* | | X | | | | Lovette & Rubenstein (2007) |
| *Dicaeum* | X | | | | | Nyári *et al.* (2009) |
| Mammals | | | | | | |
| Marsupials | | | | | | |
| *Ailurops* | X | | X | | | Kirsch & Wolman (2001) |
| Placentals | | | | | | |
| *Crocidura* | X | | | | | Ruedi Auberson & Savolainen (1998)<br>Esselstyn & Brown (2009) |
| Pteropidae | X | X | | | | Campbell *et al.* (2007) |
| *Macaca* | | X | | | | Ziegler *et al.* (2007) |
| *Macrogalidia* | | X | | | | Bininda-Emonds Gittleman & Purvis (1999) |
| *Babyrousa* | X | X | | | | Price *et al.* (2005)<br>Funk *et al.* (2007) |
| Anoa | X | X | | | | Pitra F?rbass & Seyfert (1997)<br>Schreiber *et al.* (1999) |
| Sciuridae | X | X | | | | Mercer & Roth (2003) |

APPENDIX 2

List of Philippine taxa, category membership (see Fig. 5, definitions as defined previously) and references.

| | Primitive | Asian | Australasian | Reference |
|---|---|---|---|---|
| **Plants** | | | | |
| *Hoya* | | | X | Wanntorp *et al.* (2006) |
| Annonaceae | X | | | Su & Saunders (2009) |
| **Invertebrates** | | | | |
| Molluscs | | | | |
| *Paracrostoma* | X | | | Köhler & Glaubrecht (2007) |
| Crustacea | | | | |
| *Macrobrachium* | | | X | de Bruyn Wilson & Mather (2004) |
| Insects | | | | |
| Spiders | | | | |
| *Nephila* | X | | | Su *et al.* (2007) |

APPENDIX 2 *Continued*

| | Primitive | Asian | Australasian | Reference |
|---|---|---|---|---|
| **Vertebrates** | | | | |
| Fish | | | | |
| Odacini | X | | X | Clements *et al.* (2004) |
| Amphibians | | | | |
| *Limnonectes* | | X | | Evans *et al.* (2003b) |
| *Platymantis* | X | | X | Köhler *et al.* (2008) |
| Reptiles | | | | |
| *Trimeresurus* | X | X | | Malhotra & Thorpe (2004) |
| *Tropidophorus* | | | | Honda *et al.* (2005) |
| *Varanus* | | | | McGuire & Heang (2001) |
| *V. olivaceus* | X | | X | Schulte *et al.* (2003) |
| *V. salivator* | | X | | |
| Birds | | | | |
| Cockatoos | | | X | Brown & Toft (1999) |
| *Campiphagidae* | | | X | Jønsson *et al.* (2008) |
| *Rhabdornis* | X | | | Lovette & Rubenstein (2007) |
| | | | | Zuccon *et al.* (2006) |
| *Dicaeum* | X | | | Nyári *et al.* (2009) |
| Mammals | | | | |
| Placentals | | | | |
| Pteropidae | X | X | | Campbell *et al.* (2007) |
| Tamaraw | X | X | | Schreiber *et al.* (1999) |
| Murids | X | X | X | Jansa *et al.* (2006) |
| | | | | Steppan *et al.* (2005) |

# Attribution

**Michaux, B. 2010. Biogeology of Wallacea: Geotectonic models, areas of endemism, and natural biogeographical units. *Biological Journal of the Linnean Society* 101:193–212.**

First published in 2010 in *Biological Journal of the Linnean Society*, 101(1):193–212, doi.org/10.1111/j.1095-8312.2010.01473.x. Reprinted with permission from Oxford University Press.

# *chapter eight*

# *Natural areas*

## *Natural groups in systematics*

While the definition of natural as 'not made or caused by humankind' is clear enough, it is hardly of operational utility, and this vagueness of meaning with respect to identifying natural groupings in the world around us has led many biologists to shun its use. Wiley (1981) and Wiley and Lieberman (2011) discussed the concept of naturalness in taxonomy, and suggested that it has changed with our changing views of nature and natural processes. According to these authors, the taxonomy of Linnaeus, where natural groups were defined by an essence that all members of the group possessed, gave rise to a more inclusive, phenetic approach where a natural group was composed of members that were more similar to each other than to anything outside the group, as opposed to a phylogenetic view where a natural group consisted of an ancestor and all its descendants, that is, a monophyletic group.

Sloan (1972) contended that the break with essentialist thinking in taxonomy occurred a generation prior to Linnaeus at the end of the seventeenth century as a result of a controversy that raged about how to classify plants. The opposition was between those who favoured a logical basis for classification (by essence) and those who favoured an empirical grouping (phenetic approach). Müller-Wille (2013) argued that Linnaeus was unmoved by such arguments because he thought both approaches would lead to artificial groupings. For Linnaeus – who was the first to distinguish between 'natural' and 'artificial' – natural systems (groups) could only be produced when natural characters were used to classify them. According to Müller-Wille (2013), Linnaean natural characters, which could be used to erect natural genera, for example, would be present in all species within the genus. Such characters could only be identified when all species within the genus had been examined and a complete description of their character states given. Once this had been accomplished, a natural classification could be erected. Linnaeus regarded this as a task for the future and argued that all classification systems to date, including his own, were artificial to varying degrees.

Linnaeus thought of natural characters as relational, unifying a genus rather than separating it from other genera as dichotomous keys and artificial characters do. According to Linnaeus, a natural classification system would result in a stable nomenclature because the discovery of a new genus, for example, would not disrupt existing genera. As Linnaeus commented, 'Natural orders have their value with respect to the nature of plants; artificial ones in the diagnosis of plants' (quoted in Müller-Wille 2013:314). In a modern context, I take this to mean that natural groupings allow one to investigate the ordering process, while artificial ones (for example a dichotomous key) allow you to identify a taxon and name it.

Müller-Wille's (2013) more nuanced interpretation of Linnaeus' ideas showed not only that the development of different concepts of what is natural has not been a straightforward linear progression, but also that the twin problems of what constitutes a natural group and how to discover them have had a long – and continuing – history. Linnaeus' discussion of what is natural clearly identified the importance of using natural groupings in systematics; at a practical level, it results in a stable nomenclature and at a theoretical level it allows process to be inferred. Wiley and Lieberman (2011) discussed the characteristics of natural groups and concluded that because they originated through the operation of natural processes, they must exist independently of human cognisance and, therefore, are discovered rather than invented. In phylogenetic systematics, a natural group is discovered by establishing its monophyly. Both para- and polyphyletic groups are artificial and their progressive elimination from taxonomy through cladistic analysis has been one of the great advances in modern systematics.

## *Natural areas in biogeology*

Ebach and Michaux (2017) suggested that the concept of monophyly should be applied to areas in an analogous way to its application in taxonomy. They discussed the advantage of employing a stable area 'taxonomy' in biogeography, which could be achieved if the area units used in any analysis were monophyletic (natural). While the authors stressed the utility of this approach, it is also clear that any attempt to understand natural processes requires the identification of natural areas. However, there is the problem of identifying areas of endemism – natural or otherwise – when both 'area' and 'endemism' are problematic concepts (Wiley and Lieberman 2011).

Henderson (1991) discussed the vagueness of area of endemism definitions and two decades later Wiley and Lieberman (2011:272) admitted that 'The concept of *area of endemism* is not mature'. Operational definitions of areas of endemism depend on the overlap of a number of species' ranges (e.g. Parenti and Ebach 2009; Wiley and Lieberman 2011), which ducks the question of their ontological status but allows them to be identified and used in analyses. According to Wiley and Lieberman (2011), part of the problem with the concept of area in biogeography is that two different concepts are conflated – what they called geologic areas and areas of endemism. According to these authors, geologic areas are individuated by geological processes and have an independent history from other geologic areas, but share varying degrees of relatedness with them. Areas of endemism are defined (albeit rather vaguely) by biology. Sometimes these two types of areas are congruent and a cause-and-effect relationship exists between geological processes and biological history. In such cases, biogeographic patterns of endemic clades allow the history of geologic areas to be inferred. In other cases, the history of geologic areas and areas of endemism don't coincide. Wiley and Lieberman (2011) gave an example of disjunct distributions of angiosperm between China and eastern North America, which have been interpreted as remnant populations of a once continuous boreal forest fragmented by climate change and extinction. In such cases, it is not possible to infer geological history (i.e. a composite China-eastern North America geological entity) from a biological relationship.

Understanding the historical development of Wallacea has been a recurring research theme of mine and in Michaux (2010) I returned again to this most geologically and biologically complex area. The paper starts with a discussion of areas of endemism and what constitutes a natural area, and covers similar ground to Wiley and Lieberman (2011). I too was interested in teasing out the relationships between geology and biology inherent in the concept of areas of endemism. My biogeological interest in islands and archipelagos stems partly from the fact that areas of endemism self-define in such regions because biological boundaries tend to be prescribed geographically. Although individual islands with endemic taxa are easily enough identified, they are not necessarily natural because they may be biologically composite. Using an area with a polyphyletic biota in any analysis of distributions will lead to spurious or ambiguous results as was demonstrated by the analysis of the New Zealand Subantarctic avifauna described in Chapter 6.

As my primary interest in Wallacea and other island groups peripheral to cratonic Gondwana has been to use biology (the relationships between monophyletic groups) to help reconstruct geological history, I've always depended on natural biogeographical areas. The key questions for me were what exactly are they and how might one identify them. Figure 2 of Michaux (2010) was an attempt at answering these key questions. Natural areas were defined as those that had an endemic biota with a shared geological history and were termed Wallacean Biogeographical units to distinguish them from areas of endemism that were geologically composite. Wallacean Biogeographical units are equivalent to geologic areas that are congruent with areas of endemism (Wiley and Lieberman 2011). Figure 8.1 represents a refinement of these ideas in the light of the preceding discussion and identifies three different types of areas of endemism, two of which are nominal and only one natural. Artificial areas of

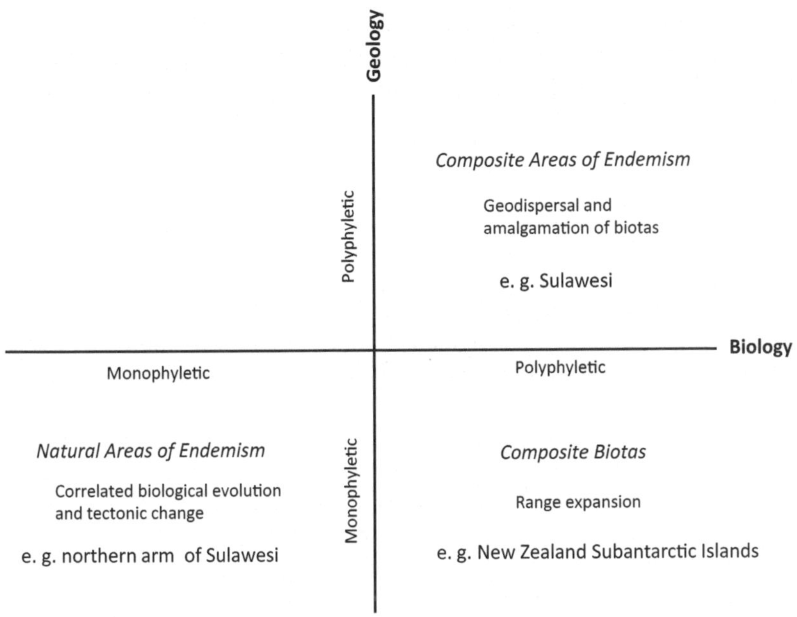


***Figure 8.1*** Different types of area of endemism based on the concept of monophyly and the correlation of biology with geology.

endemism are termed nominal because the area is a construct and the name, often that of a country, is simply a label.

## *Nominal areas of endemism*

### (i) Composite areas of endemism

Composite areas of endemism are biologically and geologically polyphyletic. Sulawesi is used as an example of a composite area of endemism because it has a mixed Asian and Australasian biota and is composed of terranes derived from different source areas during different time periods. Using Sulawesi as an area in any biogeological analysis can only lead to artificial results because Sulawesi is itself a biogeological artefact. Composite areas of endemism are formed by multiple geodispersal events, and while some endemics may still be confined to their original terranes, many can be expected to have dispersed more widely following terrane amalgamation. A phylogenetic analysis of individual endemics should establish what area clade they are part of, irrespective of their present distribution, allowing any multiple area relationships that exist to be described and related to tectonic history. If the geological and biological relationships are congruent (*sensu* Nelson and Platnick (1981); Wiley and Lieberman (2011)), then composite areas of endemism can be subdivided to yield natural areas of endemism that can be used in the construction of a general areagram.

(ii) Composite biotas

Composite biotas represent a type of area of endemism that are biologically polyphyletic but geologically monophyletic. The New Zealand Subantarctic Islands are an example of such an endemic area because the Campbell Plateau is geologically monophyletic but the biota is polyphyletic, a mix of older West Gondwanan and younger East Gondwanan endemics. The process that produces composite biotas is range expansion, the movement of whole or partial biotas in response to the removal of barriers to dispersal, either physical (such as a marine barrier) or climatic. In the case of the New Zealand Subantarctic Islands, range expansion was promoted by the amalgamation of the Campbell Plateau with the Melanesian Rift that created new habitat for taxa to expand into (Michaux and Leschen 2005). Areas of endemism with composite biotas should not be used as area units in biogeographic analyses because different parts of the biota will link it to different external areas, and the areas' placement in a general areagram will depend on what taxa are used. Once again, any area analysis should start by partitioning species on the basis of what area clade they belong to, allowing each grouping of taxa to be analysed separately.

In Figure 8.1, I have left blank the quadrant that has a monophyletic biota but polyphyletic geology. Such areas could be formed by geodispersal without biotic dispersal, in other words, if incoming terranes carried no biota or it became extinct and made no contribution to the modern biota. For example, while many terranes on the southern flanks of Eurasia are Gondwanan (Chapter 10), the modern biota is not. While such areas are theoretically nominal in the sense that biological and geological histories are not congruent, for practical purposes, they are natural areas of endemism in the sense that they can be used as inputs into an area analysis without confounding or obscuring area relationships.

## *Natural areas of endemism*

Natural areas of endemism have monophyletic geologies and biologies. These areas were called Wallacean areas of endemism by Michaux (2010) and are equivalent to congruent geologic areas and areas of endemism of Wiley and Lieberman (2011). Natural areas of endemism are formed through the natural processes of biological evolution and tectonic change, and as such are natural rather than artificial. Identification and use of natural areas of endemism are prerequisites for any biogeological analysis whose aim is to infer geological history (Wiley and Lieberman 2011). An example of a natural area of endemism is the northern arm of Sulawesi, originally an island arc carrying its own endemic taxa and with a separate tectonic history to the other terranes that make up the modern island of Sulawesi. Even if all the original endemics associated with the island arc had dispersed onto other Sulawesian terranes following docking, it would still be possible to identify original elements of the biota by showing congruence between biological and geological patterns, that is, when cladistic relationships between taxa are congruent with the geological history of the terranes. Original members of the biota (or clade if speciation had occurred post amalgamation) would be more closely related to species found on other parts of the original arc system than to other Sulawesian taxa.

## The Banda arcs

Michaux (2010) proposed a number of natural areas of endemism within Wallacea including south Maluku (Michaux 2010, Figure 4:2b), western Nusa Tenggara (Michaux 2010, Figure 4:4) and Timor (Michaux 2010, Figure 4:3), which are collectively referred to as the Banda arcs (Figure 8.2).

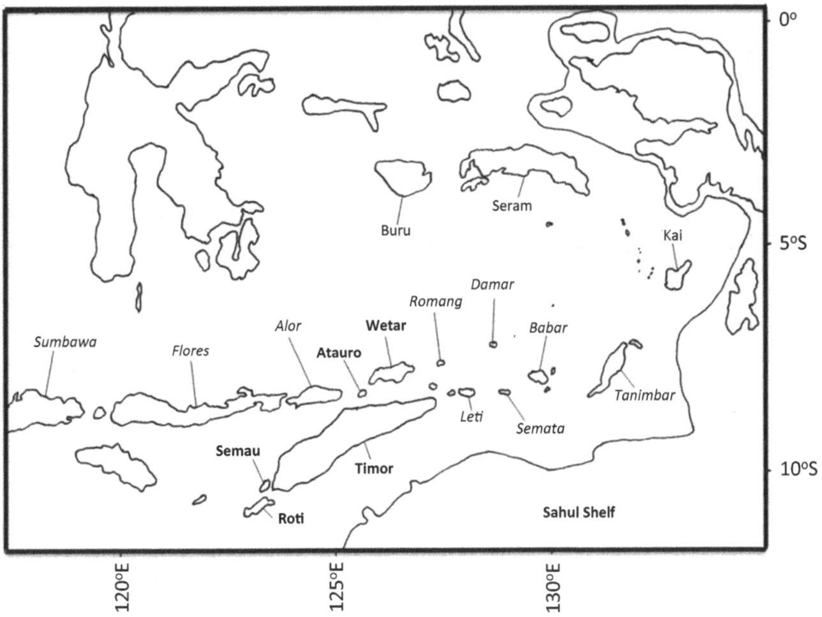


*Figure 8.2* Locality map of the Banda arc. Islands mentioned in the text are named. Wetar collision zone islands in bold.

The islands of the Banda arcs consist of two parallel, arcuate chains extending from the Sunda Shelf to the island of Buru in south Maluku, a distance of about 1500 km. These two chains are quite distinct: the northern or inner arc is young, volcanically active and structurally simple; the southern or outer arc is much older, non-volcanic and structurally very complex. The purpose of this section is to re-examine Michaux's (2010) proposal in more detail, and in the process demonstrate how one might analyse the biogeology of the Banda arcs and identify natural areas.

## *Geology of the inner Banda arc*

The islands of the inner Banda arc extend from Lombok and Sumbawa through Flores, Alor, Atauro, Wetar and Romang to Banda Island just south of Seram. The oldest volcanic rocks are found on the most westerly islands; Lower Miocene age for Lombok (Foden 1979) and Middle Miocene (15 to 12 Ma) for Flores (Otake et al. 2002; Ota and Kaneko 2010; Subarsyah and Rahardiawan 2016) and Wetar (Sewell and Wheatley 1994). The inner arc has been and continues to be built directly on oceanic crust (Sewell and Wheatley 1994; Otake et al. 2002; Ely et al. 2011). The petrology of these volcanoes is highly variable ranging from silica-rich rhyolites and dacites to intermediate andesites and silica-poor basalts (Foden 1979).

One section of the inner arc – the Wetar Zone of Ely et al. (2011) – is now extinct with volcanic activity ceasing on Atauro at 3.3 Ma, Wetar at 3 Ma, Alor at 1.4 Ma and Romang at 1.7 Ma (Ely et al. 2011). This zone is also seismically quiescent to intermediate depths down to 300 km (Sandiford 2008). The Wetar gap is interpreted as the collision zone between the Australian plate and the inner Banda arc. When Australian continental crust entered the subduction zone, it caused the subducting slab to become detached and volcanic activity to cease (e.g. Ely et al. 2011). Evidence for the presence of Australian continental crust underlying this sector of the Banda arcs comes from Pb-isotope studies reported in Elburg et al. (2004), who found that the volcanic rocks of the collision zone showed higher ratios of $^{204}Pb/^{206}Pb$, which the authors concluded was the result of contamination of melts by Australian continental crust. Uplift of limestone terraces by hundreds of metres on Atauro (Ely et al. 2011) and other islands in the collision zone (Roosmawati and Harris 2009) is interpreted in terms of the isostatic rebound of lighter, under-thrust Australian crust following slab detachment at about 3Ma. Outside the collision zone, the inner arc is separated from the outer arc by deep water basins: in the east by the Weber Deep that has a depth of 7 km and formed between 3.5 Ma and 1 Ma (Hall et al. 2017), and in the west by the Savu Basin that is several kilometres deep and thought to have formed by rapid subsidence in the Middle Miocene (Rigg and Hall 2011). Neither basin is underlain by oceanic crust and both are thought to be extensional features caused by slab rollback.

## *Geology of the outer Banda arc*

While the inner Banda arc is a classic volcanic chain that formed some 150 km above a subduction zone where oceanic crust was consumed and magmas generated, the outer Banda arc has had a complex and multiphased history, the details of which are still debated. The outer Banda arc consists of the islands of Sumba, Sawu, Roti, Semau, Timor, Leti, Semata, Babar, Tanimbar, Kai, Seram and Buru. These islands are separated from Australia and New Guinea by a series of deep troughs (Hall et al. 2017). Villeneuve et al. (2010) recognised Buru and Seram as a distinct tectonic block from the Timor-Tanimbar sector of the outer arc. Charlton (2001) proposed that the Buru-Seram block had been derived from the northwest margin of Australia, but Villeneuve et al. (2010) regarded it as a westwards extension of the Bird's Head, New Guinea, as did Hall (2012) who included Buru and Seram as part of the Sula Spur.

Villeneuve et al. (2010) recognised three tectonostratigraphic units within Timor that they traced eastwards as far as Tanimbar:

1. A para-autochthonous unit of unmetamorphosed Permian to Palaeogene sediments that form much of central Timor. They regarded this unit as a proto-Timor and suggested it was originally attached to Greater India rather than to Australia. This unit is equivalent (in part) to the Unmetamorphosed Continental Shelf Sedimentary (UCSS) unit of Ota and Kaneko (2010), although these later authors suggested an Australian origin.
2. Kalobano unit of imbricated Permian to Pliocene sediments found on the south coast and interpreted as the former leading edge of the Australian shelf.
3. An allochthonous unit found in the north of Timor and interpreted as an obducted Eurasian/Sundaic terrane.

According to Villeneuve et al. (2010), proto-Timor was translated northwards to collide with the Eurasian/Sundaic margin resulting in the emplacement of the allochthonous unit during the Upper Eocene (c 37 Ma), before moving south again to collide with the Australian margin during the Pliocene (c 5 Ma) when the Kalobano unit was sutured to Timor. Timor has been attached to Australia and moved in concert with it since the Pliocene. Similar scenarios have also been suggested by a number of other authors.

Zimmermann and Hall (2014) analysed detrital zircons from Sumba, east Timor and Tanimbar and concluded, on the basis of their zircon age profiles, that these islands were part of Australia's continental shelf during the Permian and much of the Mesozoic before they were rifted away in the Late Jurassic to collide with the Sundaic margin in the Late Cretaceous.

Standley and Harris (2009) reported that detrital zircon age profiles derived from the allochthonous unit of north Timor were typical of what they termed the Great Indonesian Arc (discussed below) and quite distinct from typical Australian zircon age profiles. Ota and Kaneko (2010) proposed that ophiolite obduction on the Timor block was a consequence of a collision with Sundaland rather than Australia because K-Ar ratios in micas from associated metamorphics indicated collision age estimates ranging from 40 Ma for Laibobar (a satellite island of Tanimbar), 37 Ma for west Timor, 11 Ma for Leti to 6–5 Ma for East Timor, all of which predate Timor's collision with Australia. Standley and Harris (2009) reported a peak of metamorphism in the allochthonous unit at 45 Ma, which they suggested was a result of the collision between the Timor block and Sundaland.

Harris (2006) referred to the allochthonous unit as the Banda terrane. According to Harris (2006), this terrane can be recognised in southeast Kalimantan (Meratus terrane), southwest Sulawesi and Flores as well as Timor. He calculated a minimum metamorphic age of 32 Ma, and like Ota and Kaneko (2010) concluded that the metamorphism pre-dated the collision with Australia. A similar minimum age of 35 Ma was derived for the volcanic melts. According to Harris (2006), the Banda terrane was part of a Cretaceous to early Tertiary 'Great Indonesian Arc' that stretched from India to New Guinea. This arc system became fractured in the Late Eocene and Oligocene forming arc/sea-floor fragments that were emplaced in various places such as Kalimantan and Sulawesi. One such fragment – the Banda terrane – was transported south with the opening of the South Banda Sea during the Late Miocene (10 Ma) to eventually collide with the Australian margin in the Pliocene (5 Ma). While there are alternative views about the history of Timor – for example, see Audley-Charles (2004) and Boger et al. (2017) – it seems that there is substantial evidence that the allochthonous unit of Timor, Leti, Sermata and Laibobar had its origin in an island arc environment that was close to the Sundaland margin during the Eocene.

## *Natural areas of endemism in the Banda arcs*

In Michaux (2010) I drew attention to the apparent close biological relationship of Timor to Wetar, its biological distinctiveness from Lombok, Sumbawa and Flores, and its mixed Australian and Asian biota. Timor has been a challenging place for biologists due to the logistics of working in an isolated region, and the rugged terrain made more difficult to travel in by the destruction of basic infrastructure as a result of the recent and prolonged armed conflict that made research impossible (Kaiser et al. 2011). However, things have improved since Timor-Leste gained independence in 2002, and recent studies of the birds of Timor (Trainor et al. 2007) and the ants of Timor and surrounding islands (Trainor and Andersen 2010) have shed light on the biogeography of the region as a whole and Timor in particular. Both the avifauna and antfauna of Timor are a mixture of Asian and Australasian taxa. According to Trainor et al. (2007), there are 168 native resident birds in Timor-Leste of which 44 species are Wallacean endemics (26%), 35 have Asian relatives (21%) and 32 Australasian relatives (19%). The ant faunas of Southeast Asia and Australia are quite distinct (Trainor and Andersen 2010) and the Timor antfauna is dominated by Asian taxa with 76% being related to tropical forest taxa of Southeast Asia and only 18% related to Australian savannah woodland species.

A PAE analysis of Timorese birds was carried out. The distributional data are from Bird Life International (http://datazone.birdlife.org/eba/factsheet/164;http://datazone.birdlife.org/eba/factsheet/165), White (1984), White and Bruce (1986), Coates and Bishop (1997), Trainor and Soares (2004), Trainor (2005a, b) and Eaton et al. (2016). Single island endemics were removed from the data set prior to analysis and the resulting data matrix is shown in Figure 8.3. The data were analysed with the phylogenetic package TNT (Goloboff et al. 2008) using the implicit enumeration option that is guaranteed to find the most parsimonious solution(s). A single most parsimonious tree (length = 87 steps) was obtained and is also shown in Figure 8.3. Three clades were recovered by the analysis: a basal clade of south Maluku (Buru/Seram) and Kai that is the sister area to all other Banda arc islands; within this latter clade Timor (and satellite islands) plus Atauro and Wetar are a sister area to all the remaining islands from Sumbawa to Tanimbar.

The most obvious feature of this biologically based area grouping is that it does not separate inner and outer arc islands into different clades; it is not congruent with geology. The basal clade contains the islands of Buru, Seram and Kai and links islands that are geologically similar to western New Guinea but isolated from it by the Seram Trough – an aseismic linear feature up to 2 km in depth (Hall et al. 2017). Further support for linking the Kai Islands with Buru and Seram comes from a PAE analysis of the endemic birds of Maluku (Coates and Bishop 1997: Appendix 2), the result of which is shown in Figure 8.4. Single island endemics were removed from the data set before analysis. The matrix derived from the endemic bird distributions is also shown in Figure 8.4 and was also analysed using the implicit enumeration (*ienum*) option in TNT (Goloboff et al. 2008). A single most parsimonious tree of length 35 steps was produced. The tree separates north and south Maluku into sister clades, links Obi with north Maluku as previously suggested by Michaux (2010) and also links Kai with Buru and Seram. The complicated trough structure to the immediate south of the Kai Islands (Hall et al. 2017: Figure 1) appears to have isolated this northern section from the rest of the outer Banda arc.

**A**

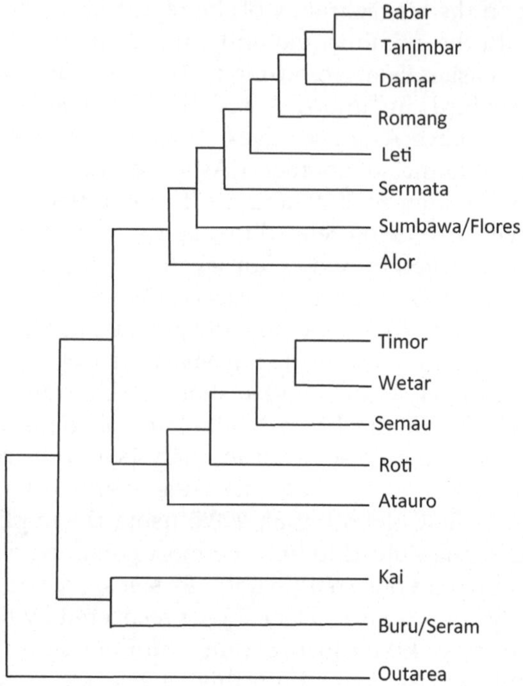


**B**

| | |
|---|---|
| Outarea | 00000000000000000000000000000000000000000 |
| Sumbawa/Flores | 1000000100111000001110000000000000000000000 |
| Roti | 11111111000000000000000000000000000000000 |
| Semau | 11011101111100000000000000000000000000000 |
| Atauro | 10000011010010000000000000100000000000000000 |
| Timor | 11111111111111111111111110000000000000000000 |
| Wetar | 01000111011001111111111110000000000000000000 |
| Romang | 10000000001000000011110111111000000000000000 |
| Leti | 10000000001000000011100011110000000000000000 |
| Sermata | 00000000001000000011000011010000000000000000 |
| Babar | 10000000001000000000111111011011111100000000000 |
| Damar | 10000000001000000000111011111100100000000000000 |
| Tanimbar | 00000000001000000010100110111111111111100000 |
| Kai | 00000000000100000000000000101110001100011111 |
| Buru/Seram | 00000000000000000000000001000000011111111101 |
| Alor | 00000000000000000011000000000000000000000000 |

***Figure 8.3*** PAE analysis of restricted range Banda arc birds. A: Single most parsimonious tree length = 87. See text for details. B: Data matrix.

A

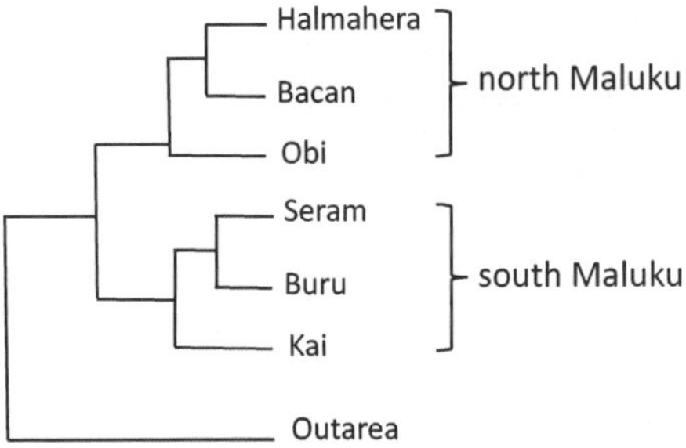


B

| | |
|---|---|
| Outarea | 000000000000000000000000000000 |
| Halmahera | 110111100111111111011110011111 |
| Bacan | 111111100111111111011110011111 |
| Obi | 0111101001?0101010111100010011 |
| Bur | 010000011000000000100001110010 |
| Seram | 010000011000000001100001110000 |
| Kai | 000000001000000000000000110000 |

***Figure 8.4*** PAE analysis of the endemic avifauna of Maluku. See text for details. A: Single most parsimonious tree, length = 35. B. Data matrix.

The other islands of the Banda arcs are arranged into two clades (Figure 8.3). While not sorting into outer and inner arc groupings, the Timor clade does contain islands found within the Wetar collision zone. The sister-area relationship confirms the close biological similarity between Timor and Wetar discussed in Michaux (2010). Further confirmation of both the Timor-Wetar sister-area relationship and the separation of collision zone islands from the rest of the Banda arc come from a PAE analysis of the Asian ant fauna of Timor described in Trainor and Andersen (2010). Single-island endemics were removed from the data matrix before analysis using TNT (Goloboff et al. 2008) using the exact search algorithm (*ienum*), which yielded two most parsimonious trees (length = 72 steps). One of these trees and the data matrix is shown in Figure 8.5. Once again, the islands of Timor, Wetar and Atauro form a clade that is a sister group to Panay and Alor of western Nusa Tenggara.

## *Biogeology of the Banda arcs*

Figure 8.6 is a summary of the geological and biological history of the Banda arcs. In Figure 8.6, proto-Timor is placed next to the Great Indonesian Arc system during the Eocene. This arc system is adjacent to the Sula Spur, an Australian continental extension that included southern Maluku (Buru and Seram) and Kai (Hall 2012). While it is unlikely that the Sula

**A**

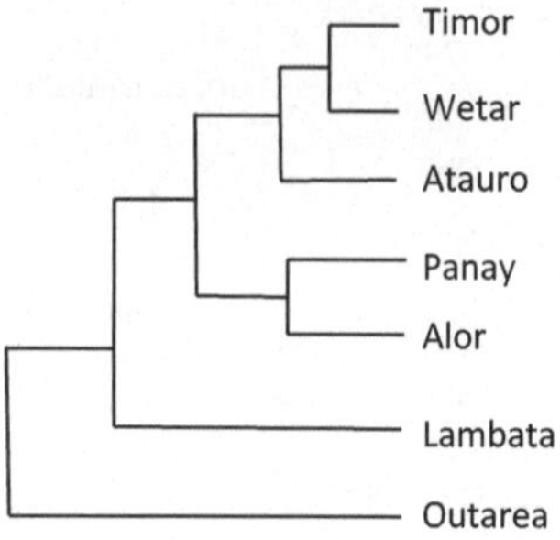


**B**

| | |
|---|---|
| Outarea | 00000000000000000000000000000000000000000000 |
| Timor | 11111111111000011111111101111101111111111111 |
| Wetar | 00110110110100011001011101110011111111110101 |
| Atauro | 00000011011000001000000011010001100000001001 |
| Panay | 11000011101011110110100001011101000000000001 |
| Alor | 11101011110011111100100011010011000000010011 |
| Lambata | 00000000100100001000100001010010000000001111 |

*Figure 8.5* PAE analysis of Timore-Leste ant fauna. A: One of two most parsimonious trees, length = 72. B: Data matrix.

Spur was physically connected to the Great Indonesian Arc, both units shared a similar history involving collision with the Sunda margin, followed by fragmentation and subsequent tectonic dispersal. The presence of an artiodactyl *Anthracotherium* in the Eocene of Timor (Lithoreau and Ducrocq 2007) is indicative of a more general range expansion of Asian taxa into newly created terrestrial environments following the obduction of the Banda terrane onto proto-Timor.

According to Harris (2006), this Great Indonesian Arc became fragmented during the Oligocene due to extensional collapse of the arc orthogonally to the strike of the trench. One of these fragments – the Timor Block – was tectonically dispersed southwards until its collision with the Australian margin in the Late Miocene. I have suggested that the Australian elements of the Timor biota, such as the Australasian-derived murid rodents (Musser 1981) and *Eucalyptus* (Payn et al. 2007), originated from this time. The Late Miocene collision of Timor with Australia resulted in Timor moving in concert with Australia northwards to collide with the inner Banda arc in the Pliocene. I interpret both the occurrence of pygmy stegodont fossils in the Pliocene/Pleistocene of Timor and the origin of the 'collision zone' clade linking Timor with Weta and Atauro (Figures 8.3 and 8.5) as a consequence of this collision. The opening of extensional basins has subsequently fragmented this region and biologically isolated individual islands (Figure 8.6).

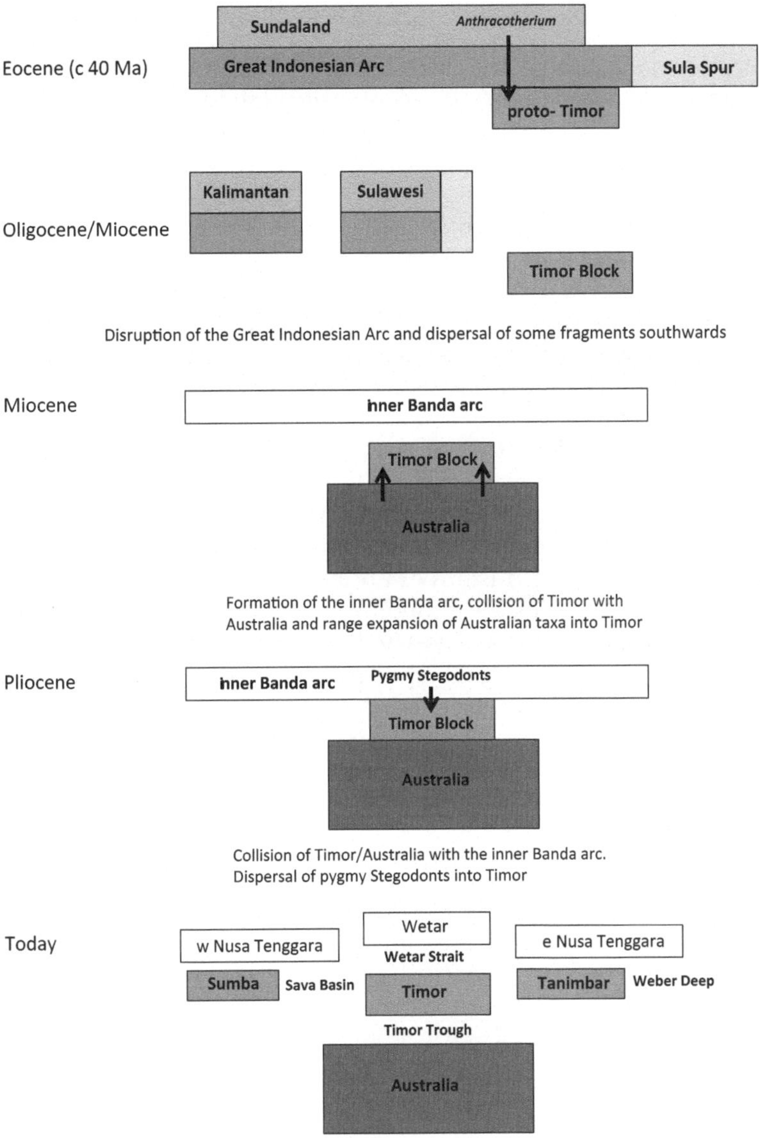


***Figure 8.6*** Cenozoic history of the Banda arc. Eocene: Amalgamation phase. Oligocene/Miocene: Fragmentation phase. Miocene: Collision of Timor block and Australia; growth of volcanic inner Banda arc. Pliocene: Collision of Timor/ Australia with the inner Banda arc. Today: Fragmentation phase. See text for details.

## *Interpretation of the Banda arc biota*

According to the outline presented here, the biota of Timor (and by extension the Banda arcs in general) is composite both in terms of source area (Asia and Australia) and time (Eocene Asian and Pliocene Asian). The following endemic areas are provisionally recognised:

1. Buru, Seram and Kai form a clade (south Maluku) that is sister area to Halmahera, Obi and Bacan (north Maluku). Maluku as a whole is a nominal area of endemism because it is geologically and temporally composite. North Maluku is probably best regarded as a composite area of endemism, even though it has a polyphyletic geology and biology. The distinction between a Melanesian/Papuan and Australian biota is a fine one and the biota of North Maluka could be regarded as Australasian. South Maluku is a natural area of endemism.
2. The Banda arc area of endemism includes all inner and outer islands apart from Kai, Seram and Buru. The Banda arc as a whole is a composite area of endemism. Within this larger endemic area there are two other monophyletic groupings: what I would term the collision zone clade of Timor and its satellite islands plus Wetar and Atauro, and a clade containing all other islands. Neither of these are natural areas of endemism.

**CSIRO** PUBLISHING

*Australian Systematic Botany*, 2016, **29**, 424–439
http://dx.doi.org/10.1071/SB16032

# A comprehensive vicariant model for Southwest Pacific biotas

*V. Ung*[A], *B. Michaux*[B,D] *and R. A. B. Leschen*[C]

[A]CNRS UMR 7205 (CNRS-UPMC-MNHN), 57 Rue Cuvier CP43, F-75005 Paris, France.
[B]Private Bag, Kaukapakapa 0843, New Zealand.
[C]Landcare Research, New Zealand Arthropod Collection, Private Bag 92170, Auckland, New Zealand.
[D]Corresponding author. Email: bjmichaux@gmail.com

**Abstract.** In the present paper, we develop a new biogeographic model for the biota of the Southwest Pacific, using 76 published phylogenies for a range of island endemics or near-endemic organisms. These phylogenies were converted to areagrams by substituting distributions for taxa. Paralogy-free subtrees (3-item statements) were derived from these areagrams and used as input data into LisBeth that uses compatibility analysis and an exhaustive branch and bound algorithm to produce optimal trees. A general areagram is derived from all three-item statements common to the optimal trees. The results of the analysis show that the Melanesian Rift is not a natural biogeographic area; the islands of the Southwest Pacific are more closely related to each other than they are to Australia; and New Caledonia has had a long history of biological isolation. There is support for a general period of mobilism during the mid-Cenozoic when the biota as a whole expanded its range in response to regional uplift. By comparing the general areagram with what is known about the tectonic development of the region, it is possible to both calibrate the nodes of the areagram, and to identify points of conflict between the geological and biological data.

Received 16 August 2016, accepted 7 October 2016, published online 11 May 2017

## Introduction

The Southwest Pacific Ocean is dotted with islands of varying sizes, ranging from small atolls or eroded volcanic remnants, such as Lord Howe Island, to considerable landmasses such as New Zealand. The floras and faunas of the islands are disharmonic in composition and often depauperate, although a minority of clades are highly diverse, with many of the taxa occupying basal positions in their phylogenies. The largest islands, New Zealand and New Caledonia, are renowned for their primitive taxa and endemic biotas, and even the smaller islands exhibit high degrees of endemicity. Many of the islands are isolated today by considerable stretches of open-water, but this was not always the case because they were once part of, or adjacent to, a now submerged continent that we will refer to as the Melanesian Rift (Fig. 1).

There has been a long history of biogeographic inquiry into how the present Southwest Pacific biota has evolved through time to attain its present distribution and characteristics. There were speculative hypotheses concerning the modern New Zealand biota included in both the correspondence of Darwin (Darwin 1903) and in the writings of Wallace (Wallace 1876), and in the publications from the earliest days of European scientific research by prominent New Zealand scientists such as Hutton (Hutton 1872) and Oliver (Oliver 1925). These hypotheses all sought to explain present distributions in terms of dispersal routes over a stable globe. By the 1970s, when the idea that the Earth's surface was a truly dynamic place gained common acceptance, biogeographic hypotheses changed to incorporate original elements rafted from Gondwana, which were added to by chance dispersals through time. For example, in Charles Fleming's influential book, *The Geological History of New Zealand and its Life* (Fleming 1979), he talked about an archaic biotic element that has always been in New Zealand, and that had been passively transported as the Tasman Sea opened at the end of the Cretaceous. In Fleming's view, this original element was added to periodically from different source areas, interspersed with times of biological isolation and increasing endemicity.

The history of biogeographic research within the Southwest Pacific became more complicated during the three decades that followed the publication of Fleming's book. There were two principal reasons for this, namely, the advent of two new biogeographic research programs (panbiogeography and vicariance biogeography), and the development of molecular methods. These developments stimulated a great deal of research that continues to the present (Heads 2014*a*); however, unfortunately, little real progress has been made in our understanding of the evolution of the biota as a whole and how this relates to the geological development of the Southwest Pacific. Too much of the research based on molecular data is collected primarily for systematic purposes and, like most vicariant studies, tends to be based on a single taxon (or a small

 

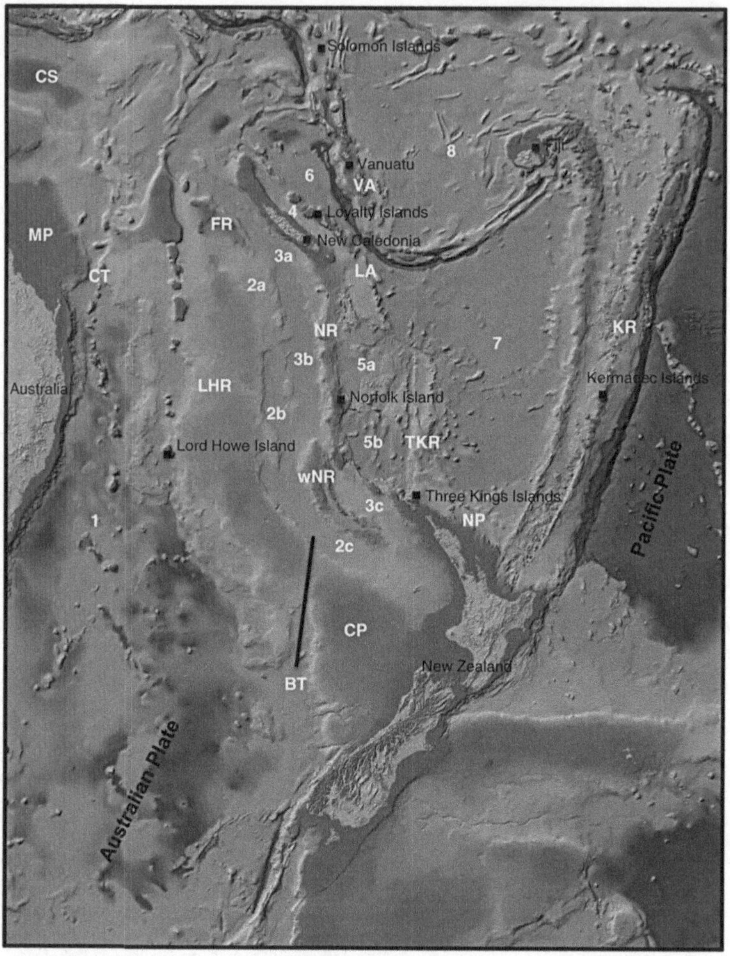


**Fig. 1.** Locality map, showing structural features discussed in the text. BT, Bellona Trough; CP, Challenger Plateau; CS, Coral Sea; CT, Cato Trough; FR, Fairway Rise; KR, Kermadec Ridge; LA, Loyalty Arc; LHR, Lord Howe Rise; MP, Marion Plateau; NP, Northland Plateau; NR, Norfolk Ridge; wNR, western Norfolk Ridge; VA, Vityaz Arc. 1, Tasman Sea; 2a, Fairway Basin; 2b, central Fairway Basin; 2c, Taranaki Basin; 3a, northern New Caledonian Basin; 3b, central New Caledonian Basin; 3c, Reinga Basin; 4, South Loyalty Basin; 5a, northern Norfolk Basin; 5b, southern Norfolk Basin; 6, North Loyalty Basin; 7, South Fiji Basin; 8, North Fiji Basin.

number of taxa), which limits the generalisations that can be drawn from such studies, whereas the exhaustive approach of panbiogeoegraphy, which gives a broad overview (see, for example, Croizat 1964), lacks the specificity to reconstruct history.

As a consequence of this lack of biogeographic focus, there are many conflicting hypotheses concerning area relationships. For example, whether New Caledonia is more closely related to Australia (Ladiges *et al.* 2003, Ladiges and Cantrill 2007; de Queiroz 2014) or New Zealand (Swenson *et al.* 2001; Arensburger *et al.* 2004*a*), or to both (Heads 2008)? Or whether it is perhaps most closely related to New Guinea (Swenson *et al.* 2001)? Similar ambiguity exists for the area-relationships shown by New Zealand, Lord Howe Island and Loyalty Islands. Despite these problems, we argue that the islands of the Southwest Pacific are ideal for examining the interplay between evolution and tectonic history because of the geographically circumscribed nature of the individual floras and faunas, the high degree of endemism shown by individual islands or island groups, and because good-quality biological and geological data have become increasingly available over the past decade. We review the tectonic history of the Melanesian Rift and compare this to a general areagram derived from an analysis of three-taxon statements for taxa in the South Pacific region. There are six key questions we wish to address, and they are as follows:

(1) Is the Melanesian Rift a natural biogeographic entity; in other words, does it require biologeographical explanation?
(2) What are the relationships among Melanesian Rift areas and what relationship do they have to Australia?
(3) Is New Caledonia an ancient island that has been biologically isolated for tens of millions of years?
(4) Are the Loyalty Islands more closely related to the islands of the Vityaz Arc rather than to New Caledonia, as suggested by Heads (2008)?
(5) Is there evidence for biological mobilism (range expansion of biotas as opposed to chance dispersal of individual taxa) associated with periods of uplift of the Melanesian Rift?
(6) Is it possible to date any internal nodes through tectonic calibration?

## Tectonic history of the Melanesian Rift

During the later stages of Gondwana's Mesozoic break up, a large, now mostly submerged, continental fragment was rifted from East Gondwana and transported away from Australia as the Tasman Sea opened. This event was unusual because tensional forces were not concentrated at a single rift zone as, for example, in the cases of India, South America, or Africa, but were distributed across a region some 300 km or more wide. One possible reason why rifting east of Australia was not confined to a single zone was that eastern East Gondwana had been subject to high-heat flows before rifting, which presumably plasticised basement rocks, weakening the structural integrity of the crust. Westward subduction of the Pacific plate under East Gondwana is thought to have occurred from early in the Cretaceous (*c.* 120 million years ago; e.g. Sdrolias *et al.* 2004; Schellart *et al.* 2006; Matthews *et al.* 2011), but may have started much earlier (e.g. Rapela *et al.* 2005). The crust of eastern Australia and the Lord Howe Rise underwent a long period of extension with eruption of felsic magmas between 120 and 95 million years ago, long before any active rifting started (Bryan *et al.* 2012). Uplift and erosion of these felsic rocks produced significant volumes of volcano-clastic sediments, filling basins within and adjacent to the Lord Howe Rise (Crawford *et al.* 2003).

The Tasman Sea opened asymmetrically, with new ocean crust being first formed in the south (84 million years ago), before rifting propagated northward, achieving a final separation of the Melanesian Rift from East Gondwana when the Cato Trough (Fig. 1, CT) opened between 63 (Sdrolias *et al.* 2001) and 55 (Schellart *et al.* 2006) million years ago. Prior to this date, the Melanesian Rift was still attached to East Gondwana via the Marian Plateau (Fig. 1, MP). Schellart *et al.* (2006) suggested rollback of the Pacific plate as a likely mechanism for the formation of the Tasman Sea. Normally, a change from compressional (subduction) to tensional (rifting) regimes is achieved by back-arc spreading, but the asymmetric opening of the Tasman Sea makes this unlikely. During rollback, the subducting plate sinks faster than its subduction rate, causing the lower plate margin to migrate towards the ocean. This results in extension of the upper plate as it tries to adjust to the receding subduction zone. This model implies that the Pacific plate was subducting to the west, and that many of the basin structures on the Melanesian Rift originated with sea floor spreading in the Tasman Sea.

## Structural elements of the Melanesian Rift

### *Dampier–Monawai Ridge*

The Dampier–Monowai Ridge is a disrupted slice of continental crust, which thins to the north. It is between 100 and 80 km wide and was isolated from the Lord Howe Rise when marginal basins formed along its eastern margin between 79 and 71 million years ago (although Crawford *et al.* (2003) favoured a later separation at 64 million years ago) and from Australia at 73 million years ago (Gaina *et al.* 1998; Stagg *et al.* 1999). The basins to the east are underlain by stretched continental crust and may be thought of as parts of the Western Rift Province of the Lord Howe Rise (Keene *et al.* 2008).

### *Lord Howe Rise*

The Lord Howe Rise (Figs 1, 2, LHR), the major feature of the Melanesian Rift at ~2000 km long and up to 400 km wide, is divided into three north–south structural elements (Gaina *et al.* 1998). The Challenger Plateau (Fig. 1, CP) and western coast of the South Island, New Zealand, form the southern section, which is separated from the central Lord Howe Rise by the Bellona Trough (Fig. 1, BT), an early continental rift formed before sea-floor spreading at 83 million years ago in the Tasman Sea (Gaina *et al.* 1998). Wood and Woodward (2002) estimated that there was 30% crustal thinning in the Bellona Trough during this rifting phase. The Challenger Plateau is composed of Palaeozoic and Mesozoic granites and metamorphosed sediments equivalent to those found on the western coast of the South Island, New Zealand (Wood 1991). A granite sample dredged from the western margin had been hydrothermally altered at 95 million years ago (Tulloch

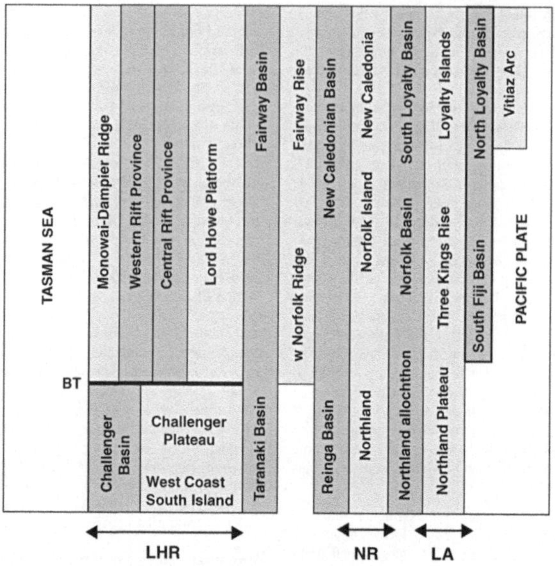


**Fig. 2.** Schematic map showing the relationships among the structural features shown in Fig. 1. BT, Bellona Trough; LA, Loyalty Arc; LHR, Lord Howe Rise; NR, Norfolk Ridge.

*et al.* 1991), indicating that the pre-rift volcanism described earlier extended to the southern parts of the Lord Howe Rise. The Challenger Plateau is only marginally thinned, with an estimated thickness of 24–25 km (Wood and Woodward 2002; Grobys *et al.* 2008). The Challenger Basin is found on the western margin of the Challenger Plateau, and is probably a lateral equivalent of the Western Rift Province (Fig. 2). The central and northern parts of the Lord Howe Rise are separated by a transform fault that corresponds to negative gravity anomalies between Australia and the Bass Straight, and which was also interpreted by Gaina *et al.* (1998) as an early, pre-continental breakup feature.

The northern and central sections of the Lord Howe Rise are structurally differentiated from west to east (Keene *et al.* 2008). The Western Rift Province basins contain up to 6 km of Cretaceous to Cenozoic sediments. The Lord Howe chain of hotspot volcanoes, found on the western margin, has been active from *c.* 30 million years ago (Nova Bank) to Recent (Lord Howe Island). All the islands are thought to have been subaerial while over the stationary hot spot (Quilty 1993). As the Australian plate moved northward, the islands became eroded and submerged. The Central Rift Province basins (Fig. 2) contain up to 4 km of Cretaceous to Cenozoic sediments. The basement is continental crust between 18 and 20 km thick. The easternmost division recognised by Keene *et al.* (2008) is the Lord Howe Platform (Fig. 2), a relatively non-stretched ribbon of continental crust that has only a thin covering of sediments. Klingelhoefer *et al.* (2007) estimated the crust of the Lord Howe Platform to be 23 km thick. Stretching of the Lord Howe Rise crust thus appears to increase to the west and north, which makes sense given the geometry of its detachment.

Deep-sea drill sites on the Lord Howe Rise have given interesting insights into its history. The occurrence 94 million years ago of rhyolite lavas at Deep Sea Drilling Project (DSDP) site 207 just north of the Bellona Trough (Mortimer 2004) confirms the involvement of the Lord Howe Rise in the extensive pre-rift volcanism discussed earlier. Two unconformities recognised in the sedimentary sequence at DSDP 207 also occur at sites to the north (DSDP 206 and 208), showing that these unconformities are regional. The older unconformity is between Late Palaeocene (55 million years ago) and Early Eocene (50 million years ago) at all three sites. The younger unconformity starts in the Late Eocene (40 million years ago) at all sites, but ends earlier in the north during the mid-Oligocene (30 million years ago) and later in the south during the mid-Miocene (15 million years ago). It is likely that much of the Lord Howe Rise was above water during these times and that subsidence proceeded from north to south. Sdrolias *et al.* (2003) suggested that the latter unconformity is a result of a major tectonic event and noted its synchronicity with the emplacement of the New Caledonian ophiolite (see below). The earlier period of uplift and erosion is consistent with westward dipping subduction along the eastern margin of the Lord Howe Rise during the Palaeocene.

### *Fairway Basin and Fairway Rise*

The Fairway Basin (Fig. 1, 2a) contains up to 4 km of sediment and is underlain by thinned continental crust some 12–15 km thick (Klingelhoefer *et al.* 2007). Lafoy *et al.* (2005*a*) considered the Fairway Basin to be a lateral equivalent of the Taranaki Basin (Fig. 1, 2c), which would make this structure a major feature along the entire eastern margin of the Lord Howe Rise. The Fairway Rise (Fig. 1, FR) is continental with crust estimated to be 22 km thick (Klingelhoefer *et al.* 2007). The ridge underwent uplift and probable subaerial exposure from the mid-Eocene (*c.* 50 million years ago) to early Oligocene (35 million years ago) in response to the obduction of the New Caledonian ophiolite (Klingelhoefer *et al.* 2007; Collot *et al.* 2008; Sutherland *et al.* 2010). Lafoy *et al.* (2005*a*) suggested that the Fairway Rise extends to the south-east as the West Norfolk Ridge (Fig. 1, wNR), which is of similar continental composition (Herzer *et al.* 1999) and thickness (Grobys *et al.* 2008).

### *New Caledonian Basin*

This enigmatic feature extends from New Caledonia into the New Zealand region (Fig. 1, 3). Whereas most constructions show the New Caledonian Basin merging into the Taranaki Basin, Lafoy *et al.* (2005*a*) suggested that its southern equivalent is the Reinga Basin (Fig. 1, 3c). There is little consensus as to its age, mechanism of formation, or the nature of the underlying crust. Age estimates range from early Cretaceous (e.g. Lafoy *et al.* 2005*b*) to mid-Eocene–Oligocene (Sutherland *et al.* 2010). Mechanisms of formation include continental rifting (e.g. Crawford *et al.* 2003), back-arc spreading (e.g. Schellart *et al.* 2006), or reactivation of older rifts and subsidence during plate boundary rearrangements (Sutherland *et al.* 2010). The underlying crust has been interpreted as stretched continental (e.g. Schellart *et al.* 2006), back-arc basin (e.g. Collot *et al.* 2008) or oceanic (e.g. Sutherland 1999).

An emerging view is that the New Caledonian Basin is a composite structure, and parts may well differ in age, composition and mode of formation. Collot *et al.* (2011) provided a summary of this multi-history view, in which they recognised three phases of development. An initial phase consisted of mid-Cretaceous (*c.* 90 million years ago) rifting of stretched continental crust in the northern New Caledonian Basin (Fig. 1, 3a), forming a relatively shallow trough. Collot *et al.* (2011) cited the volume of sediment-fill (Lafoy *et al.* 2005*b* estimated <8 km) and 10 km thick crust (Klingelhoefer *et al.* 2007) in support of the view that the underlying crust is not oceanic. In the second phase, the northern New Caledonian Basin deepened in response to compression associated with the westward obduction of the New Caledonian ophiolite, which caused uplift of the Fairway Ridge and rapid subsidence in the eastern part of the New Caledonian Basin at *c.* 40 million years ago. The third phase is associated with the regional subsidence in the Oligocene (*c.* 30 million years ago), during which the central parts of the New Caledonian Basin (Fig. 1, 3b) subsided to bathyl depths (see Sutherland *et al.* (2010) for a possible mechanism). According to Pelletier (2007), the central region has a basement of oceanic crust with magnetic lineations indicating an age of 62–56 million years. This view was supported by Lafoy *et al.* (2005*b*) who suggested a crustal thickness of 8–9 km in this central region, but not by Grobys *et al.* (2008) whose model indicated a uniform thickness of 14 km apart from thinner (11 km) crust between 35°S and 37°S.

### *Norfolk Ridge*

This continental ribbon (Fig. 1, NR) stretches ~1500 km from the d'Entrecasteaux Reef to northern New Zealand and varies between 70 and 100 km in width (Pelletier 2007). Grobys *et al.* (2008) model suggested crustal thickness of 21 km in the north, with thinner crust in the central sector. The island of New Caledonia (Fig. 1, NC) in the northern sector of the Norfolk Ridge has a Mesozoic geology, similar to that of the North Island, New Zealand (Collot *et al.* 2011). Norfolk Island (Fig. 1, NI) is found at the southern end of the northern Norfolk Ridge. It and the nearby Phillip Island are eroded remnants of Pliocene basaltic volcanoes active some 3 million years ago.

### *Loyalty Basin and Loyalty Arc*

The South Loyalty Basin (Fig. 1, 4) is interpreted as a back-arc structure formed over a westward dipping subduction zone east of the Loyalty Arc (Crawford *et al.* 2003; Sdrolias *et al.* 2004; Schellart *et al.* 2006; Collot *et al.* 2008, 2011). The oceanic crust of the South Loyalty Basin lacks any magnetic lineations to date it; however, on the basis of micropalaeontology and geochemistry (Collot *et al.* 2008, 2011), is thought to have formed during the Upper Cretaceous to Palaeocene, or perhaps into the earliest Eocene. Schellart *et al.* (2006) estimated that it may have been up to 750 km wide in the early Eocene, before plate boundary reorganisation in the mid-Eocene (*c.* 45 million years ago) and initiation of eastward subduction of the basin crust beneath the western margin of the Loyalty Arc. The subduction of south Loyalty Basin crust eventually brought the Loyalty Arc into contact with the Norfolk Ridge (see below).

In the reconstructions of Crawford *et al.* (2003), Sdrolias *et al.* (2004) and Schellart (2006), the South Loyalty Basin extended north into the New Guinea area and south into the New Zealand region. The remnants of the basin are now found as dispersed plateau regions in the Norfolk Basin (Fig. 1, 5a,b; Sdrolias *et al.* 2004; DiCaprio *et al.* 2009), and as obducted oceanic crust of the Northland Allochthon (Schellart 2006). Schellart *et al.* (2009) interpreted a lower mantle high-velocity anomaly as evidence of this fossil north-east dipping subduction zone stretching 2500 km between New Caledonia and New Zealand.

The Loyalty Arc extends both northward as the South d'Entrecasteaux Ridge (Schellart *et al.* 2006) and to the south as the Three Kings Rise (Fig. 1, TKR) and the Northland Plateau (Fig. 1, NP; Schellart 2006). The Loyalty Arc is now disrupted by the Cook and Vening–Meinesz fault zones, which are found to the north and south of the Three Kings Rise respectively. If one accepts the models of Crawford *et al.* (2003), Sdrolias *et al.* (2004) and Schellart *et al.* (2006), then the Loyalty Arc originated in the Cretaceous above a westward dipping subduction zone along the margin of north-eastern Gondwana. Evidence of Cretaceous continental basement has been deduced from geochemical signatures of lavas for the Three Kings Islands (Nicholson *et al.* 2008) and New Caledonia (Nicholson *et al.* 2011), and from dredged material from the Northland Plateau (Mortimer *et al.* 2008). According to DiCaprio *et al.* (2009), there is also Mesozoic continental crust trapped within the Norfolk Basin. An Eocene arc was built on this earlier structure as South Loyalty Basin crust was subducted eastward beneath it. Arc volcanism ceased as the Loyalty Arc collided with the Norfolk Ridge. This collision was asymmetric, starting earlier in the north at the end of the Eocene (40–34 million years ago), on the basis of the age of ophiolite emplacement in New Caledonia, during the Oligocene (30–23 million years ago) in the Three Kings Ridge region, on the basis of the dating of a collision orogeny recognised on the Three Kings Rise by Meffre *et al.* (2006), and in the Miocene (23–21 million years ago) in the south as the Northland Plateau collided with northern New Zealand, coincidental with the emplacement of the Northland Allochthon (Ballance and Spörli 1979, Mortimer *et al.* 2008). Following the collision of the Loyalty Arc with the Norfolk Ridge, subduction polarity switched to a westerly direction and the subduction zone migrated southward as the South Fiji Basin (Fig. 1, 7) and the modern Norfolk Basin opened as a result of back-arc extension (Herzer *et al.* 2011).

### *Vityaz Arc*

The Vityaz Arc (Fig. 1, VA) is composed of the Solomon Islands, Vanuatu and Fiji. This Eocene (45 million years ago) arc formed along the north-eastern margin of the Melanesian Rift (Crawford *et al.* 2003), as the westward subduction of Pacific plate resumed once more, following the collision of the Loyalty Arc with the Norfolk Ridge. This subduction eventually detached the proto-Vityaz Arc from the northern Loyalty Arc when the back-arc North Loyalty Basin (Fig. 1, 6) opened. As the North Loyalty Basin widened, the Vityaz Arc was translated eastward until the Solomon Islands collided with the Ontong Java Plateau (Seton *et al.* 2012). This collision caused a series of readjustments, leading to the opening of the North Fiji Basin (Fig. 1, 8) and clockwise rotation of Vanuatu back towards the Melanesian Rift at 12–10 million years ago (Greene *et al.* 1994). The Vityaz Arc became fragmented at this time, as Fiji became isolated and Vanuatu was translated westward. The present tectonic setting of Vanuatu provides a modern analogue of the tectonic history of the Loyalty Arc, and it too is destined to eventually collide with the Melanesian Rift margin as the remaining crust of the North Loyalty Basin is subducted beneath it. The Vityaz Arc is part of the Melanesian Arc, which extends westward into the New Guinea region and south, via the Kermadec Ridge, to the North Island, New Zealand (Coleman 1997).

## Biogeographic implications

The extent and distribution of emerged land is a critical factor in any biogeographic analysis of terrestrial biota. The key question, therefore, is whether the present configuration of limited land and general submergence to bathyl depths occurred soon after detachment? Rapid subsidence to bathyl depths of extensive areas of continental crust can be achieved only in passive tectonic environments and when the crust is stretched and thinned, otherwise it retains too much buoyancy. The evidence suggests that thinning of Melanesian Rift crust intensified to the north-west (i.e. towards the final detachment hinge) and areas such as the Challenger Plateau, Lord Howe Platform, and the ridge systems to the east remained essentially unaltered. So, although it is clear that there was a rapid and wide-scale subsidence to depth in western and north-western parts of the Melanesian Rift during the Palaeocene, lack of sediment cover, crustal thicknesses in excess of 20 km in southern and eastern areas, and an active subduction system along its eastern margin suggest that other parts of the Melanesian Rift did not experience regional subsidence. Uplift of the Lord Howe Rise in the Late Palaeocene to Early Eocene (53.5–50 million years ago) would have provided new land for surviving terrestrial biota to expand into, and promoted a period of biotic mobilisation.

The second period of regional uplift on the Lord Howe Rise occurred in the mid-Eocene at *c.* 45 million years ago and was related to the collision of the Loyalty Arc with the Norfolk Ridge. Once again, extensive areas of land would have become available for terrestrial organisms to colonise. The interesting aspect of this second period of regional uplift was that the subsequent subsidence occurred much earlier in the north (*c.* 30 Ma) than in the south (*c.* 20 Ma). One possible effect of this asymmetry was that the terrestrial biota would have been channelled southward as New Zealand moved northward as part of the Australian plate. The other area affected by the collision of the Loyalty Arc with the Norfolk Ridge was the Three Kings Rise, which experienced a collision-related orogeny that formed a large emergent island on its eastern flank between 38 and 21 million years ago.

Whereas uplift and basin closure would have promoted range expansion and the mixing of floras and faunas, the opening and deepening of sedimentary basins would have promoted biological isolation and speciation. The timing of these opening and subsidence events coincided with periods of uplift, which probably affected biogeographic patterning because deep basins could remain barriers to dispersal even in times of regional uplift. To complicate the matter still further, the geometry of Melanesian Rift basins, which are subparallel to the rift, would have hindered west–east range expansion while promoting north–south dispersal.

## Methods

The difficulties of constructing general areagrams have been extensively discussed in the literature (e.g. Nelson and Ladiges 1996; van Veller *et al.* 2000; Donoghue and Moore 2003; dos Santos 2011; Wiley and Lieberman 2011). In real-world situations, individual areagrams are unlikely to be congruent with each other and the problem, therefore, arises as to how best to deal with this incongruency. The main sources of incongruency are the occurrence of widespread taxa (multiple areas on a single terminal branch, or MASTs, for short), redundant areas (resulting in taxonomic paralogy), missing areas and inadequate methods of analysis (dos Santos 2011). Redundancy, the repeated occurrence of the same area in different branches on the areagram is nigh on universal and results in paralogous nodes. Such nodes represent a serious problem because they are non-random because of their increasing frequency towards the root of the areagram, yield no information about area relationships and obscure the real relationship between areas (Nelson and Ladiges 1996). Using only non-paralagous nodes removes both the noise caused by redundant areas and the inherent bias of paralogy. There has also been much debate about the methodology of dealing with MASTs and missing taxa (e.g. van Veller *et al.* 2000). Ebach *et al.* (2005) argued that the procedures for resolving paralogy and for solving distribution patterns should be separated because the implementation of Assumption 2 (Nelson and Platnick 1981) may itself produce paralagous nodes. These authors suggested a refinement of Assumption 2, termed the transparent method, for overcoming this problem. We present here a three-item analysis based on paralogy-free subtrees combined with the transparent method of Ebach *et al.* (2005).

A literature search for phylogenies involving taxa from Lord Howe Island, Norfolk Island, New Caledonia, Loyalty Island and Three Kings Islands produced a total of 78 phylogenies, the details of which are shown in Table 1, including their three- or four-letter acronyms. The phylogenies covered a wide range of taxa including plants (ferns, gymnosperms, monocots and dicots), invertebrates (molluscs, arachnids, myriapods, insects and crustacea) and vertebrates (reptiles, fish and birds). These phylogenies (or the relevant monophyletic clades) were then converted into areagrams by substituting areas for taxa. So as to increase the resolving power of the analysis and to work within the constraints imposed by an exhaustive algorithm for finding optimal trees, the number of areas used had to be limited. Large areas such as Australia, New Zealand and New Guinea were treated as single areas of endemism, despite the probability that they are not. Future studies are needed to refine area relationships within these areas to other Melanesian Rift and wider Pacific regions. The Solomon Islands, Vanuatu and Fiji were coded as 'Vityaz Arc', a geological structure that these islands were part of before the arc became disrupted. All other Polynesian or Micronesian distributions were coded as 'PAC', and any areas outside of the Pacific (e.g. Indian Ocean, Indonesia, Asia) were coded as 'OUT'. OUT-coded areas were pruned before the final analysis, because a preliminary analysis showed OUT to be basal to the general areagram, which does not need rooting with an outarea. This reduced the number of phylogenies by two (highlighted in Table 1) because they were reduced to two-area statements.

LisBeth ver. 1.3 (Bagils *et al.* 2012; http://infosyslab.fr/downloadlisbeth/LisBeth.exe, accessed 8 December 2016), which was used to analyse the data, has been developed to analyse three-item statements for biogeographic studies. LisBeth translates cladograms of taxa to taxon-area cladograms (TACs) by replacing terminal taxa with the areas of endemism where representatives are found. Lisbeth applies the paralogy-free subtree analysis (Nelson and Ladiges 1996) before the transparent method of Ebach *et al.* (2005), to resolve paralogy from the TACs, and extracts biogeographic three-item statements. Three-item statements are rooted trees that represent area relationships directly rather than in a matrix form. Optimal trees are constructed from all the three-item statements employing an exhaustive branch and bound algorithm using compatibility analysis. An intersection tree is formed from all the three-item statements common to the optimal trees, and is equivalent to a consensus tree.

## Results

The 76 phylogenies (Table 1) yielded 105 mast- and paralogy-free subtrees that were input as characters into LisBeth. These subtrees produced 166 three-item statements that resulted in the following two optimal trees:

$$((((((\mathrm{NZ}, \mathrm{TKI}), \mathrm{KER}), (\mathrm{LHI}, \mathrm{NOR})), \mathrm{PNG}), (\mathrm{PAC}, \mathrm{VIT})), (\mathrm{LOY}, \mathrm{NC})), \mathrm{AUS}) \text{ and}$$

$$(((((((\mathrm{NZ}, \mathrm{TKI}), \mathrm{KER}), (\mathrm{LHI}, \mathrm{NOR})), \mathrm{PAC}), \mathrm{PNG}), \mathrm{VIT}), (\mathrm{LOY}, \mathrm{NC})), \mathrm{AUS})$$

These two optimal trees yielded the intersection tree shown in Fig. 3. The intersection tree had a retention index of 0.58 and a completeness index of 0.53. In LisBeth, the retention index measures the proportion of three-item statements deduced from the primary hypothesis of homology that fit the intersection tree (Bagils *et al.* 2012). The completeness index is the proportion of three-item statements present in the intersection tree of the number of three-item statements from the intersection tree that are also present in the characters. These values showed that half of the primary homologies were found in the intersection tree, i.e. half of the characters supported the cladogram. The codes for the original phylogenies that support each node are also shown in Fig. 3.

## Discussion

### *Is the Melanesian Rift a natural biogeographic area?*

In vicariance biogeography, natural areas are defined by their endemic biotas (Rosen 1978; Parenti and Ebach 2009) and shared history (Michaux 2010; Ung *et al.* 2016). The results of the present study showed that the Melanesian Rift is not a natural biogeographic unit (i.e. it is not circumscribed by a common geological history and has its own endemic flora and fauna), but is a paraphyletic area because areas external to it (VIT, PAC, PNG) are nested within it. As such, it is not a real biogeographic entity in the sense that a paraphyletic taxon is not a real evolutionary entity. When Fig. 3 is converted to a geogram (East Gondwana (northern Norfolk Ridge (Vityaz Arc (New Guinea (central Melanesian Rift (Kermadec Ridge (southern Melanesian Rift)))))), some natural areas within it are apparent, including the northern section of the Norfolk Ridge represented by the islands of New Caledonia and the Loyalty Islands, the central section of the Melanesian Rift represented by Norfolk and Lord Howe islands and the southern section of New Zealand and the Three Kings Islands.

### *Area relationships*

Australia (East Gondwana) is the sister area to all other areas. Within the ingroup, significant aspects of the intersection tree include the following:

(1) the basal position of New Caledonia,
(2) the inclusion of the Vityaz Arc within a clade that includes New Guinea and the southern and central sections of the Melanesian Rift rather than with the Loyalty Islands,

**Table 1. Details of phylogenies used in the study**

NZ, New Zealand; TKI, Three Kings Islands; KER, Kermadec Islands; LHI, Lord Howe Island LOY, Loyalty Islands NC, New Caledonia; NOR, Norfolk Island; PAC, Polynesia + Micronesia; AUS, Australia + Tasmania, PNG, New Guinea; OUT, areas outside Australia, Pacific, and Melanesian Rift; VIT, Vityaz Arc (Solomons + Vanuatu + Fiji)

| Taxon | Code | Clade size | Area cladogram | Reference |
|---|---|---|---|---|
| Plants | | | | |
| Ferns | | | | |
| *Grammitis* | GRAM | 7 spp. | ((NC,LHI)(AUS(AUS(NZ(NZ+AUS,TAS))))) | Papadopulos *et al.* (2011) |
| *Hymenophyllum* | HYM | 3 spp. | (NC(AUS,LHI)) | Papadopulos *et al.* (2011) |
| *Hymenophyllum* | HYM | 3 spp. | (LHI(AUS,NZ))) | Papadopulos *et al.* (2011) |
| *Polystichum* | POL | 5 spp. | ((AUS(AUS(LHI,AUS)))NZ) | Papadopulos *et al.* (2011) |
| *Polystichum* | POL | 3 spp. | ((NZ,LHI)AUS) | Perrie *et al.* (2003) |
| Tree ferns | | | | |
| Cyatheaceae | CYA | 5 spp. | ((PNG(LHI,NOR)OUT)NZ) | Korall *et al.* (2007) |
| Cyatheaceae | CYA | 3 spp. | (NC(LHI,AUS)) | Korall *et al.* (2007) |
| Cyatheaceae | CYA | 4 spp. | ((PNG,NZ)(NC,AUS)) | Korall *et al.* (2007) |
| Cyatheaceae | CYA | 4 spp. | (((AUS,NZ)LHI)NC) | Korall *et al.* (2007) |
| Cyatheaceae | CYA | 7 spp. | (NC(OUT(OUT,PNG))(LHI(AUS,NZ))) | Korall and Pryer (2014) |
| Cyatheaceae | CYA | 8 spp. | (NC(OUT(OUT(NZ+TKI+VIT+PAC((LHI,NOR)OUT, PNG))))) | Korall and Pryer (2014) |
| Cyatheaceae | CYA | 3 spp. | ((LHI,AUS)NC) | Korall and Pryer (2014) |
| Gymnosperms | | | | |
| *Agathis* | AGA | 5 spp. | ((VIT+AUS(AUS,OUT)NC)NZ) | Stöckler *et al.* (2002) |
| *Araucaria* | ARA | 4 spp. | (AUS(NOR,NC)) | Givnish and Renner (2004) |
| *Araucaria* | ARA | 3 spp. | ((OUT,AUS)((NC,NOR)AUS+PNG)) | Stefenon *et al.* (2006) |
| Monocots | | | | |
| *Dendrobium* | DEN | 7 spp. | (AUS((AUS,LHI)(AUS(AUS(AUS,AUS+NC+LHI))))) | Papadopulos *et al.* (2011) |
| *Howea* | HOW | 5 spp. | (LHI(NC(OUT+PAC,VIT))) | Papadopulos *et al.* (2011) |
| *Howea* | HOW | 4 spp. | ((AUS(PNG,AUS))(LHI,AUS)) | Papadopulos *et al.* (2011) |
| Eudicots | | | | |
| Casuarinaceae | CAS | 8 spp. | (OUT(AUS(AUS(NC(AUS,OUT))))(NC(OUT,AUS))) | Steane *et al.* (2003) |
| Chloranthaceae | CHL | 6 spp. | (OUT(NC,OUT(NZ(VIT,PAC)))) | Zhang and Renner (2003) |
| *Coprosma* | COP | 3 spp. | (NZ(LHI,PAC)) | Papadopulos *et al.* (2011) |
| *Coprosma* | COP | 8 spp. | (VIT(LHI((NZ(NZ,AUS))(NZ(PAC,AUS))))) | Papadopulos *et al.* (2011) |
| *Dracophyllum* | DRA | 5 spp. | (((NZ,NC)TAS)(LHI,AUS)) | Wagstaff *et al.* (2010) |
| *Ficus* | FIC | 7 spp. | ((AUS(PNG(AUS,VIT)))(AUS(AUS,LHI))) | Rønsted *et al.* (2008) |
| Gesneriaceae | GES | 6 spp. | (((((OUT,AUS)OUT)(AUS(LHI,NC)))(NC,NZ))OUT) | Woo *et al.* 2011 |
| *Ixora* | IXO | 6 spp. | (OUT(NC(VIT(NC(PAC(NC,OUT)))))) | Mouly *et al.* (2009) |
| *Korthalsella* | KOR | 6 spp. | (AUS+PNG(NZ(OUT(PAC(LHI,LHI+AUS))))) | Papadopulos *et al.* (2011) |
| *Macropiper* | KOR | 6 spp. | (LHI+NOR+NZ+TKI(LHI((PAC,TKI)(VIT,PAC)))) | Papadopulos *et al.* (2011) |
| *Melicope* | MELI | 3 spp. | (LHI(PAC,NZ)) | Papadopulos *et al.* (2011) |
| *Melicope* | MELI | 6 spp. | (LHI(AUS,AUS+OUT,OUT,PNG,AUS+PNG+VIT) | Papadopulos *et al.* (2011) |
| *Melicytus* | MEL | 16 spp. | ((((NZ+TAS,NZ,AUS+LHI,NZ)(NZ,NOR))NZ)((NOR (KER(NZ,TKI)))((NZ,VIT)NZ)(NC,AUS))) | Mitchell *et al.* (2009) |
| *Metrosideros* | MET | 7 spp. | (NZ(PAC+VIT(PAC(NZ(LHI(LHI,TKI+NZ)))))) | Papadopulos *et al.* (2011) |
| *Metrosideros* | MET | 5 spp. | (PAC,VIT(LHI(NZ,KER))) | Wright *et al.* (2000) |
| *Metrosideros* | MET | 3 spp. | (PAC(NZ,LHI)) | Percy *et al.* (2008) |
| *Myrsine* | MYR | 8 spp. | ((LHI,LHI+OUT)((AUS,OUT)(NZ((TKI,KER)NZ)))) | Papadopulos *et al.* (2011) |
| *Nothofagus* | NOT | 6 spp. | ((PNG(PNG(PNG,NC))(NZ,AUS,OUT)) | Heenan and Smissen (2013) |
| *Olearia* | OLE | 3 spp. | ((NZ,LHI)AUS) | Papadopulos *et al.* (2011) |
| *Olearia* | OLE | 5 spp. | (AUS(TAS((AUS,LHI)AUS))) | Papadopulos *et al.* (2011) |
| Pennantia | PEN | 4 spp. | (AUS(TKI(NZ,NOR) | Landcare Research (2015) |
| Pittosporum | PIT | 6 spp. | (((((PAC,NC)NC)NZ,PAC)OUT)AUS) | Gemmill *et al.* 2002 |
| Sapotaceae | SAP | 16 spp. | (((AUS(AUS,LHI))(((NZ,AUS)(AUS,NC))(OUT(VIT, PAC)))))(PNG((VIT,PAC)(PNG(PNG,PAC))))) | Swenson *et al.* (2013) |
| *Santalum* | SAN | 8 spp. | (AUS((PAC(PNG(AUS, PAC)))(AUS(NC+LOY,VAN)))) | Harbaugh and Baldwin (2007) |
| *Vitex* | VIE | 5 spp. | ((NC((NZ,AUS)NC))(OUT(OUT,AUS))) | Pratt 2013 |
| *Xylosoma* | XYL | 3 spp. | (OUT(LHI(NC,LHI))) | Papadopulos *et al.* (2011) |
| Invertebrates | | | | |
| Mollusca | | | | |
| Pectinidae | PEC | 5 spp. | ((((PAC,LOY)VAN+LOY)LOY)AUS) | Puslednik and Serb (2008) |
| Pectinidae | PEC | 7 spp. | (((LOY,PNG)LOY)OUT)((LOY,PAC)PAC) | Puslednik and Serb (2008) |

(*continued next page*)

**Table 1.** (*continued*)

| Taxon | Code | Clade size | Area cladogram | Reference |
|---|---|---|---|---|
| *Placostylus* | PLA | 4 spp. | (TKI(NC(LHI,NZ))) | Ponder *et al.* (2003) |
| Oligochaeta | | | | |
| Megascolecidae | MEG | 10 spp. | ((((AUS,NI)TAS)NZ)(NZ(AUS,TKI))(OUT(NC,AUS))) | Buckley *et al.* (2011) |
| Megascolecidae | MEG | 7 spp. | ((NC,NZ)(NZ,AUS)) | Buckley *et al.* (2011) |
| Arachnida | | | | |
| Thanisidae | THA | 4 spp. | (AUS(TKI,NZ))) | Sirvid *et al.* (2013) |
| Myriapoda | | | | |
| *Henicops* | HEN | 3 spp. | (((NZ+AUS)LHI)NC) | Edgecombe *et al.* (2006) |
| Insecta | | | | |
| Orthoptera | | | | |
| Grylloidea | GRY | 3 spp. | (PNG,NC(LOY,VIT)) | Nattier *et al.* (2011) |
| Grylloidea | GRY | 4 spp. | ((VAN(LOY,VIT))NC) | Nattier *et al.* (2011) |
| Hemiptera | | | | |
| Anostostomatidae | ANO | 4 spp. | (AUS(NZ(NC,NZ))) | Pratt *et al.* (2008) |
| Cicadidae | CIC | 4 spp. | ((NZ,NC)(NZ,AUS)) | Buckley *et al.* (2002) |
| *Kikihia* | KIK | 4 spp. | (((NZ,KER)NOR)NZ) | Arensburger *et al.* (2004*b*) |
| Peloridiidae | PEL | 6 spp. | (((NZ,NC)NZ)(OUT(AUS,AUS+LHI))) | Burckhardt (2009) |
| Phasmatodea | | | | |
| Dytiscidae | DYS | 8 spp. | ((OUT(VIT(AUS+PNG,VIT)))(OUT((NC,PNG)(NC, PAC)))) | Balke *et al.* (2007) |
| Eurycanthinae | PHA | 10 spp. | (((AUS,OUT)(AUS(AUS,LHI)))((AUS,VIT(NC(NC, NZ))))) | Buckley *et al.* (2009) |
| Lanceocercata | LAN | 10 spp. | ((AUS+PAC,AUS)((AUS((AUS,VIT)(AUS,PNG)))(AUS (AUS,LHI)))) | Buckley *et al.* (2010) |
| Trogossitidae | TRO | 9 spp. | (AUS(NC(NZ(LHI(NC(TKI(NZ(NZ(TKI,OUT)))))))) | Leschen and Lackner (2013) |
| Diptera | | | | |
| *Simulium (Inseliellum)* | SIM | 3 spp. | (VIT(NC(PAC))) | Craig *et al.* (2001) |
| Hymenoptera | | | | |
| *Leptomyrmex* | LEP | 6 spp. | ((PAC,AUS)((PNG,AUS)(AUS,NC))) | Lucky (2011) |
| *Encarsia* | ENC | 5 spp. | ((AUS,NZ)(PAC(PAC,AUS))) | Schmidt *et al.* (2006) |
| Lepidoptera | | | | |
| Micropterigidae | MIC | 7 spp. | (((((NC,NZ)NZ)NZ)OUT)(AUS(NZ,NC))) | Gibbs 2010 |
| Crustacea | | | | |
| *Paratya* | PAR | 5 spp. | (NZ((NC(LHI,NOR))AUS)) | Page *et al.* (2005) |
| Vertebrates | | | | |
| Fish | | | | |
| Gobiidae | GOB | 17 spp. | ((((PAC,NC)(NC,VIT))(VIT,PAC)((PAC,VIT)(PAC, NC))))((OUT,VIT)(OUT(PAC,VIT)(VIT,NC)))) | Keith *et al.* (2011) |
| Reptiles | | | | |
| Elapidae | ELA | 10 spp. | ((NC+LOY,VIT)((PNG,VIT),PAC,OUT))(((NC(LOY, VIT))VIT)OUT)) | Lane and Shine (2011) |
| Geckonidae | GEC | 4 spp. | ((AUS,NC)(AUS,NZ)) | Nielsen *et al.* (2011) |
| Geckonidae | GEC | 5 spp. | ((NC,AUS)(AUS(PNG,AUS))) | Gamble *et al.* (2008) |
| Scincidae | SCI | 5 spp. | ((NC,NC(LHI,NZ))NC) | Smith *et al.* (2007) |
| Scincidae | SCN | 4 spp. | (AUS(NC(LHI,NZ))) | Chapple *et al.* (2009) |
| Birds | | | | |
| Columbidae | | | | |
| *Hemiphaga* | HEM | 5 spp. | (((NZ,NOR)NZ)(PNG,AUS)) | Goldberg *et al.* (2011) |
| Psittacidae | | | | |
| *Cyanoramphus* | CYAN | 4 spp. | (((NZ,NOR)NZ)NC) | Boon *et al.* (2001) |
| Platycercinae | PLT | 7 spp. | (AUS((AUS(VIT(NZ(LOY,NC))))AUS)) | Schweizer *et al.* (2013) |
| Strigidae | | | | |
| *Ninox* | NIN | 4 spp. | (AUS(TAS(NOR,NZ) | Wink *et al.* (2008) |
| Aegothelidae | | | | |
| *Aegotheles* | AEG | 8 spp. | ((PNG(OUT(PNG(PNG,AUS)))PNG)(NC,NZ)) | Dumbacher *et al.* (2003) |
| Acanthidsidae | | | | |
| *Gerygone* | GER | 12 spp. | ((PNG,AUS)((OUT,PNG)(AUS(OUT(AUS(VIT(AUS(AUS (PNG(NZ,NOR))))))))) | Nyári (2011) |

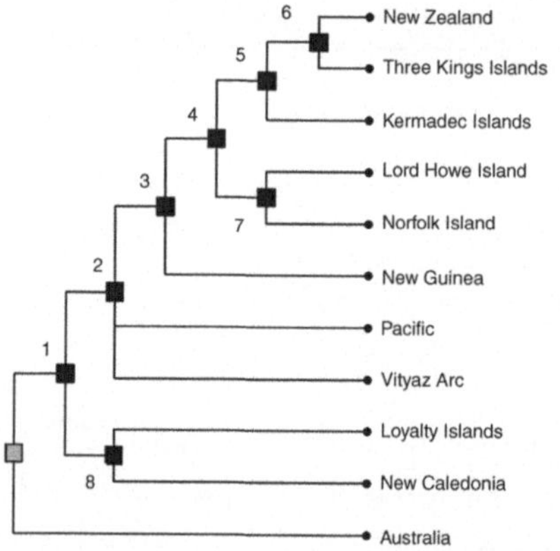


**Fig. 3.** General areagram of the Melanesian Rift and adjacent areas. RI = 0.58; completeness index = 53%. Taxa supporting each node with codes as in Table 1. 1, TRO, GRAM, PAR, DYS, ARA, AEG, SCN, MIC, PIT, GES, PLT, NOT, DRA, SAN, PEC, PEL, ANO, PHA; 2, SAP, ELA, CYA, PEC, HOW, CHL, PIT, MEL, PLT, KOR; 3, VIE, CYA, NOT, ARA; 4, TRO, POL, SCI, CYA, HEM, CYAN, NIN, SCN, GER, PEN, MEL, MET, KOR, COP, OLE, PLA, HEN; 5, GRAM, MYR, MEL, DRA, KIK, MET, NIN, ; 6, TRO, PEN, MYR, MEL, MET, THA; 7, CYA, PAR; 8, ELA, PLT.

(3) New Guinea as sister area to southern and central sections of the Melanesia Rift, and
(4) the sister area relationship between central and southern areas of the Melanesian Rift.

The general areagram showed that all Pacific islands marginal to the Australian craton are more closely related to each other than they are to Australia. The northern Melanesian Rift, represented by the modern island of New Caledonia, is basal to all other ingroup areas, suggesting that it has had a long and unique geological history, and that its highly endemic flora and fauna are ancient as previously suggested by several biogeographers (e.g. Haase and Bouchet 1998; Jaffre *et al.* 1997; Ladiges and Cantrill 2007; Heads 2008; Swenson *et al.* 2013). The Loyalty Islands are more closely related to New Caledonia than to the Vityaz Arc, despite the earlier separation of the Loyalty Islands from New Caledonia (opening of the South Loyalty Basin) than the separation of the Vityaz Arc from the Loyalty Islands (opening of the North Loyalty Basin), and despite the present separation from New Caledonia by a deep, trench-like oceanic structure.

### *Dating nodes of the general areagram*

The general areagram represents relationships among areas based on the biological relationships deduced from their flora and faunas. How one interprets any general areagram, which is a necessary prerequisite for comparison of relationships among the same areas on the basis of geology, is not obvious. In a traditional cladogram, synapomorphies are attached to branches and nodes represent common ancestors (or later the hypothetical taxonomic unit). What do the branches and nodes of an areagram represent? There appear to be two alternative interpretations of areagram nodes in the literature. One interpretation holds them to represent ancestral areas, echoing Hennig's idea of node as the common ancestor (e.g. Bremer 1992; Ronquist 1994; Hausdorf 1998), the other interprets them as a vicariant event that splits the ancestral area, giving rise to two sister areas (e.g. Hovenkamp 1997). It appears to us that these are two mutually exclusive interpretations. In our view, the nodes in Fig. 3 are best interpreted as range-splitting events, such as basin formation, and branches between nodes as ancestral ranges. This allows the tectonic history described earlier to be mapped onto the areagram and a comparison between the two sets of data to be made, i.e. how many nodes conform to any proposed tectonic history.

We agree with the analysis of Heads (2014*b*) that all potential dating methods are problematic. Published ages of molecular-based phylogenies consistently underestimate clade ages. These ages are usually calculated using one of two general methods, namely, extrapolation of known base substitution rates or using the age of the oldest fossil, often combined with the idea of clock-like change, although some substitution models now allow for different rate change over different parts of the phylogeny. Both of these general methods are built on such unrealistic assumptions that neither can be expected to yield accurate estimates of divergence times. Using substitution rates derived from modern taxa and then applying them over evolutionary time, often to groups only distantly related, is not justifiable, no matter how sophisticated the analysis might be. The same can be said for dating phylogenies using the age of the oldest fossil, which, despite giving only a minimum age for divergence, becomes a maximum estimate by proxy (Heads 2014*b*). Using tectonic events to calibrate phylogenies is not without its problems either, including the uncertainty of age estimates for any proposed tectonic event, the probable episodic nature of such processes and the difficulty of applying tectonic dating to individual phylogenies because any cladistic signal will become clearer as multiple phylogenies are processed and combined into general areagrams. Potentially, it is possible to date all internal nodes and then observe whether they are consistent or not; i.e. dates of more derived nodes are younger than older nodes.

The least ambiguous node to date is Node 3 in Fig. 3, which shows New Guinea as sister area to the central and southern Melanesian Rift. New Guinea has long been known as a geologically composite area (Pigram and Davies 1987), with the terranes of the Central Highlands and the South-east Peninsula being linked to the Melanesian Rift (Croizat 1964; Michaux 1989, 1994, 1998; Heads 2002, 2014*b*). The vicariant event that disrupted the ancestral area Central Highlands and the South-east Peninsula + central and southern Lord Howe Rise was the opening of the Coral Sea (Fig. 1, CS) between 61 and 52 million years ago (Baldwin *et al.* 2012). This date for Node 3 sets a minimum age for the earlier nodes and a maximum for later nodes.

Although the basal position of New Caledonia + Loyalty Islands with respect to other areas of the Melanesian Rift supports an early opening of the north New Caledonia Basin (Fig. 1, 3a) and suggests that the Norfolk Ridge (Fig. 1, NR) to the south of New Caledonia was also submerged at an early date, dating of Node 1 is not possible because of the uncertain tectonic history of the New Caledonian Basin. However, its age is constrained to a lower limit of 85 and 63 million years ago (Fig. 3, basal node = opening of the Tasman Sea) and an upper limit set by the age of Node 2.

Node 2 in Fig. 3 is an anomaly. It forms a trichotomy between the Vityaz Arc, Polynesia–Micronesia (PAC), and the southern and central sections of the Melanesian Rift. The areas coded as 'PAC' are almost certain to be polyphyletic and their relationships to areas within the Melanesian Rift or to the Vityaz Arc are beyond the resolution power of the present study. The Vityaz Arc was formed at 45 million years ago, setting a maximum date for Node 2 that is younger than the date established for Node 3. If the geology is wrong, then there must have been an earlier vicariance event, which would be evinced by the presence of both older continental basement beneath the Eocene arc and oceanic crust that pre-dates the opening of the North Loyalty Basin (Fig. 1, 6). Geological hypotheses also indicate that the Vityaz Arc was formed adjacent to New Caledonia, but the biology suggests that the flora and fauna were derived from other areas of the Melanesian Rift. Alternatively, the position of the Vityaz Arc in the areagram may be incorrect and future studies incorporating fresh data will resolve the anomaly.

The branch linking Nodes 4 and 7 represents the ancestral area of Lord Howe plus Norfolk Island (central Melanesian Rift). Both islands are young and neither existed even 3 million years ago; however, the structures they are part of, namely, the Norfolk Ridge (Fig. 1, NR) and central Lord Howe Rise (Fig. 1, LHR), did. There are three possible times when the ancestral area of central Melanesian Rift existed. It could be primitive, that is dating from pre-rift times, or a result from regional uplift experienced between *c.* 55 and 50 million years ago and again between *c.* 45 and 30 million years ago. The earliest date can be ruled out because it would be incompatible with the age of Node 3 (61–52 million years ago). Node 7 is associated with the opening (or deepening) of the central part of the Fairway Basin (Fig. 1, 2b) between the Lord Howe Rise and the west Norfolk Ridge (Fig. 1, wNR). No date is available for the opening of this part of the Fairway Basin; however, it is predicted to date from either *c.* 50 million years ago or *c.* 30 million years ago, depending on which of the uplift events were responsible. The results of the present study lend support to a period of biotic mobilism on the Melanesian Rift during the Cenozoic uplift of the central and southern sections of the Melanesian Rift. Node 4 is linked to marine transgressions within the Taranaki and Reinga basins (Fig. 1, 2c and 3c) that had probably extended far enough to the south by *c.* 55–50 million years ago to isolate the southern and central parts of the Melanesian Rift.

Node 5 is associated with the opening of the South Fiji Basin (Fig. 1, 7) between 31 and 25 million years ago and Node 6 with movement along the Vening–Meinesz Fracture Zone between 28 and 25 million years ago (Herzer *et al.* 2011) or 26–22 million years ago (Mortimer *et al.* 2007). This latter event separated the Three Kings Ridge (Fig. 1, TKR) from northern North Island, New Zealand, and more specifically from the Northland Plateau (Fig. 1, NP). Whereas it was not possible to investigate relationships between areas of endemism within New Zealand and its offshore islands in this study, there is a small but growing literature that could form the basis of such a comprehensive biogeographic analysis. The Three Kings Islands, the only part of the Three Kings Rise (Fig. 1, TKR) above water, have been the subject of a recent study by Buckley and Leschen (2013). These authors looked at the phylogeny of five beetle and one stick insect lineages that are found on the Three Kings Islands and elsewhere in New Zealand. They concluded that the Three Kings Islands show two types of relationship; one a sister area to the North Cape (northern tip of the North Island, New Zealand) and the other as the outarea to New Zealand as a whole. Age estimates for the separation of the Three Kings Islands from mainland New Zealand, on the basis of the molecular phylogeny of the beetle *Brachynopus scutellaris* by using previously calculated substitution rates, ranged from 24 to 2.24 million years ago. Even though the authors used complex substitution models to help overcome some of the problems inherent in using fixed substitution rates over multiple taxa for long time periods, we favour the older estimate because it coincides with the activation of the Vening–Meinez Fracture Zone and, therefore, is congruent with a tectonic calibration. Similar biogeographic patterning has also been found for geckos (Nielsen *et al.* 2011), skinks (Chapple *et al.* 2009), earthworms (Buckley *et al.* 2011), land snails (Spencer *et al.* 2006) and cicadas (Marshall *et al.* 2012).

Node 8 dates the split between New Caledonia and the Loyalty Islands. There are two possible dates for Node 8, depending on whether the vicariant event can be traced to either the opening of the South Loyalty Basin (*c.* 70–60 million years ago) or following the obduction of the New Caledonian ophiolites (40–34 million years ago) when the Loyalty Arc collided with New Caledonia.

## Conclusions

A large micro-continent was rifted from the north-eastern margin of East Gondwana during the Upper Cretaceous, which probably became biologically isolated (*c.* 70 million years ago) before complete separation (*c.* 63 million years ago) from the Australian sector of East Gondwana. Regional tension resulted in a series of subparallel rift basins forming barriers to dispersal and promoting local endemism, for example the early (*c.* 70 million years ago) isolation of the northern section of the Norfolk Ridge. The north-western region of the micro-continent was probably submerged completely because of the substantially thinned crust and distance from any tectonically active boundary, but there was little crustal thinning in the east and south and there was an active subduction zone along the eastern margin of the micro-continent. Land area on the Lord Howe Platform, Norfolk Ridge and Challenger Plateau would have been reduced as the micro-continent moved away from East Gondwana. Land probably existed as a series of isolated islands in the east and south until *c.* 53 million years ago when the micro-continent experienced a short period (3 million years) of regional uplift. This event, probably related to change in subduction rate, would have increased available land area and promoted mobilism and mixing of previously differentiated floras and faunas. How extensive this mixing was depended on the efficacy of the subparallel rift basins to act as barriers to dispersal.

Subsequent regional subsidence was again arrested at *c.* 45 million years ago when a second period of regional uplift was recorded, this time as a consequence of the collision of the Loyalty Arc with the Norfolk Ridge. The modern island of New Caledonia dates from this time, forming a refuge for an archaic northern Melanesian Rift flora and fauna that have survived to the present day. Subduction of the South Loyalty Basin beneath the Norfolk Ridge was oblique and, as the collision zone moved south, land began to subside once more, starting at 30 million years ago in the north and finishing at *c.* 20 million years ago in the south. This pattern of subsidence would have resulted in flora and fauna being funnelled southward, a pattern undoubtedly enhanced by the geometry of the then flooded rift basins. As tectonic activity migrated southward, the central and northern sections of the Melanesian Rift subsided to their present depths. Hot-spot activity along the western margin of the micro-continent from 30 million years ago produced land to preserve a central Melanesian Rift flora and fauna as that area foundered, as did the volcanic activity from 3 million years ago that formed Phillip and Norfolk Islands on the Norfolk Ridge.

Fundamental rearrangement of plate boundaries from 20 million years ago resulted in further fragmentation of the southern sector of the micro-continent as the South Fiji and Norfolk basins opened, and the Cook and Vening–Meinesz fracture zones became active. Future work will have to untangle the precise relationship between the Three Kings Rise, other sectors of the Loyalty Arc and northern New Zealand.

## Acknowledgements

V. Ung is supported by the Centre National de la Recherche Scientifique (CNRS). R. A. B. Leschen was supported in part by the New Zealand Ministry for Science and Innovation through the 'Defining New Zealand's Land Biota' programme.

## References

Arensburger P, Buckley TR, Simon C, Moulds M, Holsinger KE (2004*a*) Biogeography and phylogeny of the New Zealand cicada genera (Homoptera: Cicadidae) based on nuclear and mitochondrial DNA data. *Journal of Biogeography* **31**, 557–569. doi:10.1046/j.1365-2699.2003.01012.x

Arensburger P, Simon C, Holsinger K (2004*b*) Evolution and phylogeny of the New Zealand cicada genus *Kikihia* Dugdale (Homoptera: Auchenorrhyncha: Cicadidae) with special reference to the origin of the Kermadec and Norfolk islands' species. *Journal of Biogeography* **31**, 1769–1783. doi:10.1111/j.1365-2699.2004.01098.x

Bagils RZ, Ung V, Grand A, Vignes-Lebbe R, Cao N, Ducasse J (2012) LisBeth: new cladistics for phylogenetics and biogeography. *Comptes Rendus. Palévol* **11**, 563–566. doi:10.1016/j.crpv.2012.07.002

Baldwin SL, Fitzgerald PG, Webb LE (2012) Tectonics of the New Guinea Region. *Annual Review of Earth and Planetary Sciences* **40**, 495–520. doi:10.1146/annurev-earth-040809-152540

Balke M, Wewalka G, Alarie Y, Ribera I (2007) Molecular phylogeny of Pacific Island Colymbetinae: radiation of New Caledonian and Fijian species (Coleoptera, Dytiscidae). *Zoologica Scripta* **36**, 173–200.

Ballance PF, Spörli KB (1979) Northland allochthon. *Journal of the Royal Society of New Zealand* **9**, 259–275. doi:10.1080/03036758.1979.10419416

Boon WM, Kearvell JC, Daugherty CH, Chambers GK (2001) Molecular systematics and conservation of kakariki (*Cyanoramphus* spp.). *Science for Conservation* **176**, 1–46.

Bremer K (1992) Ancestral areas: a cladistic reinterpretation of the center of origin concept. *Systematic Biology* **41**, 436–445. doi:10.1093/sysbio/41.4.436

Bryan SE, Cook AG, Allen CM, Siegel C, Purdy DJ, Greentree JS, Tonguc IU (2012) Early–mid Cretaceous tectonic evolution of eastern Gondwana: from silicic LIP magmatism to continental rupture. *Episodes* **35**, 142–152.

Buckley TR, Leschen RAB (2013) Comparative phylogenetic analyses reveals long-term isolation of lineages on the Three Kings Islands, New Zealand. *Biological Journal of the Linnean Society. Linnean Society of London* **108**, 361–377. doi:10.1111/j.1095-8312.2012.02009.x

Buckley TR, Arensbburger P, Simon C, Chambers GK (2002) Combined data, Bayesian phylogenetics, and the origin of the New Zealand cicada genera. *Systematic Biology* **51**, 4–18. doi:10.1080/106351502753475844

Buckley TR, Attanayake D, Bradler S (2009) Extreme convergence in stick insect evolution: phylogenetic placement of the Lord Howe Island tree lobster. *Proceedings. Biological Sciences* **276**, 1055–1062. doi:10.1098/rspb.2008.1552

Buckley TR, Attanayake D, Nylander JAA, Bradler S (2010) The phylogenetic placement and biogeographical origins of the New Zealand stick insects (Phasmatodea). *Systematic Entomology* **35**, 207–225. doi:10.1111/j.1365-3113.2009.00505.x

Buckley TR, James S, Allwood J, Bartlam S, Howitt R, Prada D (2011) Phylogenetic analysis of New Zealand earthworms (Oligochaeta: Megascolecidae) reveals ancient clades and cryptic taxonomic diversity. *Molecular Phylogenetics and Evolution* **58**, 85–96. doi:10.1016/j.ympev.2010.09.024

Burckhardt D (2009) Taxonomy and phylogeny of the Gondwanan moss bugs or Peloridiidae (Hemiptera: Coleorrhyncha). *Deutsche Entomologische Zeitschrift* **56**, 173–235. doi:10.1002/mmnd.200900019

Chapple DG, Ritchie PA, Daugherty CH (2009) Origin, diversification and systematics of the New Zealand skink fauna (Reptilia: Scincidae). *Molecular Phylogenetics and Evolution* **52**, 470–487. doi:10.1016/j.ympev.2009.03.021

Coleman PJ (1997) Australia and the Melanesian arcs: a review of tectonic settings. *AGSO Journal of Australian Geology & Geophysics* **17**, 113–125.

Collot J, Géli L, Lafoy Y, Vially R, Cluzel D, Klingelhoefer F, Nouzé H (2008) Tectonic history of northern New Caledonian Basin from deep offshore seismic reflection: relation to late Eocene obduction in New Caledonia, Southwest Pacific. *Tectonics* **27**, TC6006. doi:10.1029/2008TC002263

Collot J, Lafoy Y, Géli L (2011) 'Explanatory Notes of the Structural Provinces of the Southwest Pacific Map.' (Geological Society of New Caledonia: Nouméa, New Caledonia)

Craig DA, Currie DC, Joy DA (2001) Geographical history of the central-western Pacific black fly subgenus *Inseliellum* (Diptera: Simuliidae: *Simulium*) based on a reconstructed phylogeny of the species, hot-spot archipelagos and hydrological considerations. *Journal of Biogeography* **28**, 1101–1127. doi:10.1046/j.1365-2699.2001.00619.x

Crawford AJ, Meffre S, Symonds PA (2003) 120 to 0 Ma tectonic evolution of the Southwest Pacific and analogous geological evolution of the 600 to 220 Ma Tasman Fold Belt System. In 'Geological Society of Australia Special Publication 22', pp. 377–397. (Geological Society of Australia: Sydney, NSW, Australia)

Croizat L (1964) 'Space, Time, Form. The Biological Synthesis.' (Self-Published: Caracus, Venezuela)

Darwin C (1903) 'More Letters of Charles Darwin. Vol.1.' (Ed. F. Darwin) (John Murray: London, UK)

de Queiroz A (2014) 'The Monkey's Journey: how Improbable Journeys Shaped the History of Life.' (Basic Books: New York, NY, USA)

DiCaprio L, Müller RD, Gurnis M, Goncharov A (2009) Linking active margin dynamics to overriding plate deformation. Synthesizing geological images with geological data from the Norfolk Basin. *Geochemistry Geophysics Geosystems* **10**, Q01004. doi:10.1029/2008GC002222

Donoghue MJ, Moore BR (2003) Toward an integrative historical biogeography. *Integrative and Comparative Biology* **43**, 261–270. doi:10.1093/icb/43.2.261

dos Santos CMD (2011) On the role of assumptions in cladistic biogeographical analyses. *Papéis Avulsos de Zoologia* **51**, 295–305. doi:10.1590/S0031-10492011001900001

Dumbacher JP, Pratt TK, Fleischer RC (2003) Phylogeny of the owlet-nightjars (Aves: Aegothelida) based on mitochondrial DNA sequence. *Molecular Phylogenetics and Evolution* **29**, 540–549. doi:10.1016/S1055-7903(03)00135-0

Ebach MC, Humphries CJ, Newman RA, Williams DM, Walsh SA (2005) Assumption 2: opaque to intuition? *Journal of Biogeography* **32**, 781–787. doi:10.1111/j.1365-2699.2005.01283.x

Edgecombe GD, Colgan DJ, Sharkey D (2006) Phylogeny and biogeography of the Australasian centipede *Henicops* (Chilopoda: Lithobiomorpha): a combined morphological and molecular approach. *Insect Systematics & Evolution* **37**, 241–256. doi:10.1163/187631206788838590

Fleming CA (1979) 'The Geological History of New Zealand and its Life.' (University of Auckland Press: Auckland, New Zealand)

Gaina C, Roest WR, Müller RD, Symonds P (1998) The opening of the Tasman Sea: a gravity anomaly animation. *Earth Interactions* **2**, 1–23. doi:10.1175/1087-3562(1998)002<0001:TOOTTS>2.3.CO;2

Gamble A, Bauer AM, Greenbaum E, Jackman TR (2008) Evidence for Gondwanan vicariance in an ancient clade of gecko lizards. *Journal of Biogeography* **35**, 88–104.

Gemmill CE, Allan GJ, Wagner WL, Zimmer EA (2002) Evolution of insular Pacific *Pittosporum* (Pittosporaceae): origin of the Hawaiian radiation. *Molecular Phylogenetics and Evolution* **22**, 31–42. doi:10.1006/mpev.2001.1019

Gibbs GW (2010) Micropterigidae (Lepidoptera) of the southwestern Pacific: a revision with the establishment of five new genera from Australia, New Caledonia and New Zealand. *Zootaxa* **2520**, 1–48.

Givnish TJ, Renner SS (2004) Tropical intercontinental disjunctions: Gondwana breakup, immigration from the Boreotropics, and transoceanic dispersal. *International Journal of Plant Sciences* **165**, S1–S6. doi:10.1086/424022

Goldberg J, Trewick SA, Powlesland RG (2011) Population structure and biogeography of *Hemiphaga* pigeons (Aves: Columbidae) on islands in the New Zealand region. *Journal of Biogeography* **38**, 285–298. doi:10.1111/j.1365-2699.2010.02414.x

Greene HG, Collot J-Y, Fisher MA, Crawford AJ (1994) Neogene tectonic evolution of the New Hebrides Island arc: a review incorporating ODP drilling results. *Proceedings of the Ocean Drilling Program, Scientific Results* **134**, 1946.

Grobys JWG, Gohl K, Eagles G (2008) Quantitative tectonic reconstructions of Zealandia based on crustal thickness estimates. *Geochemistry Geophysics Geosystems* **9**, 1–18. doi:10.1029/2007GC001691

Haase M, Bouchet P (1998) Radiation of crenobiontic gastropods on an ancient continental island: the *Hemistomia*-clade in New Caledonia (Gastropoda: Hydrobiidae). *Hydrobiologia* **367**, 43–129. doi:10.1023/A:1003219931171

Harbaugh DT, Baldwin BG (2007) Phylogeny and biogeography of the sandalwoods (*Santalum*, Santalaceae): repeated dispersals throughout the Pacific. *American Journal of Botany* **94**, 1028–1040.

Hausdorf B (1998) Weighted ancestral area analysis and a solution of the redundant distribution problem. *Systematic Biology* **47**, 445–456. doi:10.1080/106351598260815

Heads M (2002) Birds of paradise, vicariance biogeography, and terrane tectonics in New Guinea. *Journal of Biogeography* **29**, 261–283. doi:10.1046/j.1365-2699.2002.00667.x

Heads M (2008) Panbiogeography of New Caledonia, South-west Pacific: basal angiosperms on basement terranes, ultramafic endemics inherited from volcanic island arcs and old taxa endemic to young islands. *Journal of Biogeography* **35**, 2153–2175. doi:10.1111/j.1365-2699.2008.01977.x

Heads M (2014*a*) Biogeography by revelation: investigating a world shaped by miracles. *Australian Systematic Botany* **27**, 282–304. doi:10.1071/SB14038

Heads M (2014*b*) 'Biogeography of Australasia.' (Cambridge University Press: Cambridge, UK)

Heenan PB, Smissen RD (2013) Revised circumscription of *Nothofagus* and recognition of the segregate genera *Fuscospora*, *Lophozonia*, and *Trisyngyne* (Nothofagaceae). *Phytotaxa* **146**, 1–31. doi:10.11646/phytotaxa.146.1.1

Herzer RH, Sykes R, Killops SD, Funnell RH, Burggraf DR, Townend J, Raine JI, Wilson GJ (1999) Cretaceous carbonaceous rocks from the Norfolk Ridge system, Southwest Pacific: implications for regional petroleum potential. *New Zealand Journal of Geology and Geophysics* **42**, 57–73. doi:10.1080/00288306.1999.9514831

Herzer RH, Barker DH, Roest WR, Mortimer N (2011) Oligocene–Miocene spreading history of the northern South Fiji Basin and implications for the evolution of the New Zealand plate boundary. *Geochemistry Geophysics Geosystems* **12**, Q02004. doi:10.1029/2010GC003291

Hovenkamp P (1997) Vicariance events, not areas, should be used in biogeographical analysis. *Cladistics* **13**, 67–79. doi:10.1111/j.1096-0031.1997.tb00241.x

Hutton FW (1872) On the origin of the fauna and flora of New Zealand. *Transactions and Proceedings of the New Zealand Institute* **5**, 227–256.

Jaffre T, Bouchet P, Veillon J-M (1997) Threatened plants of New Caledonia: is the system of protected areas adequate? *Biodiversity and Conservation* **7**, 109–135. doi:10.1023/A:1008815930865

Keene J, Potter A, Baker C, Tran M (2008) Sedimentology and geomorphology of the East Marine Region: a spatial analysis. (Geoscience Australia Publications) Available at http://horizon.documentation.ird.fr/exl-doc/pleins_textes/pleins_textes_7/b_fdi_53-54/010020262.pdf [Verified October 2016]

Keith P, Lord C, Lorion J, Watanabe S, Tsukamoto K, Couloux A, Dettai A (2011) Phylogeny and biogeography of Sicydiinae (Teleostei: Gobiidae) inferred from mitochondrial and nuclear genes. *Marine Biology* **158**, 311–326. doi:10.1007/s00227-010-1560-z

Klingelhoefer F, Lafoy Y, Collot J, Cosquer E, Géli L, Nouzé H, Vially R (2007) Crustal structure of the basin and ridge system west of New Caledonia (Southwest Pacific) from wide-angle and reflection seismic data. *Journal of Geophysical Research* **112**, B11102. doi:10.1029/2007JB005093

Korall P, Pryer KM (2014) Global biogeography of scaly tree ferns (Cyatheaceae): evidence for Gondwanan vicariance and limited transoceanic dispersal. *Journal of Biogeography* **41**, 402–413. doi:10.1111/jbi.12222

Korall P, Conant DS, Metzgar JS, Schneider H, Pryer KM (2007) A molecular phylogeny of scaly tree ferns (Cyatheaceae). *American Journal of Botany* **94**, 873–886. doi:10.3732/ajb.94.5.873

Ladiges PY, Cantrill D (2007) New Caledonia–Australian connections: biogeographic patterns and geology. *Australian Journal of Botany* **20**, 383–389.

Ladiges PY, Udovicic F, Nelson G (2003) Australian biogeographical connections and the phylogeny of large genera in the plant family Myrtaceae. *Journal of Biogeography* **30**, 989–998. doi:10.1046/j.1365-2699.2003.00881.x

Lafoy Y, Brodien I, Vially R, Exon NF (2005*a*) Structure of the basin and ridge system west of New Caledonia (Southwest Pacific): a synthesis.

*Marine Geophysical Researches* **26**, 37–50. doi:10.1007/s11001-005-5184-5
Lafoy Y, Géli L, Klingelhoefer F, Vially R, Sichler B, Nouzé H (2005*b*) Discovery of continental stretching and oceanic spreading in the Tasman Sea. *Eos* **86**, 101–105. doi:10.1029/2005EO100001
Landcare Research (2015) Phylogeny of New Zealand plants, Pennantiaceae. Available at http://plantphylogeny.landcareresearch.co.nz/webforms/ViewTree.aspx?ObjectID=a3ae47d8-56a1-442c-b2a6-1804dad911b6 [Verified October 2016
Lane A, Shine R (2011) Phylogenetic relationships within laticaudine sea snakes (Elapidae). *Molecular Phylogenetics and Evolution* **59**, 567–577. doi:10.1016/j.ympev.2011.03.005
Leschen RAB, Lackner T (2013) Gondwanan Gymnochilini (Coleoptera: Trogossitidae): generic concepts, review of New Zealand species and long-range Pacific dispersal. *Systematic Entomology* **38**, 278–304. doi:10.1111/j.1365-3113.2012.00661.x
Lucky A (2011) Molecular phylogeny and biogeography of the spider ants, genus *Leptomyrmex* Mayr (Hymenoptera: Formicidae). *Molecular Phylogenetics and Evolution* **59**, 281–292. doi:10.1016/j.ympev.2011.03.004
Marshall DC, Hill KBR, Marske KA, Chambers C, Buckley TR, Simon C (2012) Limited, episodic diversification and contrasting phylogeography in a New Zealand cicada radiation. *BMC Evolutionary Biology* **12**, 117–135. doi:10.1186/1471-2148-12-177
Matthews KJ, Hale AJ, Gurnis M, Müller RD, DiCaprio L (2011) Dynamic subsidence of Eastern Australia during the Cretaceous. *Gondwana Research* **19**, 372–383. doi:10.1016/j.gr.2010.06.006
Meffre S, Crawford AJ, Quilty PG (2006) Arc-continent collision forming a large island between New Caledonia and New Zealand in the Oligocene. In 'Australian Earth Sciences Convention 2006: Conference Abstracts', 2–8 July 2006, Melbourne, Vic., Australia. (Geological Society of Australia) Available at http://www.publish.csiro.au/?act=view_file&file_id=ASEG2006ab111.pdf [Verified October 2016]
Michaux B (1989) Generalized tracks and geology. *Systematic Zoology* **38**, 390–398. doi:10.2307/2992404
Michaux B (1994) Land movements and animal distributions in east Wallacea (eastern Indonesia, Papua New Guinea and Melanesia). *Palaeogeography, Palaeoclimatology, Palaeoecology* **112**, 323–343. doi:10.1016/0031-0182(94)90079-5
Michaux B (1998) Terrestrial birds of the Indo-Pacific. In 'Biogeography and Geological Evolution of SE Asia'. (Eds R Hall, JD Holloway) pp. 361–391. (Backhuys: Leiden, Netherlands)
Michaux B (2010) Biogeology of Wallacea: geotectonic models, areas of endemism, and natural biogeographical units. *Biological Journal of the Linnean Society. Linnean Society of London* **101**, 193–212. doi:10.1111/j.1095-8312.2010.01473.x
Mitchell AD, Heenan PB, Murray BG, Molloy BPJ, de Lange PJ (2009) Evolution of the south-western Pacific genus *Melicytus* (Violaceae): evidence from DNA sequence data, cytology and sex expression. *Australian Systematic Botany* **22**, 143–157. doi:10.1071/SB08042
Mortimer N (2004) Basement gabbro from Lord Howe Rise. *New Zealand Journal of Geology and Geophysics* **47**, 501–507. doi:10.1080/00288306.2004.9515072
Mortimer N, Herzer RH, Gans PB, Laporte-Magoni C, Calvert AT, Bosch D (2007) Oligocene–Miocene tectonic evolution of the South Fiji Basin and Northland Plateau, SW Pacific Ocean: evidence from petrology and dating of dredged rocks. *Marine Geology* **237**, 1–24. doi:10.1016/j.margeo.2006.10.033
Mortimer N, Dunlap WJ, Palin JM, Herzer RH, Hauff F, Clark M (2008) Ultra-fast early Miocene exhumation of Cavalli Seamount, Northland Plateau, Southwest Pacific Ocean. *New Zealand Journal of Geology and Geophysics* **51**, 29–42. doi:10.1080/00288300809509848
Mouly A, Razafimandimbison SG, Khodabandeh A, Bremer B (2009) Phylogeny and classification of the species-rich pantropical showy genus *Ixora* (Rubiaceae-Ixoreae) with indications of geographical monophyletic units and hybrids. *American Journal of Botany* **96**, 686–706. doi:10.3732/ajb.0800235
Nattier R, Robillard A, Desutter-Grandcolas L, Coloux A, Grandcolis P (2011) Older than New Caledonian emergence? A molecular phylogenetic study of the eneopterine crickets (Orthoptera: Grylloidea). *Journal of Biogeography* **38**, 2195–2209. doi:10.1111/j.1365-2699.2011.02563.x
Nelson G, Ladiges PY (1996) Paralogy in cladistic biogeography and analysis of paralogy-free subtrees. *American Museum Novitates* **3167**, 1–58.
Nelson G, Platnick NI (1981) 'Systematics and Biogeography: Cladistics and Vicariance.' (Columbia University Press: New York, NY, USA)
Nicholson KN, Black PM, Sporli KB (2008) Cretaceous–Oligocene multiphase magmatism on Three Kings Islands, northern New Zealand. *New Zealand Journal of Geology and Geophysics* **51**, 219–229. doi:10.1080/00288300809509861
Nicholson KN, Maurizot P, Black PM, Picard C, Simonetti A, Stuart A, Alexander A (2011) Geochemistry and age of the Noumea Basin lavas, New Caledonia: evidence for Cretaceous subduction beneath the eastern Gondwana margin. *Lithos* **125**, 659–674. doi:10.1016/j.lithos.2011.03.018
Nielsen SV, Bauer AM, Jackman TR, Hitchmough RA, Daugherty CH (2011) New Zealand geckos (Diplodactylidae): cryptic diversity in a post-Gondwanan lineage with trans-Tasman affinities. *Molecular Phylogenetics and Evolution* **59**, 1–22. doi:10.1016/j.ympev.2010.12.007
Nyári AS (2011) Origin and evolution of the unique Australo-Papuan mangrove-restricted avifauna: novel insights form molecular phylogenetic and comparative phylogeographic analyses. PhD thesis, University of Kansas, Lawrence, KS, USA. Available at https://kuscholarworks.ku.edu/bitstream/handle/1808/7691/Nyari_ku_0099D_11510_DATA_1.pdf?sequence=1 [Verified October 2016]
Oliver WRB (1925) Biogeographical relations of the New Zealand region. *Journal of the Linnean Society Botany* **47**, 99–140. doi:10.1111/j.1095-8339.1925.tb01794.x
Page TJ, Baker AM, Cook BD, Hughes JM (2005) Historical transoceanic dispersal of a freshwater shrimp: the colonization of the South Pacific by the genus *Paratya* (Atyidae). *Journal of Biogeography* **32**, 581–593. doi:10.1111/j.1365-2699.2004.01226.x
Papadopulos AST, Baker WJ, Crayn D, Buttin RK, Kynast RG, Hutton I, Savolainen V (2011) Speciation with gene flow on Lord Howe Island. *Proceedings of the National Academy of Sciences of the United States of America* **108**, 13188–13193. doi:10.1073/pnas.1106085108
Parenti LR, Ebach ME (2009) 'Comparative Biogeography.' (University of California Press: Oakland, CA, USA)
Pelletier B (2007) Geology of the New Caledonia region and its implications for the study of the New Caledonian biodiversity. In 'Compendium of Marine Species of New Caledonia'. (Eds C Payri, B Richer de Forges) pp. 19–32. (IRD: Nouméa, New Caledonia)
Percy DM, Garver AM, Wagner WL, James HF, Cunningham CW, Miller SE, Fleischer RC (2008) Progressive island colonization and ancient origin of Hawaiian *Metrosideros* (Myrtaceae). *Proceedings. Biological Sciences* **275**, 1479–1490. doi:10.1098/rspb.2008.0191
Perrie LR, Brownsey PJ, Lockhart PJ, Brown EA, Large MF (2003) Biogeography of temperate Australasian *Polystichum* ferns as inferred from chloroplast sequence and AFLP. *Journal of Biogeography* **30**, 1729–1736.
Pigram CJ, Davies HL (1987) Terranes and accretion history of the New Guinea orogeny. *BMR Journal of Australian Geology and Geophysics* **10**, 193–211.
Ponder WF, Colgan DJ, Gleeson DM, Sherley GH (2003) Relationships of *Placostylus* from Lord Howe Island: an investigation using the mitochondrial cytochrome c oxidase 1 gene. *Molluscan Research* **23**, 159–178. doi:10.1071/MR03001

Pratt SJ (2013) Evolution of the genera *Vitex* (Lamiaceae) and *Zygogynum* (Winteraceae) on New Caledonia. MSc thesis, University of Waikato, Hamilton, New Zealand. Available at http://researchcommons.waikato.ac.nz/handle/10289/7570 [Verified October 2016]

Pratt RC, Morgan-Richards M, Trewick SA (2008) Diversification of New Zealand weta (Orthoptera: Ensifera: Anostostomatidae) and their relationships in Australasia. *Philosophical Transactions of the Royal Society of London. Series B, Biological Sciences* **363**, 3427–3437. doi:10.1098/rstb.2008.0112

Puslednik L, Serb JM (2008) Molecular phylogenetics of the Pectinidae (Mollusca: Bivalvia) and effect of increased taxon sampling and outgroup selection on tree topology. *Molecular Phylogenetics and Evolution* **48**, 1178–1188. doi:10.1016/j.ympev.2008.05.006

Quilty PG (1993) Tasmantid and Lord Howe seamounts: biostratigraphy and palaeoceanographic significance. *Alcheringa* **17**, 27–53. doi:10.1080/03115519308619487

Rapela CW, Pankhurst RJ, Fanning J, Hervé F (2005) Pacific subduction coeval with the Karoo mantle plume: early Jurassic subcordilleran belt of northwestern Patagonia. In 'Terrane Processes at the Margin of Gondwana'. (Eds APM Vaughan, PI Leat, RJ Pankhurst) Geological Society Special Publication 246, pp. 1–21. (Geological Society Publishing House: Bath, UK)

Ronquist F (1994) Ancestral areas and parsimony. *Systematic Biology* **43**, 267–274. doi:10.1093/sysbio/43.2.267

Rønsted N, Weiblen GD, Savolainen V, Cook JM (2008) Phylogeny, biogeography and ecology of *Ficus* section *Malvanthera* (Moraceae). *Molecular Phylogenetics and Evolution* **48**, 12–22. doi:10.1016/j.ympev.2008.04.005

Rosen DE (1978) Vicariant patterns and historical explanation in biogeography. *Systematic Zoology* **27**, 159–188. doi:10.2307/2412970

Schellart WP (2006) Fitting Northland, New Caledonia and d'Entrecasteaux geology into the Late Cretaceous–Cenozoic Southwest Pacific tectonic framework. In 'AESC Extended Abstracts 2006', 2–8 July 2006, Melbourne, Vic., Australia. (Geological Society of Australia) Available at http://www.publish.csiro.au/ex/ASEG2006ab157 [Verified October 2016]

Schellart WP, Lister GS, Toy VG (2006) A Late Cretaceous and Cenozoic reconstruction of the Southwest Pacific region: tectonics controlled by subduction and slab rollback processes. *Earth-Science Reviews* **76**, 191–233. doi:10.1016/j.earscirev.2006.01.002

Schellart WP, Kennett BLN, Spakman W, Amaru M (2009) Plate reconstructions and tomography reveal a fossil lower mantle slab below the Tasman Sea. *Earth and Planetary Science Letters* **278**, 143–151. doi:10.1016/j.epsl.2008.11.004

Schmidt S, Driver F, de Barro P (2006) The phylogenetic characteristics of three different 28S rRNA gene regions in *Encarsia* (Insecta, Hymenoptera, Aphelinidae). *Organisms, Diversity & Evolution* **6**, 127–139. doi:10.1016/j.ode.2005.07.002

Schweizer M, Güntert M, Hertwig ST (2013) Out of the Bassian province: historical biogeography of the Australasian platycercine parrots (Aves, Psittaciformes). *Zoologica Scripta* **42**, 13–27. doi:10.1111/j.1463-6409.2012.00561.x

Sdrolias M, Müller RD, Gaina C (2001) Plate tectonic evolution of eastern Australian marginal ocean basins. *Eastern Australian Basin Symposium* **2001**, 227–237.

Sdrolias M, Müller RD, Gaina C (2003) Tectonic evolution of the Southwest Pacific using constraints from backarc basins. In 'Evolution and Dynamics of the Australian Plate' (Eds RR Hillis, D Muller) Geological Society of Australia Special Publication 22 and Geological Society of America Special Paper 372, pp. 343–359. (Geological Society of America: Boulder, CO, USA)

Sdrolias M, Müller RD, Mauffret A, Bernardel G (2004) Enigmatic formation of the Norfolk Basin, SW Pacific: a plume influence on back-arc extension. *Geochemistry Geophysics Geosystems* **5**, Q06005. doi:10.1029/2003GC000643

Seton M, Flament N, Müller RD (2012) Subduction history in the Melanesian border region, SW Pacific. In 'East Australasian Basins Symposium IV', 10–14 September 2012, Brisbane, Qld, Australia. (Ed. T Mares) (Petroleum Exploration Society of Australia: Brisbane, Qld, Australia) Available at http://www.earthbyte.org/Resources/Pdf/Seton_Melanesian_borderlands_subduction_history_EABS4_2012.pdf [Verified October 2016]

Sirvid PJ, Moore NE, Chambers GK, Prendergast K (2013) A preliminary molecular analysis of phylogenetic relationships of New Zealand Thomisidae (Areneae) using a multilocus approach. *Invertebrate Systematics* **27**, 655–672.

Smith SA, Sadlier RA, Bauer AM, Austin C, Jackman T (2007) Molecular phylogeny of the scincid lizards of New Caledonia and adjacent areas: evidence for a single origin of the endemic skinks of Tasmantis. *Molecular Phylogenetics and Evolution* **43**, 1151–1166. doi:10.1016/j.ympev.2007.02.007

Spencer HG, Brook FJ, Kennedy M (2006) Phylogeography of Kauri snails and their allies from Northland, New Zealand (Mollusca: Gastropoda: Rhytididae: Paryphantinae). *Molecular Phylogenetics and Evolution* **38**, 835–842. doi:10.1016/j.ympev.2005.10.015

Stagg HMJ, Borissova I, Alcock M, Moore AMG (1999) Tectonic provinces of the Lord Howe Rise. *AGSO Research Newsletter* **31**, 31–32. http://www.agso.gov.au/information/publications/resnews/

Steane DA, Wilson KL, Hill RS (2003) Using *matK* sequence data to unravel the phylogeny of Casuarinaceae. *Molecular Phylogenetics and Evolution* **28**, 47–59. doi:10.1016/S1055-7903(03)00028-9

Stefenon VM, Gailing O, Finkeldey R (2006) Phylogenetic relationship within genus *Araucaria* (Araucariaceae) assessed by means of AFLP fingerprints. *Silvae Genetica* **55**, 45–52.

Stöckler K, Daniel IL, Lockhart PJ (2002) New Zealand kauri (*Agathis australis* (D.Don) Lindl., Araucariacea) survives Oligocene drowning. *Systematic Biology* **51**, 827–832. doi:10.1080/10635150290102474

Sutherland R (1999) Basement geology and tectonic development of the greater New Zealand region: an interpretation from regional magnetic data. *Tectonophysics* **308**, 341–362. doi:10.1016/S0040-1951(99)00108-0

Sutherland R, Collot J, Lafoy Y, Logan GA, Hackney R, Stagpoole V, Uruski C, Hashimoto T, Higgins K, Herzer RH, Wood R, Mortimer N, Rollet N (2010) Lithosphere delamination with foundering of lower crust and mantle caused permanent subsidence of New Caledonian Trough and transient uplift of Lord Howe Rise during Eocene and Oligocene initiation of Tonga–Kermadec subduction, western Pacific. *Tectonics* **29**, TC2004. doi:10.1029/2009TC002476

Swenson U, Hill RS, McLoughlin S (2001) Biogeography of *Nothofagus* supports the sequence of Gondwana break-up. *Taxon* **50**, 1025–1041. doi:10.2307/1224719

Swenson U, Nylinder S, Munzinger J (2013) Sapotaceae biogeography supports New Caledonia being an old Darwinian island. *Journal of Biogeography* **40**, 797-809. doi:10.1111/jbi.12246

Tulloch AJ, Kimbrough DL, Wood RA (1991) Carboniferous granite basement dredged from a site on the southwest margin of the Challenger Plateau, Tasman Sea. *New Zealand Journal of Geology and Geophysics* **34**, 121–126. doi:10.1080/00288306.1991.9514449

Ung V, Zaragüeta-Bagils R, Williams DM (2016) Comparative biogeography of Southeast Asia and the West Pacific region. *Biological Journal of the Linnean Society. Linnean Society of London* **117**, 372–385. doi:10.1111/bij.12670

van Veller MGP, Kornet DJ, Zandee M (2000) Methods in vicariance biogeography: assessment of the implementations of assumptions 0, 1, and 2. *Cladistics* **16**, 319–345. doi:10.1006/clad.2000.0135

Wagstaff SJ, Dawson MI, Venter S, Munzinger J, Crayn DM, Steane DA, Lemson KL (2010) Origin, diversification and classification of the Australasian genus *Dracophyllum* (Richeeae: Ericaceae). *Annals of the Missouri Botanical Garden* **97**, 235–258. doi:10.3417/2008130

Wallace AR (1876) 'The Geographical Distribution of Animals.' (MacMillan: London, UK)

Wiley EO, Lieberman BS (2011) 'Phylogenetics: the Theory of Phylogenetic Systematics', 2nd edn. (Wiley-Blackwell: Hoboken, NJ, USA)

Wink M, Heidrich P, Sauer-Gürth H, Elsayed A-A, Gonzales J (2008) Molecular phylogeny and systematics of owls (Strigiformes). In 'Owls of the World'. pp. 42–63. (Helm: London, UK)

Woo VL, Funke JF, Smith PJ, Lockhart PJ, Garnock-Jones PJ (2011) New World origins of Southwest Pacific Gesneriaceae: multiple movements across and within the South Pacific. *International Journal of Plant Sciences* **172**, 434–457. doi:10.1086/658183

Wood RA (1991) Structure and seismic stratigraphy of the western Challenger Plateau. *New Zealand Journal of Geology and Geophysics* **34**, 1–9. doi:10.1080/00288306.1991.9514433

Wood R, Woodward D (2002) Sediment thickness and crustal structure of offshore western New Zealand from 3D gravity modelling. *New Zealand Journal of Geology and Geophysics* **45**, 243–255. doi:10.1080/00288306.2002.9514971

Wright SD, Yang CG, Dawson JW, Whittaker DJ, Gardner RC (2000) Riding the ice age El Niño? Pacific biogeography and evolution of *Metrosideros* subg. *Metrosideros* (Myrtaceae) inferred from nuclear ribosomal DNA. *Proceedings of the National Academy of Sciences of the United States of America* **97**, 4118–4123. doi:10.1073/pnas.050351197

Zhang L-B, Renner S (2003) The deepest splits in Chloranthaceae as resolved by chloroplast sequences. *International Journal of Plant Sciences* **164**, S383–S392. doi:10.1086/376882

Handling editor: Malte Ebach

## Attribution

**Ung, V., Michaux, B., Leschen, R.A.B. 2016. A comprehensive vicariant model for Southwest Pacific Biotas. *Australian Systematic Botany*, 29:424–439.**

First published in 2017 in *Australian Systematic Botany*, 29(6):424–439, doi:10.1071/SB16032. Reproduced with permission from CSIRO Publishing.

*chapter nine*

# *The final piece of the jigsaw puzzle*

Having written extensively on the geology and biogeography of Wallacea, New Guinea and the Campbell Plateau, it was a natural progression to fill the gap and tackle the biogeology of New Zealand. Rich Leschen had suggested for some time that we should write about the 'Melanesian Rift', in part because of his entomological interests in the islands off northern New Zealand (e.g. Buckley and Leschen 2013) and partly to complement our Campbell Plateau paper (Chapter 6), thereby completing a comprehensive treatment of New Zealand biogeography. We teamed up with Theary Ung to undertake an analysis of endemic or restricted range species found on Lord Howe Island, Norfolk Island, New Caledonia, New Zealand and other areas adjacent to it such as eastern Australia, Vanuatu and Fiji. As we do not live in the best of all possible worlds (Chapter 5), a literature search for phylogenies that happened to include species distributed in the areas of interest had to serve. Paralogy-free trees were produced and three-item statements derived and combined to form an optimal, general areagram.

Judging by some of the reviews we received, there is a belief that combining individual areagrams into general areagrams is not valid because individual phylogenies on which areagrams are based are unique historical occurrences that have to stand by themselves, or that three-item statements are passé and the subject has moved on. We argued that you can indeed combine individual areagrams, that there is an extensive literature that has explored the issue conceptually, and computational methods have been developed to do so (Platnick and Nelson 1978; Nelson and Platnick 1981; Nelson and Ladiges 1991; Morrone and Carpenter 1994; Morrone and Crisci 1995; Page 1994; Ebach et al. 2003, Bagils et al. 2012). Nobody questions the validity of combining different character or gene trees representing different evolutionary pathways into a single most parsimonious phylogeny, which is an analogous process to combining three-area statements that imply different area relationships into a single optimal tree. I go further and suggest that you *have* to combine individual areagrams if you want to interpret the biogeology of areas because individual studies cannot be assessed for their validity and are analogous to anecdote (Chapter 4). The issue of relevance is irrelevant. The ontological basis of the method is robust, and efficient computational methods have been developed to implement it.

Our analysis highlighted a number of features of the Melanesian Rift including its structural complexity and heterogeneous tectonic history. To treat this area as a uniform block with an identical history is too simplistic. We believe our comprehensive model describing periods of vicariance, geodispersal and range expansion accounts for what we can observe today, is consistent with tectonic and palaeontological data and provides, in conjunction with Michaux and Leschen (2005), a coherent explanation for the origin of New Zealand's biota. In the following sections, I explore the history of New Zealand's biota in greater detail using three examples – New Zealand temperate rain forests, in particular Northland's kauri forests, Cenozoic Mollusca and the alpine flora and fauna.

## *Origin of New Zealand's forests*

### *Kauri forests of northern New Zealand*

Once, not so long ago, much of the upper North Island of New Zealand was covered in kauri (*Agathis australis*) forest (Figure 9.1). New Zealand kauri is a magnificent emergent tree that has a columnar trunk and spreading canopy when mature, and produces timber of exception quality, quantity and durability (Plate 9.1). It formed the basis of a major North Auckland timber industry during the early days of European settlement from the mid-1800s and, together with the collection of kauri gum (resin), provided the initial impetus for the economic development of the whole of Northland. By the early 1900s, much of this forest had been felled and turned into farmland and what remained was further reduced by a push to fell remaining stands of mature trees in the 1920s and 30s. Only remnants of this once extensive forest cover remain today and this is now under threat by a water-borne fungus-like mould *Phytophthora agathidicide* that is attacking and killing even mature trees (www.teara.govt.nz/en/kauri-forest).

Despite its occurrence in temperate areas of northern New Zealand, kauri forest has a number of characteristics of tropical rainforest – many leaves of plants growing in the understory have drip tips; trunks of the canopy trees are often buttressed; lianas, epiphytes, perching orchids and climbers are common; and soils are thin and infertile. A tropical character is also apparent in the distribution of its constituent species. The genus *Agathis* exemplifies a typical spatial pattern for many species found in kauri forest. The Catalogue of Life (www.catalogueoflife.org) lists 17 extant species in the genus that are found on what might be termed Gondwanan terrane fragments (Figure 9.2): Australia (3 spp.), New Zealand (1 sp.), New Caledonia (5 spp.), Melanesia (2 spp.), New Guinea (2 spp., one shared with Australia), and Wallacea, the Greater Sundas, Peninsula Malaysia (6 spp.). Such a distribution indicates a Mesozoic origin for *Agathis* because the

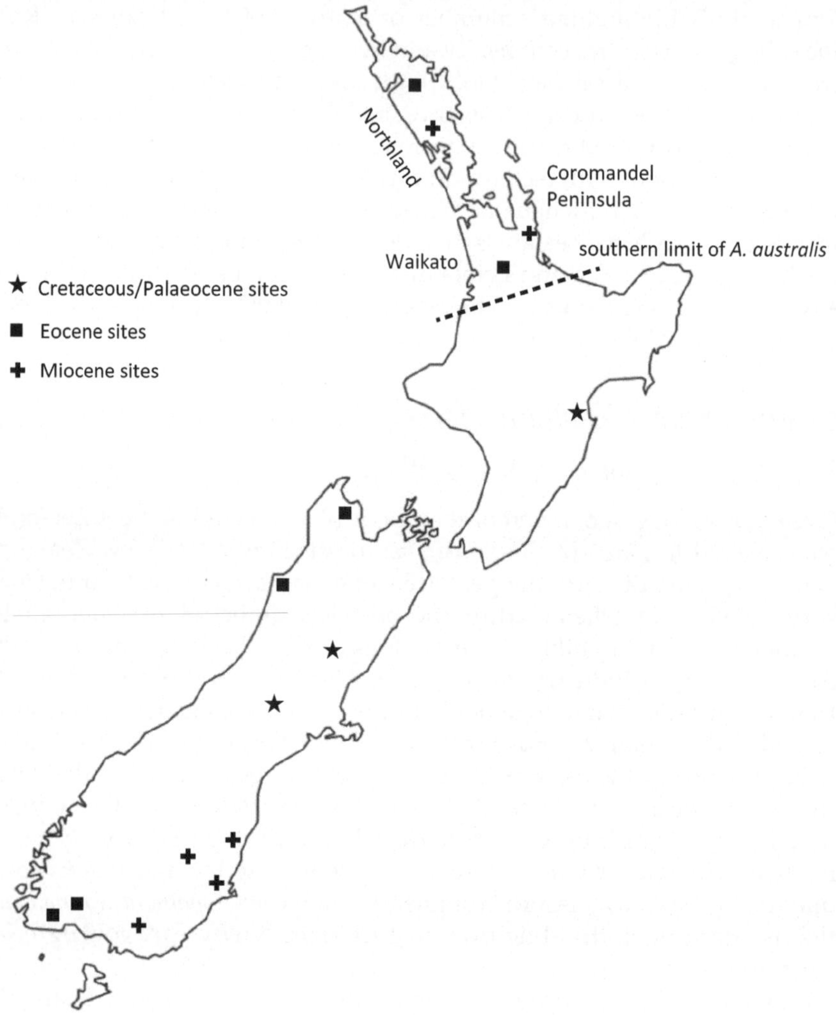


***Figure 9.1*** Locality map of New Zealand Late Cretaceous and Cenozoic plant fossil localities.

Gondwanan terranes found in mainland Southeast Asia and the Greater Sundas were derived from Gondwana and accreted to a Palaeozoic Asian core more or less continuously throughout the Mesozoic until the mid-Cretaceous (e.g. Metcalfe 2009; Hall 2011).

*Agathis* is a well-supported monophyletic genus and sister taxon to the monotypic *Wollemia*. *Agathis australis* is the basal taxon to all other extant *Agathis* species (Stockler et al. 2002; Knapp et al. 2007; Liu et al. 2009; Biffin et al. 2010; Escapa and Catalano 2013; Kranitz et al. 2014). While

***Plate 9.1*** Mature kauri (*Agathis australis*) showing columnar trunk and spreading crown, scale approximately 10 m to first branch. Note nikau palms (*Rhopalostylis sapida*) in understory.

fossil evidence for Middle Cretaceous *Agathis* fossils in New Zealand (Stockler et al. 2002) is disputed (e.g. Biffin et al. 2010), there is a range of ages derived from molecular studies that, despite a consistent underestimation these methods produce, put the origin of *Agathis* back into the Cretaceous. Knapp et al. (2007) gave a range of dates for the origin of *A. australis* from 56 ± 11 Ma to 92 ± 26 Ma depending on the method of estimation, and concluded that they couldn't falsify the hypothesis of its in situ evolution within New Zealand. Liu et al. (2009) estimated that the genus *Agathis* originated at 61 ± 15 Ma but unfortunately didn't include the basal *A. australis* in their study. They dated the divergence of *Wollemia* from *Agathis* at 118 Ma. Kranitz et al. (2014) gave an age range of 91–55 Ma for the split between *Wollemia* and *Agathis*. I note that even the sceptical Biffin et al. (2010), whose paper reported large error bars around the age of origin of *A. australis*, had a lower limit divergence age near the base of the Early Cretaceous in one of the models they presented.

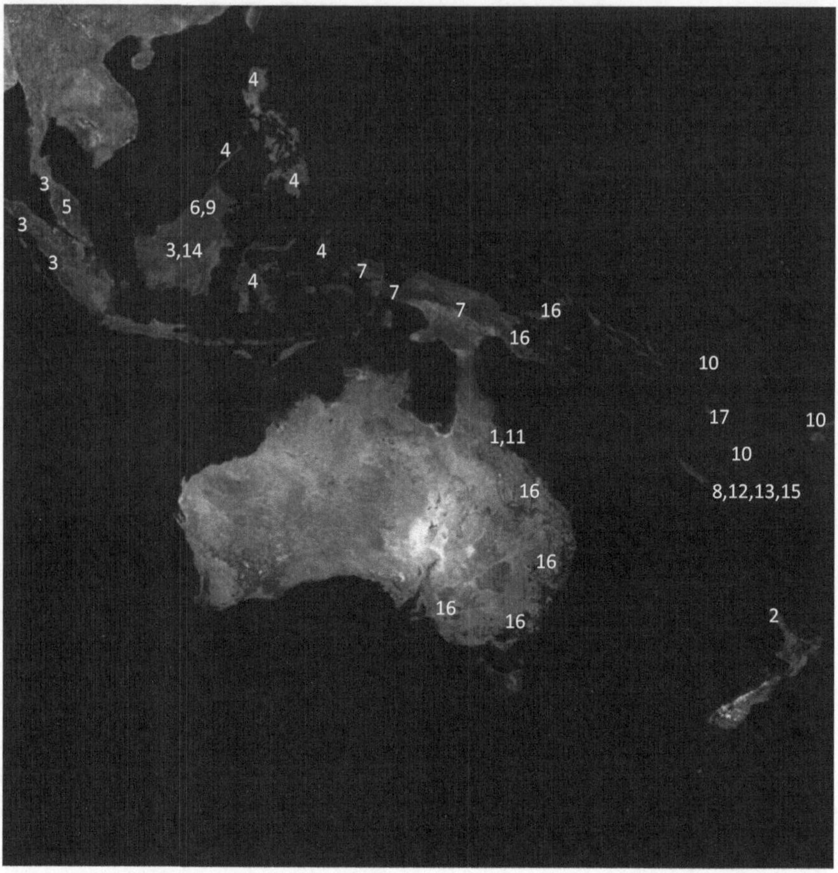


***Figure 9.2*** Distribution of *Agathis* in the Indo-Pacific. 1 = *A. atropurpurea* (ne Queensland); 2 = *A. australis* (n North Island); 3 = *A. borneensis* (Malay Peninsula, nw Sumatra, Borneo); 4 = *A. dammara* (Sulawesi, n Maluku, Mindanao, Luzon, Palawan); 5 = *A. flavescens* (Malay Peninsula – moist montane forest); 6 = *A. kinabaluensis* (n Borneo, ne Sarawak); 7 = *A. labillardieri* (w New Guniea); 8 = *A. lanceolata* (New Caledonia); 9 = *A. lenticula* (n Borneo, ne Sarawak); 10 = *A. macrophylla* (Fiji, s Vanuatu, s Solomon Islands); 11 = *A.microstachya* (ne Queensland); 12 = *A. montana* (New Caledonia); 13 = *A. moorei* (New Caledonia); 14 = *A. orbicula* (Borneo – highlands); 15 = *A. ovata* (New Caledonia); 16 = *A. robusta* (e and s Australia, Southwest Peninsula, New Guinea, New Britain); 17 = *A. silbae* (Vanuatu – Espiritu Santo). n = north, w = west, e = east, s = south.

Other species that are important components of the distinctive kauri-podocarp-hardwood forest and that share this Gondwana terrane-distributional pattern include the podocarp *Dacrydium* (rimu – *D. cupressinum*), *Melicope* (wharangi – *M. ternate*), *Dysoxylum* (kokhekohe – *D. spectabile*) and *Alectryon* (titoki – *A. excelsus*). Terrane taxa restricted to the tropical Pacific include *Coprosma, Olearia, Hedycarya* (porokaiwhiri *H. arborea*), *Knightia* (rewarewa – *K. excelsa*), *Carpodetus* (putaputaweta – *C. serratus*), *Quintia, Streblus* and *Corynocarpus* (karaka – *C. laevigatus*). In addition, more widespread tropical genera common in kauri forest include *Beilschmiedia* (taraire – *B. tarairi*), *Macropiper* (kawakawa – *M. excelsum*) and *Pittosporum*.

## *New Zealand forests through time*

It is somewhat of an enigma that a rainforest should be growing in modern New Zealand because the area that was to become New Zealand was at high polar latitudes prior to its separation from Gondwana. Although there were no ice caps during the Late Cretaceous and the climate was mild even at high latitudes (Vajda and Raine 2010), there would still have been protracted periods of complete darkness making it difficult for the type of evergreen angiosperm species growing in the modern kauri forests of Northland to survive, although perhaps not impossible (Read and Francis 1992). Mobile animals, such as Cretaceous New Zealand dinosaurs, could have simply migrated in and out of the area with the changing seasons, although Vajda and Raine (2010) have suggested that some dinosaurs may also have been able to overwinter in the milder conditions that prevailed then.

Plant macrofossils and palynological research have provided an insight into what these high-latitude polar forests, which were widespread and ranged over southern South America, West Antarctica, southeast Australia and New Zealand, were like (Hill 2004; Lee et al. 2016). The estimated mean annual temperature (MAT) for the South Island of New Zealand during the Late Cretaceous ranged from 7–11°C in the south to 12–16°C in the north (Kennedy et al. 2002; Kennedy 2003). As a comparison, the MAT of northern New Zealand today is about 16°C (Chappell 2013). Figure 9.1 shows the location of Cenozoic plant fossil localities in New Zealand referred to in the text. The macroflora of New Zealand's South Island was dominated by Araucariaceae genera such as *Aruacaria* and *Otakauia* in the Late Cretaceous, which largely disappeared in the Palaeocene to be replaced by other conifer genera (Pole and Vajda 2009). Pollen analysis of Late Cretaceous sediments in the South Island indicated a temperate rainforest composed of podocarps (40% of pollen), tree ferns (40%) and Araucariaceae (6%). Angiosperm pollen consisted predominantly of *Nothofagus* and Proteaceae (Vajda and Raine 2003). Similar Late Cretaceous forests also grew at sites in the North Island of New Zealand (Vajda and Raine 2010). The mass extinction event at the end of the Cretaceous resulted in a proliferation of fern spores followed by the gradual recovery of gymnosperms and then angiosperms as the climatic effects of the Yucatan asteroid strike moderated (Vajda and Raine 2003). According to Poole and Cantrill (2006), modern analogues of the

Cretaceous polar forests of West Antarctica occur today in the temperate rainforests of South America, and New Zealand Cretaceous forests are suggestive of modern mixed podocarp-beech forests of the North and South Islands of New Zealand.

There was also a similarity between Early to Middle Eocene forests of southern South America, southeast Australia, New Zealand (Wilf et al. 2009) and Wilkes Land, East Antarctica (Contreras et al. 2013). These forests were characterised by abundant *Casuarina* and Proteaceae (Pocknall 1989) and the presence of palms (Arecaceae) (Conran et al. 2015). Contreras et al. (2013) likened the Wilkes Land forests to those now growing in New Caledonia. The subtropical to tropical temperature estimates of Carpenter et al. (2012) and Hollis et al. (2012) for high-latitude Australasian sites during the Eocene, while consistently higher than those obtained from oxygen isotope studies, are consistent with a proposed Early and Middle Eocene subtropical or tropical biota in New Zealand (e.g. Adams et al. 1990; Beu and Maxwell 1990).

Hollis et al. (2012) recorded declining temperatures in the Late Eocene, which coincided with the replacement of *Casuarina* and Proteaceae forests of the South Island by *Nothofagus* (*fusca*-group) dominated assemblages, although in the Waikato (northern New Zealand) *Nothofagus* spp. were found in more mixed communities that included *Casuarina*, Proteaceae and Araucariaceae (Pocknall 1989), and evidence of continued warm to subtropical temperatures as far as 50°S is provided by the presence of *Calamus*-like palms (Hartwich et al. 2010). Continued cooling resulted in the replacement of *fusca*-group species with *brassie*-group species during the Oligocene. Pocknall (1989) suggested that the closest modern analogues of these forests are those of the central mountains of New Guinea, which grow in regions with a MAT of 13–18°C and high rainfall.

The Lower Miocene climate of the southern South Island was warmer than today with a MAT in the range 17–20°C and high rainfall (Pole 2014; Reichgelt et al. 2015). Forests in the South Island at this time were characterised by a high diversity of podocarps that were taxonomically close to (or identical with) extant New Zealand species (Jordan et al. 2011), treeferns and angiosperms such as *Nothofagus*, Casuarinaecea, Myrtaceae, Proteaceae, *Eucalyptus*, *Metrosideros*, Lauraceae and palms (Pole et al. 2003; Reichgelt et al. 2015). An Early Miocene site in eastern Southland has yielded fossils of *Corynocarpus* (karaka) and *Avicennia* (mangrove), *Pomaderris* (kumarahou) and *Pouteria* (tawapou) that are now restricted to the northern parts of the North Island (Campbell 2002). A Lower Miocene site on the Hukatere Peninsula, Kaipara Harbour, yielded macrofossils/pollen from *Agathis* (kauri), *Avicennia* (mangrove), *Beilschmedia* (tawa), *Coprosma*, *Leptospermum* (now *Kunzea*; manuka and kanuka), *Myrsine*, (matipou) *Rhopalostylus* (nikau palm) and *Vitex* (puriri) as well as many other species that characterise modern, mixed podocarp-broadleaved-kauri forests (Jones 1970; Sutherland 2003). A Middle Miocene locality from the Coromandel Peninsula dated at 15–13 Ma indicted that kauri made up 10–20% of canopy trees there (Moore and Wallace 2000).

Some caution is required in making comparisons between reconstructed ancient forest types and their modern equivalents, because it is often not possible to say that fossil taxa are the same as modern species; pollen can travel a long distance, and local extinctions can alter forest composition. The identification of kauri (*Agathis australis*) illustrates the problem of identifying extant plant species in the fossil record. *Agathis* pollen is difficult to distinguish from pollen of other Araucariaceae and macrofossils often lack diagnostic characters. As a consequence, the evidence for the presence of *A. australis* much earlier than the Miocene is contested (Pole 2008). Modern New Zealand forests also differ from earlier forests because of extinctions. For example, no native *Eucalyptus* or *Casuarina* are now found in New Zealand, presumably a consequence of the sharp decline in temperature after about 12 Ma (Pole 2014), and there are only two Proteaceae species found in New Zealand now, despite the family having had a much higher diversity in the Palaeogene.

However, it is also apparent that some modern austral forests have deep in situ historical roots. Modern Podocarp-*Nothofagus* forest clearly had its origin back in the Late Mesozoic. This floral association was widespread in southeast Australia, New Zealand, West Antarctica and southern South America during the Cretaceous. The widespread occurrence of *Nothofagus*, Podocarpaceae and Araucariaceae prior to the fragmentation of East Gondwana provided the base from which some of New Zealand's modern forests evolved. Mixed podocarp/*Nothofagu*s forests still grow today in high rainfall, warmer parts of the South Island (mainly west of the Southern Alps but missing from central areas) and in the upland areas of the North Island. Pure stands of *Nothofagus* are restricted to the drier parts of the South Island, mainly east of the Southern Alps, and at higher altitudes in montane regions. The exact composition of these forests through time was dependent on temperature, rainfall, seasonality and species present at the time.

Northland's kauri forests are another matter, and I'm inclined to think that Mike Pole (1994) is right and that this forest type most probably dates from the end of the Oligocene or Lower Miocene when kauri forest and associated broadleaved assemblages are found in various places within New Zealand. This late development of kauri forest long after the detachment of the Melanesian Rift from East Gondwana led Mike Pole, among others, to speculate that kauri and other plant species colonised New Zealand by long-distance dispersal over open ocean (Pole 1994).

In our original paper (Ung et al. 2016), we drew attention to two regional unconformities discovered on the Melanesian Rift during the

Deep Sea Drilling Project (sites DSDP 206–208). The first of these hiatuses lasted approximately three million years between 53 and 50 Ma (Middle Eocene). The second was more prolonged and the subsequent marine transgression was asymmetric with subsidence first occurring in the north and then proceeding southwards. This second event started at 40 Ma (Late Eocene) and lasted until 30 Ma (Late Oligocene) in the north and until 15 Ma (Middle Miocene) in the south. We stressed the potential of these events, particularly the younger one, for channelling biotas southwards along the Norfolk Ridge into northern New Zealand because of the asymmetric subsidence and north-south trending basins. Rather than relying on a series of miraculous events to explain the sudden appearance of kauri forest, perhaps it was simply growing further north on an emergent Norfolk Ridge and expanded its range southwards as subsidence and marine transgression flooded land to the north. Thus, this most iconic of northern trees might not have originated in New Zealand at all but is a relative newcomer.

## *New Zealand Cenozoic Mollusca*

According to Fleming (1965), it was Dieffenbach who collected the first fossil molluscs for scientific purposes from New Zealand, which he subsequently sent to J.E. Gray at the British Museum who described the material in 1843. This marked the start of a long and continuing research effort into New Zealand's molluscs, particularly those of the Cenozoic. Pioneer workers such as Hutton, Suter, Marshall, Finlay, Marwick and Powell (and Fleming himself) laid the groundwork for our modern understanding of this extensive fauna. As Fleming (1965) eloquently put it, 'Before paleontology can properly function either as the handmaid of stratigraphy or the historian of biology, its materials must be described and classified'. In the view of Beu and Maxwell (1990), Fleming was the greatest of these students of New Zealand Mollusca and the first to be in a position to interpret the fauna as a whole.

Apart from the Palaeocene to Early Eocene and Early Oligocene, time periods not well represented and generally poor in molluscs, the New Zealand Cenozoic fossil record is remarkably complete and rich in species. Beu and Maxwell (1990) estimated that the total Cenozoic molluscan fauna probably amounts to between four and five thousand species. Fleming (1966) produced his first interpretive synthesis of Cenozoic Mollusca in the form of a 'pine tree' diagram (reproduced here as Figure 9.3). The diagram shows the changes in number of 'Indo-Pacific' genera through time and, although he used a rather vague definition of what constituted an Indo-Pacific genus and bearing in mind the lack of molluscan fossils particularly in the early part of the Cenozoic, it is clear that there are spikes in the first occurrence of warm water genera in the Middle Eocene and

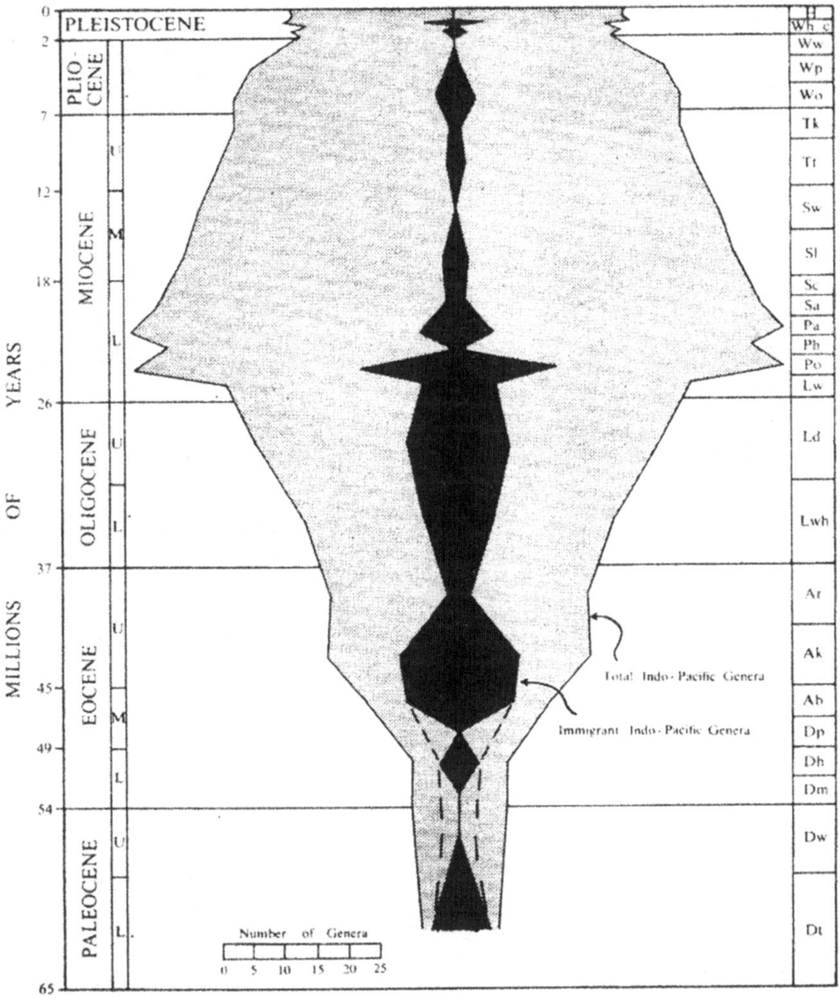


*Figure 9.3* Fleming's summary of New Zealand Cenozoic Indo-Pacific Mollusca.

Lower Miocene of New Zealand. Beu and Maxwells' (1990) monumental *Cenozoic Mollusca of New Zealand* (dedicated to Fleming who had died during the later stages of its production) still gives the most complete and detailed account, of which the following is a summary.

The general location of Late Palaeocene, Eocene and Middle Miocene fossil localities within New Zealand are shown in Figure 9.4. The Middle Eocene fossil fauna is characterised by high levels of endemism with a subtropical or tropical climate inferred from the high number of warm

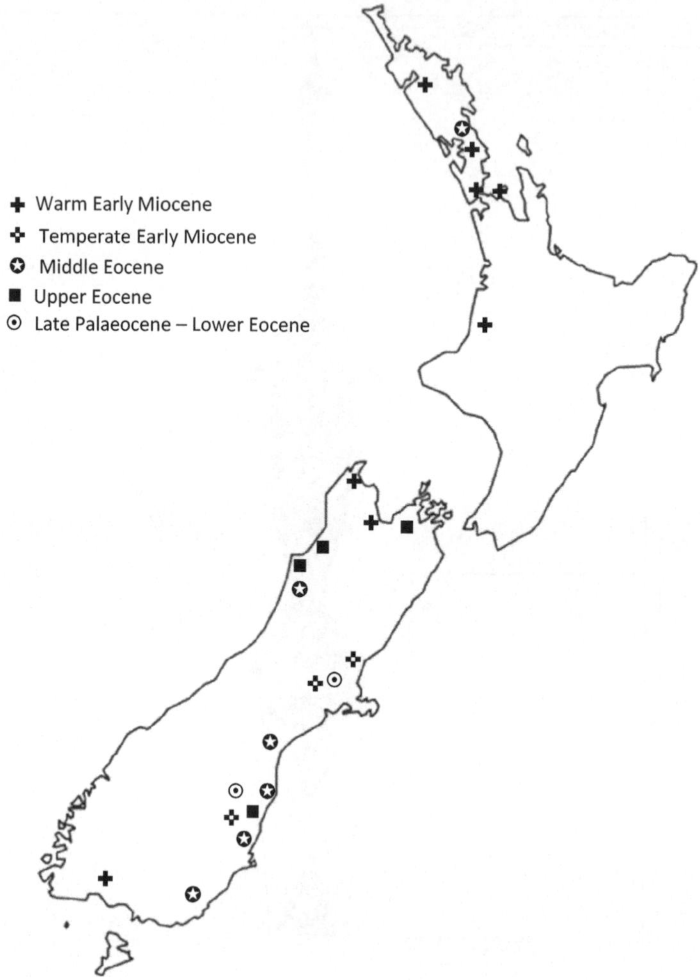


***Figure 9.4*** New Zealand Cenozoic molluscan fossil localities.

water taxa present. It is distinct from the preceding Palaeocene and Lower Eocene fauna, although this may in part be due to the deeper water environments and scarceness of molluscan fossils in these earlier rocks, and from the succeeding Upper Eocene faunas, in part at least because many of the warm water species of the Middle Eocene had died out. There is widespread evidence to support a regional unconformity at circa 50 Ma (i.e. at the beginning of the Middle Eocene). Evidence for this unconformity includes eroded older beds; abrupt lithological changes often marked by an increase in terrigenous sediment; burrowed contact zones and presence of phosphorite nodules; and reworking of microfossils. While this 50 Ma marine regression and the influx of warm water species may be coincidental, it is synchronous with a regional unconformity further to the north on the Melanesian Rift discussed in our original paper. New shallow water habitats would have been created as seas became shallower and at least some species could have expanded their ranges due to individual organisms tracking their preferred environments (Coope 1979). These new dispersal pathways from the north would, in combination with the warm Middle Eocene climate, account for the observed patterns described by Fleming (1966) and Beu and Maxwell (1990).

A Late Oligocene (28 Ma) unconformity in the South Island immediately pre-dated a dramatic increase in molluscan diversity there. The fauna of South Island localities is characterised by the first occurrence of many taxa, predominantly endemic or shared with southeast Australia with only a limited number of warm water species being recorded. While changes in oceanic circulation patterns associated with the development of the Antarctic circumpolar current or sea level falls have been suggested to account for this regional unconformity, Michaux and Leschen (2005) favoured a tectonic explanation involving the suturing of the Campbell Plateau to the Melanesian Rift. In the North Island there is no evidence of a Late Oligocene unconformity where the dramatic increase in molluscan diversity and tectonic activity occurred later in the Lower Miocene.

Beu and Maxwell (1990) divided the Lower Miocene faunas into two distinct provinces (Figure 9.4). Northland localities are characterised by many new warm water species, reef building corals and large Foraminifera (Hayward et al. 1990). This fauna is restricted to Northland during the early Middle Miocene (Otaian) but dispersed into Nelson and Southland localities in the later Middle Miocene (Altonian). The Northland fauna is distinct from that of Otago and Canterbury, which is characterised by a temperate assemblage (although probably warmer than the present). The occurrence of a distinctive Northland fauna coincided with the emplacement of the Northland Allochthon (Ballance and Spörli 1979). In our original paper, we attributed the emplacement of the allochthon to the closing of the South Loyalty Basin and the collision of the Northland Plateau with Northland (Ung et al. 2016: Figure 2). Again, the closing of the South Loyalty Basin was asynchronous, occurring earlier in New Caledonia (end of the Eocene) than in New Zealand (Lower Miocene). I interpret the distinct Northland warm water Otaian molluscs, associated reef corals and large Foraminifer as a migrant fauna that was either geodispersed into northern New Zealand along with the Northland Plateau or had expanded its range southwards along the Norfolk Ridge. As a community it was relatively short-lived with many species not surviving beyond the Lower Miocene (Fleming 1966), indicating that its presence in New Zealand was always marginal.

## *New Zealand biogeology over the past 100 Ma*

A long period of compression tectonics resulting from offshore subduction ended by 100 Ma in the New Zealand sector of East Gondwana following the culmination of the Rangitata orogeny (145–105 Ma). This orogeny was the result of the collision and amalgamation of Eastern Province terranes to Western Province rocks, resulting in the formation of a marginal mountain range. This period of mountain building and subsequent erosion would have provided many diverse habitats for Cretaceous plants and animals to expand their ranges into these newly created terrestrial environments. The occurrence of a continuous Late Cretaceous forest cover over the maritime regions of southeast Australia, New Zealand, West Antarctica and southern South America discussed earlier exemplifies this process.

Almost immediately after compression ceased, an extensional tectonic regime was initiated that formed rift systems throughout East Gondwana, resulting in the eventual fragmentation of the Australasian sector of Gondwana and the formation of a number of discrete microcontinents (Chapter 7). Figure 9.5 summarises the major events affecting the Australian-Antarctica-NZ-South American region over the past 100 Ma. These events include the isolation of Campbell Plateau-West Antarctica-South America (Figure 9.5: West Gondwana) from Australia-East Antarctica-Melanesian Rift (Figure 9.5: East Gondwana); the separation of the Campbell Plateau from West Antarctica-South America and its eventual suturing to the Melanesian Rift during the Late Oligocene resulting in a two-way biotic exchange; the separation of the Lord Howe Rise from the Norfolk Ridge and the early isolation of the Challenger Plateau and New Caledonia; two periods of regional uplift and movement of biota from north to south; and periods of volcanic activity providing refugia for meta-populations of terrestrial species on subsiding landmasses (Heads 2017).

The core components of modern New Zealand – in terms of geographical extent – are the Eastern Province terranes of the North and South Islands and the Western Province Challenger Plateau (West Coast, South Island). These two East Gondwanan units are found in the southern sectors of the Norfolk Ridge and Lord Howe Rise respectively and would have originally possessed a Cretaceous coastal biota composed of a mixture of East Gondwanan endemics and widespread taxa found in both East and West Gondwana. The Challenger Plateau was probably biologically isolated from the rest of the Lord Howe Rise because of the early, even pre-rifting opening of the Bellona Trough on its northern margin, and from the Norfolk Ridge by the opening of the Taranaki and Reinga basins. A possible consequence of these tectonic barriers is that the Challenger Plateau may not have been greatly affected by the southwards

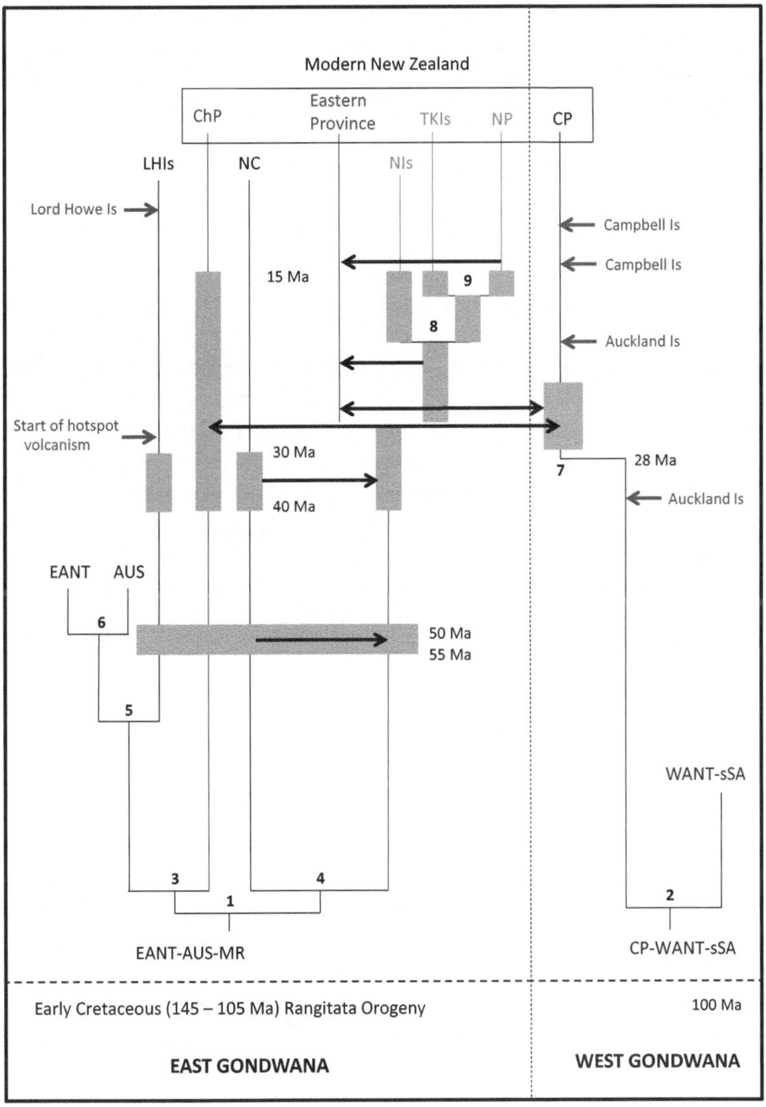


***Figure 9.5*** Biogeology of modern New Zealand. AUS = Australia, CP = Campbell Plateau, ChP = Challenger Plateau, EANT = East Antarctica, LHIs = Lord Howe Island, MR = Melanesian Rift, NC = New Caledonia, NIs = Norfolk Island. NP = Northland Plateau, sSA = southern South America, TKIs = Three Kings Islands. 1 = Fairway/New Caledonian basins, 2 = Antipodes Rift (84–80 Ma), 3 = Bellona Trough (c 80 Ma), 4 = New Caledonian Basin (early opening), 5 = Coral Sea (63–55 Ma), 6 = Southern Ocean (53 Ma), 7 = Collision of CP with MR, 8 = Norfolk Basin (23–18 Ma), 9 = Vening-Meinesz Fracture Zone (c 20 Ma). black arrows = range expansion (arrow indicates direction), grey arrow = volcanism, grey bar = marine regressions.

range expansions hypothesised to have occurred on the Melanesian Rift in the Middle Eocene and Early Miocene. If this is indeed the case, the present-day biota of the west coast of the South Island is closest, given 80 million years of evolution, to the original southern Melanesian Rift biota in terms of overall composition.

The original configuration of those Eastern Province terranes now found east of the Alpine Fault (Figure 6.1) in the Late Cretaceous is not clear but they would have been further north than their present position to account for later dextral movement along the Alpine Fault, and were presumably attached to the Australian Plate to which they were originally sutured rather than the Pacific Plate that they are now part of. The East Cape region (Figure 6.1) of the North Island, New Zealand, was also further north and was thought to have been adjacent to Northland at the time of the emplacement of the Northland allochthon (Ballance and Spörli 1979) in the Early Miocene. The Eastern Province biota may have been supplemented during the Middle Eocene regional uplift event by taxa from the north expanding their range southwards as indicated in Figure 9.5, and was substantially added to by a significant central Norfolk Ridge biota during the Early Miocene, first in Northland and subsequently in other parts of New Zealand. While the Challenger Plateau may represent a relatively unaltered biota, the biota of the Eastern Province is temporally composite.

## *The enigma of New Zealand's alpine flora and fauna*

The New Zealand alpine flora consists of about 600 highly endemic species that represent about half of all New Zealand's native dicotyledonous plants. Many alpine genera are speciose such as the mountain daisies (*Celmesia*, 48 spp.), *Hebe* (40 spp.), buttercups (*Ranunculus*, 25 spp.) and the eyebrights (*Euphrasia*, 25 spp.) (Wilson 2007). Invertebrates such as weta, stoneflies, grasshoppers, weevils, cockroaches, butterflies and moths, and cicadas also show high diversity in subalpine and alpine habitats (Buckley and Simon 2007).

The origin of this diverse and specialist flora and fauna has long been a subject of debate because subalpine and alpine habitats are geologically recent in New Zealand. Because the climate during much of the Cenozoic was warm, even subtropical (Chapter 9), and the land had little relief, it was difficult to account for a significant alpine biota. Heenan and McGlone (2013) dated alpine habitats from 1.9 Ma and as a permanent feature from as recently as 0.95 Ma. Although deformation associated with the Kaikoura orogeny started about 8 Ma, Heenan and McGlone (2013) suggested that early erosion rates were high enough to counteract uplift until 5.33 Ma, when there was also a coincidental sharp deterioration in climate. If these habitats are of recent origin, where did all these alpine specialists come from?

There have been a number of suggestions to account for the origin of New Zealand's alpine biota that can be classified into one of three broad classes of explanation (Winkworth et al. 2005):

1. Alpine lineages survived in suitable habitats within New Zealand or Antarctica.
   a. Survived in cool, wet upland habitats on infertile soils (e.g. Cockayne 1928).
   b. Lived in Antarctica and dispersed into New Zealand when the climate cooled in the Middle Miocene (e.g. Fleming 1963).
   c. Lived on the New Zealand Subantarctic Islands and dispersed into the New Zealand mainland when suitable alpine habitats formed (e.g. Wardle 1963; Michaux and Leschen 2005).

2. Alpine lineages recently dispersed into New Zealand.
   a. Dispersed from the northern hemisphere via the Malay Peninsula and New Guinea (e.g. Raven 1973).
   b. Recent dispersal from Australia (e.g. Pole 1994).

3. Evolved in situ from lowland ancestors as mountains rose.

There is evidence that some alpine species evolved from lowland ancestors. For example, Buckley and Simon (2007) presented phylogenetic evidence that the alpine species of *Maoricicada* were monophyletic and were sister group to a lowland clade, a conclusion that Fleming (1971) had come to on the basis of morphology and the occurrence of lowland species within the genus. McGlone (1985) also argued that some alpine plant species were also derived from 'pre-adapted' lowland ancestors. It seems inconceivable that some alpine species did not evolve in situ from lowland ancestors, although it would need to be established that alpine species are derived with respect to putative lowland ancestors – otherwise, one couldn't falsify the hypothesis that lowland forms recently evolved from alpine ancestors.

Michaux and Leschen (2005) suggested that the enigma of New Zealand's alpine biota might be solved if many of the alpine specialists originated on the West Gondwanan Campbell Plateau, as originally proposed by Wardle (1963). There would certainly have been suitable wet, windy, cool and infertile habitats available on the Campbell Plateau in the past, and there is a distinct austral and/or South American phylogenetic connection shown by some New Zealand alpine taxa (Wardle et al. 2001; Winkworth et al. 2005; Meudt and Simpson 2006; Wagstaff et al. 2006), that one would expect for species that evolved from Campbell Plateau palaeo-endemics, which show clear sister-group relationships to South American taxa (Michaux and Leschen 2005). The amalgamation of the Campbell Plateau with the Melanesian Rift completed the assemblage of the various components of New Zealand's biota. This amalgamation also resulted in a spatially composite biota composed of East and West Gondwanan elements. A final example will serve to demonstrate how artifactual results and misinterpretations can result from a failure to recognise composite biotas.

## *Moa, kiwi and the phylogeny of the ratites*

Recent phylogenetic analyses (Harshman et al. 2008; Phillips et al. 2010; Johnston 2011; Haddrath and Baker 2012; Smith et al. 2012; Mitchell et al. 2014) have shown that the ratites are paraphyletic and that the tinamous, previously regarded as sister group to the ratites, are nested within the ratites. Figure 9.6 shows two palaeognath phylogenies based on a multilocus nuclear gene analysis from Haddrath and Baker (2012) (Figure 9.6A) and a total molecular evidence analysis from Smith et al. (2013) (Figure 9.6B).

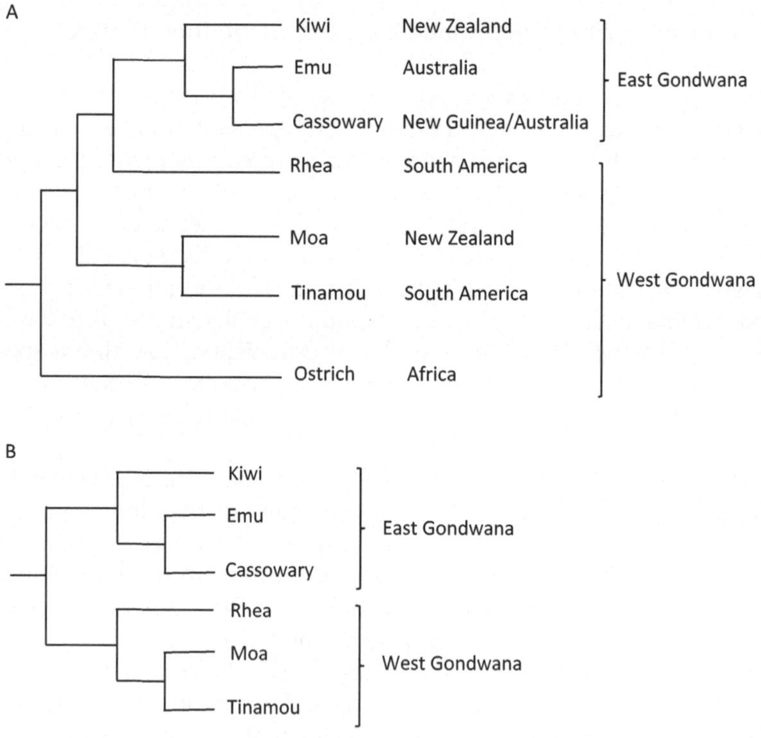


*Figure 9.6* Ratite phylogeny. A. Multilocus nuclear gene phylogeny after Haddrath and Baker (2012). B. Total molecular evidence phylogeny after Smith et al. (2013).

The elephant bird from Madagascar is missing from these phylogenies. Mitchell et al. (2014) produced a phylogeny based on mitochondrial gene sequences that placed elephant birds as sister taxon to the kiwi in the kiwi/emu/cassowary clade. Johnston's (2011) cladistic analysis of the tongue apparatus and cranial osteology placed elephant birds basally between the ostrich and all other ratites plus tinamous. The rather controversial placement of the elephant bird by Mitchell et al. (2014) needs further corroboration in light of the conflict with morphology and potential weaknesses inherent in the use of mitochondrial data. These weaknesses, outlined by Smith et al. (2013), include sensitivity to model selection and taxon sampling, and inability to detect discordance between gene and species trees. The recent discovery of hidden diversity within the elephant birds reported in Hansford and Turvey (2018) would need to be considered when taxon sampling.

The moa/tinamous clade and kiwi/emu/cassowary clades appear well supported in recently published phylogenies. The inclusion of the volant tinamous within the ratite phylogeny has cast considerable doubt on previous interpretations about the evolution of flightlessness in ratites. It now appears that flightlessness has probably evolved independently in all the ratite lineages rather than it being an ancestral condition. The alternative hypothesis that flight was independently regained by the tinamous is less likely due to the complexity of regaining flight once lost (Smith et al. 2013). The independent loss of flight has led to further inferences that are doubtful at best and ideological at worst. Harshman et al. (2008) have clearly articulated these widely held beliefs:

> Multiple loss of flight, with the implication of greater dispersal ability for ancestral paleognaths, make a strict vicariant model less compelling. (p. 13465)
>
> Dispersal of ratites is further suggested by phylogenies in which the extinct moas of New Zealand are not sister to the extant kiwis as would be predicted by strict vicariance. (p. 13465)

I've argued at some length that the 'dispersal ability' of an organism is irrelevant because chance dispersal is an extraordinary event that allows an organism to cross what is normally a barrier, and if it isn't extraordinary there is no barrier to cross and the two areas are part of the organism's normal range. While 'dispersal ability' may well influence the size of an organism's range, this is not the same as claiming it allowed the ancestors of kiwi and moa to fly across the Pacific from South America on two separate occasions. If they did so, then we could reasonably expect tinamous on all the southern continents too.

Rather than flights of fancy, I would suggest that the non-sister relationship between kiwi and moa is a result of New Zealand's spatially composite nature. Kiwi are part of a clade that contains Australian and New Guinean taxa and is clearly of East Gondwanan origin. Molecular dating indicates a New Zealand/Australia + New Guinea split sometime between 80 Ma (Haddrath and Baker 2012) and 64.5–71.6 Ma (Cooper et al. 2001), both dates of which fall within the age range for the separation of the Melanesian Rift from Australia (Chapter 9). The moa are nested within a South American clade and are equally clearly West Gondwanan. While it is possible that the ancestor of moa was a widespread species and present on the Melanesian Rift prior to its detachment from Australia, the sister-group relationship with other South American taxa rather than a more basal position strongly suggests that moa were part of the Campbell Plateau biota.

Additional evidence to support this hypothesis includes the absence of any moa fossils older than 19–16 Ma (Tennyson et al. 2010) despite the birds being large with robust bones that should readily fossilise. Bunce et al. (2009) suggested that all moa species are members of a single clade and had a South Island origin. The lack of fossils older than Miocene (19–16 Ma) suggests that the moa may not have inhabited the Melanesian Rift prior to this time, and that they were derived from a single ancestral lineage that was cold-adapted. Subsequent speciation occurred as this ancestor exploited the new habitats formed during the uplift of the Southern Alps. For instance, the alpine-specialist *Megalopteryx didinus* probably evolved in situ from a lowland ancestral form 'pre-adapted' to cool and winy habitats. This period of mountain building not only created altitudinal differences, but also significantly affected rainfall patterns, creating a mosaic of lowland/upland, wet/dry and forested/grassland habitats (Bunce et al. 2009). Fragmented habitats and the isolation of populations led to the diversification of moa, as happened to a number of other diverse alpine-specialist groups.

# *chapter ten*

# *Synthesis*

## *Breakup of Gondwana*

### *The 'classical' phase*

When Harshman et al. (2008) talked about a 'strict vicariant model' in their discussion of ratite evolution, they were referring to the breakup sequence and timing that gave rise to the southern continents as Gondwana fragmented. This sequence is well established and the major events are summarised in Table 10.1. The ages quoted are those of the oldest oceanic crust identified within each basin, but the process of continental fragmentation – from rifting to formation of new oceanic crust – can be a protracted process (discussed here). Africa was the first of the southern continents to become isolated, initially from Madagascar/India at 167 Ma with the opening of the Mozambique Channel, followed by separation from Antarctica at 147 Ma and South America at 130 Ma. India detached from both Australia and Antarctica at 120 Ma as the Indian and Southern Oceans opened. Madagascar and India separated at 85 Ma. The third arm of the India/Antarctica/Australia triple junction ran between Antarctica and Australia. This rift system had a complicated and prolonged history before sea floor spreading eventually detached Australia from Antarctica and translated it northwards at c. 45 Ma. South America was the last of the southern continents to separate from Antarctica when the Drake Passage connecting Patagonia to the Antarctic Peninsula opened at 35 Ma.

Using rifting ages to date vicariant events (i.e. to date the biological isolation of populations) is not straightforward because rifting can be a protracted process and barriers to dispersal for some organisms may form earlier than the first occurrence of oceanic crust. Cochran (1983) modelled the rifting of continental crust and rejected the idea of instantaneous rifting used in previous models, suggesting instead that the difference in time between rifting and formation of basins was anywhere between 10 and 50 million years. Cochran (1983) discussed a number of examples to support these delay intervals, and recently Brune and Autin (2013) described the opening of the Gulf of Aden between Somalia and Yemen, dating initiation of rifting to 34 Ma and the generation of oceanic crust to 17.6 Ma, a period of approximately 16 million years. There are examples of much slower rifting than the 50-million-year upper limit suggested by Cochran (1983), including

*Table 10.1* Classical breakup sequence of Gondwana. Ages are those of the oldest oceanic crust identified

| Continent | Separated from | Basin opening | Timing | Reference |
|---|---|---|---|---|
| **Africa** | Madagascar/India | Mozambique Basin | 167 Ma | Veevers (2012) |
| | Antarctica | Weddell Sea | 147 Ma | |
| | South America | Southern Atlantic Ocean | 130 Ma | |
| **India** | Australia | Indian Ocean | 120 Ma | Ali and Aitchison (2008) |
| | Antarctica | Southern Ocean | 120 Ma | |
| | Madagascar | Mascarene Basin | 85 Ma | |
| **Australia** | Antarctica | Southern Ocean | 45 Ma | Veevers et al. (1991) |
| **South America** | West Antarctica | Drake Passage | 35 Ma | Livermore et al. (2005) |

the Australia/Antarctica Rift (discussed in Chapter 7) and the opening of the North Atlantic off Newfoundland (Hopper et al. 2004), which were both intermittently active for some 70 million years before oceanic crust was generated.

## *Peeling an onion skin*

The formation of the southern continents can be thought of as the 'classical' phase of Gondwanan breakup. However, their formation occurred quite late in the breakup history, and prior to this there had been major and prolonged episodes of marginal rifting producing generally ribbon-like terranes along the northern margin of Gondwana from Australia to Africa/Arabia (Table 10.2). Robertson (2007) related the phenomenon of marginal rifting in the eastern Tethys to the effects of long-term asthenosphere flow, slab pull related to subduction beneath Laurasia and melt-induced crustal weakening associated with pulsed rifting or plume effects. Robertson (2007) rejected back-arc spreading as a plausible mechanism to explain rifting because subduction occurred beneath Eurasia not Gondwana in his reconstructions. However, back-arc spreading in combination with slab rollback has been widely used to explain basin formation in areas including the Pacific and Wallacea (e.g. Schellart et al 2006; Spakman and Hall 2010), and not all researchers agree that the southern margin of Laurasia was a long-lived Andean-type subduction zone and favour episodes of back-arc spreading and intra-ocean subduction (Zahirovic et al. 2014).

*Table 10.2* Gondwana Tethyan terranes. Dates are rifting ages

| Time | Rifted terranes | Ocean basin | Reference |
|---|---|---|---|
| **Late Devonian** | South China, North China, Tarim, Indochina, Qaidam | Palaeo-Tethys | Metcalfe (2006) |
| c 350 Ma | Asiatic Hunic terranes | Palaeo-Tethys | Stampfli and Borel (2004) |
| | European Hunic terranes (e.g. Serbia-Macedonia) | Palaeo-Tethys | Stampfli and Borel (2004) |
| **Late Permian** | Cimmerian Continent (West Cimmeria, Qiangtang, Sibumasu) | Meso-Tethys | Metcalfe 2006 |
| c 250 Ma | Eastern Mediterranean terranes | Neo-Tethys | Stampfli and Borel (2004) |
| | | Triassic Tethys | Robertson (2007) |
| **Late Jurassic** | Lhasa, Banda/wSulawesi/wBorneo (Argoland) | Ceno-Tethys | Metcalfe (2006) |
| c 150 | Opening of Central Atlantic and its extension Alpine Tethys | Neo-Tethys | Stampfli and Borel (2004) |
| | | Southern Neo-Tethys | Robertson (2007) |
| **Mid-Tertiary** | Sula Spur | Banda Sea | Hill and Hall (2003) |
| c 30 Ma | Western Mediterranean terranes | | Rosenbaum et al. (2002) |
| | | | Golonka (2004) |

The first phase of marginal rifting occurred in the Late Devonian (Table 10.2) as Asiatic, Himalayan and European terranes rifted from Gondwana's northern margin in response to the subduction of Palaeo-Pacific crust beneath northern hemisphere continental blocks. Individual elements of this initial phase of rifting included the North and South China Blocks, Tarim, Indochina, East Malaya, West Sumatra, South Tibet and Central European terranes, such as Serbia/Macedonia. As these terranes were translated northwards, new ocean crust was created behind them forming the Palaeo-Tethys Ocean (Metcalfe 2011).

A second phase of terrane generation occurred at the end of the Palaeozoic in the Late Permian with the formation of the Cimmerian Continent in the east and various East Mediterranean/Middle Eastern terranes in the west. The Palaeo-Tethys Ocean was consumed at subduction zones along its northern margin with Eurasia, and a new ocean – the

Meso-Tethys of Metcalfe (2006) or the Neo-Tethys of Stampfli and Borel (2004) – was formed outboard of Gondwana. A third phase of rifting in the Late Jurassic produced terranes now found in Southeast Asia and the Himalayas, and was coincidental with the opening of the central Atlantic and ocean basin formation in the western Mediterranean. Once again, Tethyan crust was destroyed in subduction zones along the Eurasian margin and a new Tethys ocean was formed outboard of Gondwana. The final act leading to today's world was the northward movement of South America, India, Africa and Australia, bringing these large blocks into contact with North America and Eurasia, forming the Caribbean, Mediterranean, Himalayan and Wallacean collision zones.

While I have distinguished between a late 'classical' breakup phase that produced continental-sized fragments and an earlier production of continental rift structures, these two phases are end points of a continuum. While rifted continental terranes were a feature of Gondwana while it was still part of Pangaea (until Middle Jurassic times), continental rifts continued to be produced from the Australian sector until the Late Cretaceous (Chapters 4 and 8). Neither is the size distinction absolute with structures such as the Lord Howe Rise approaching continental size while showing features of rifted continental terranes (Chapter 9). In short, the development of Gondwana through time is a complex process not usefully simplified.

## *Australia–Sunda–Pacific collision zones (Glen 2005)*

Figure 10.1 and Table 10.3 summarise the tectonic structures and biological connections of the area covered in this book. The tectonic structures include cratonic fragments derived from the classical phase of Gondwanan breakup, and Palaeozoic, Mesozoic and Cenozoic terranes and island arcs. Biological elements include 'autochthonous' taxa linked to particular cratons or terranes, and 'allochthonous' taxa that have expanded their ranges or have been geodispersed.

### *Palaeozoic terranes and microcontinents*

Palaeozoic terranes and microcontinents of Gondwanan origin are found in China and Southeast Asia. The Early Devonian (pre-rift) biotas of these terranes were Gondwanan, but by the Carboniferous (drift-phase) had evolved into a distinctive Cathaysian biota, which was replaced with a Eurasian biota in the Middle Jurassic (accretion-phase) (Metcalfe 2013: Figure 5). Their modern biotas have no Gondwanan biological signal because of their early derivation ages and are now entirely Eurasian (Table 10.2).

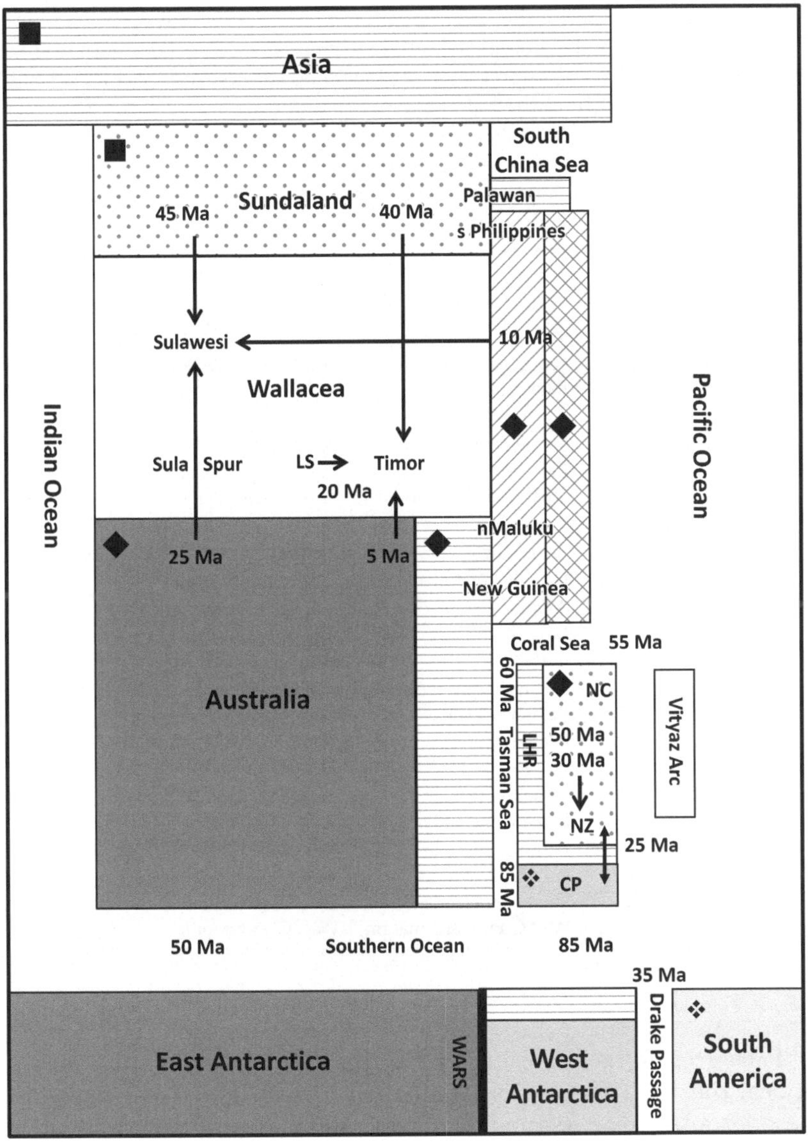


*Figure 10.1* Dark grey fill = East Gondwana cratons. Light grey fill = West Gondwana microcontinents. Horizontal lines = Palaeozoic terranes. Dense stipple = Mesozoic terranes. Medium stipple = Palaeogene arcs and continental fragments. Light stipple = Neogene arcs. Square = Eurasian biota. Diamond = East Gondwanan biota. Broken diamond = West Gondwanan biota. Arrows indicate direction of range expansions.

***Table 10.3*** Summary of the tectonic and biological composition of regions discussed in the text. Geological composition refers to terrane history: For example, Halmahera is composed of a Palaeocene island arc that collided with the Australian margin in the Late Palaeocene and on which a second arc was built in the Neogene. Halmahera is biologically composite with both cratonic Australian and Melanesian species. The Melanesian biota is temporally composite having been derived at two different periods. Detailed explanations in the text. ? = uncertainty about the presence of a Neogene arc in the Philippines and if it is present whether any biota were geodispersed with it

| | | Geological composition | Biological composition | Temporally composite |
|---|---|---|---|---|
| Melanesia | Halmahera | Neogene arc<br>Palaeogene arc<br>Australian margin (Obi) | Neogene Melanesian<br>Palaeogene Melanesian<br>Australian | Yes |
| | New Guinea | Neogene arc<br>Palaeogene arc<br>Australian margin | Neogene Melanesian<br>Palaeogene Melanesian<br>Australian | Yes |
| Wallacea | Sulawesi | Neogene arc<br>Sula Spur<br>Sundaic microcontinent | Neogene Melanesian<br>Neogene Australian<br>Palaeogene Sundaic | No |
| | Timor | Australia collision<br>Inner Banda arc collision<br>Sundaic margin collision | Neogene Australian<br>Neogene Sundaic<br>Palaeogene Sundaic | Yes |
| | Philippines | Asian margin terrane<br>Palaeogene arc<br>Neogene arc?<br>Sunda margin terrane | Palaeogene Asian<br>Palaeogene Melanesian<br>Neogene Melanesian?<br>Palaeogene Sundaic | ? |
| Southwest Pacific | North and South Islands New Zealand | Neogene uplift<br>Palaeogene collision<br>Palaeogene uplift<br>Australian margin | Neogene Norfolk Ridge<br>West Gondwanan<br>Palaeogene Norfolk Ridge<br>Australian | Yes |
| | Campbell Plateau | Palaeogene collision<br>West Gondwana margin | East Gondwanan<br>West Gondwanan | No |

Palaeozoic terranes are also found in eastern Australia, southern New Guinea, the Lord Howe Rise, Campbell Plateau and West Antarctica. These terranes were originally sutured to the Australian craton and to at least some of the microcontinental terranes that form West Antarctica. By the mid-Cretaceous, a similar flora and fauna was found along all the coastal regions of polar Gondwana, but became increasingly differentiated after 100 Ma as rifting and the opening of ocean basins intensified (Chapter 7). The primary biotic disjunction was between West and East Gondwana biotas with secondary areas of endemism developed as fragmentation and geodispersal gathered pace (Chapter 6).

## Mesozoic Terranes

Mesozoic terranes make up much of Sundaland and the Melanesian Rift (North Zealandia). In Sundaland, there was episodic accretion of terranes onto a Palaeozoic core (Indochina-East Malaya block), starting in the Middle Triassic when the Sibumasu terrane docked. Pre-Cretaceous Mesozoic sediments are rare on Sundaland, indicating it was terrestrial for much of the period following amalgamation of the Sibumasu terrane. Accretion resumed in the Early Cretaceous (Southwest Borneo – west Java – west Sumatra) and again in the mid-Cretaceous (Argoland). The biota of Sundaland has strong Asian affinities, presumably a result of range expansion by Asian species following accretion and uplift. However, some Sundaic (and Indian and Southeast Asian) taxa are distributed both east and west of Wallace's Line, and while these distributions are uniformly interpreted as evidence of chance dispersal events, some studies have recovered early dates for the origin of Asian and Australasian sister clades (e.g. Schulte et al. 2002; Brown et al. 2006). Mesozoic geodispersal of Gondwanan biotas via the Indian craton or Gondwanan rifted terranes now found in the Himalayas or Sundaland cannot be ruled out as a possible explanation for taxa whose distribution patterns ignore Wallace's Line.

Mesozoic terranes also make up much of the Melanesian Rift. The amalgamation of these Eastern Province terranes to Gondwana is recorded by the Rangitata orogeny, which was followed by a prolonged period of erosion and eventual rifting as the Tasman Sea and Coral Sea opened, isolating a Late Cretaceous/Palaeocene East Gondwanan biota on North Zealandia (Chapter 9).

## Palaeogene terranes

Palaeogene terranes include continental fragments and an island arc system, parts of which collided with New Guinea and became accreted onto the Australian foreland. Arc fragments associated with this system are also found in Halmahera (Chapter 2) and the Philippines (Table 10.3). Clockwise rotation of the Philippine Sea Plate translated arcs northwards along with a Melanesian biota recognised in Halmahera (e.g. birds-of-paradise) and the Philippines (Michaux 2010: Figure 5). Other Palaeocene terranes are predominantly continental fragments (Table 10.3): central Sulawesi/southwest peninsula microcontinent that rifted from the Sundaland margin carrying with it an Asian biota (Chapter 5); the collision of proto-Timor with the margin of Sundaland and the range expansion of Eocene Sundaic taxa onto it (Chapter 8); and the emplacement of various Australian fragments in the Central Highlands of New Guinea (Chapter 4). The collision of the Campbell Plateau with Northern

Zealandia in the late palaeogene juxtaposed East and West Gondwanan biotas, including species 'pre-adapted' to the cool, wet and windy alpine and subalpine environments that were to develop in the Late Neogene of New Zealand.

### *Neogene terranes*

Neogene tectonics in Wallacea are dominated by the collision of the Australian craton with Sundaland and the Pacific plate. The collision was initiated when the Sula Spur collided with the Sundaic margin, geodispersing an Australian biota as far west as Sulawesi (Table 10.3). Its subsequent fragmentation through the opening of marginal basins also translated Timor southwards to collide with the Australian margin. The continued northwards movement of Australia resulted in collision with Nusa Tengarra in the Wetar region (Chapter 8), and the emplacement of Neogene island arc fragments along the northern margin of New Guinea (Chapter 4). Further south along the Pacific margin, the obduction of the Northland Allochthon coincided with an influx of a warm marine biota and subtropical flora into northern New Zealand (Chapter 9).

## *Postscript*

The analysis of the spatial and temporal distribution of life in the Indonesian-west Pacific region described herein is by necessity an integrative process. It requires a synthesis of taxonomy, systematics, natural history, ecology and geology into a coherent whole. As such, it is an endeavour that can provide an antidote to the specialisation that dominates much of contemporary life sciences, which in my experience as a teacher is discouraging many of the ablest students from pursuing science degrees at university. Their perception is that life sciences are technically- rather than ideas-driven and don't provide the creative potential and big-picture thinking they seek. From my limited knowledge of the tertiary sector, there appears to be a growing feeling of frustration on the part of graduate students about the ability to pursue personally meaningful research within the academic environment. These warning signs are dangerous to ignore.

Whether you agree or disagree with the conclusions outlined in this book, the approach I've described is methodologically sound and any assumptions clearly articulated and justified, and it represents a point of view that needs airing if healthy debate is to be encouraged. Yet this book would almost certainly not have been written if I'd not been an independent, non-affiliated researcher because there are no programmes or facilities dedicated to historical biogeographical research that would have encouraged its writing. The need for a more integrated approach in the life sciences has never been more urgent as modern society grapples with global-scale environmental issues. We know that the solutions to these issues are multifactorial and immune to 'silver bullet' approaches and require big-picture thinking to contextualise the problem.

Consider the efforts to halt the alarming decline in New Zealand biodiversity since European settlement from the mid-1800s. Historically, this issue has been addressed through the creation of national parks and other protected areas to save places with specific conservation values from economic exploitation; then the emphasis switched to saving iconic species of our avifauna such as kiwi (*Apteryx* spp), kakapo (*Strigops habroptilus*), takahe (*Porphyrio hochstetteri*) and kokako (*Callaeas wilsoni*), or threatened and much-loved flora such as pohutakawa (*Metrosideros excelsa*) and more recently kauri (*Agathis australis*). Presently, the emphasis is on pest control and the creation of pest-free islands or mainland sanctuaries. The future aspirational goal is having the entire country pest-free by 2050. While these are all laudable in intention, they are reactive, ad hoc solutions: setting aside areas without an integrated pest management programme is never likely to deliver much of conservation value; saving individual species without addressing the issues of habitat modification and loss that caused their decline in the first place can never be much more than a band-aid solution; and creating a pest-free environment without addressing national air, water and land use issues will be a lost opportunity. What is really missing is a long game that might just guide conservation effort, and this is where an understanding of the history of the biota is so important.

Rather than being the flotsam and jetsam of contingent events, in a constant state of flux as new migrants arrive and resident species go extinct, biogeology teaches us that the New Zealand biota, like many found on isolated islands, has had a long coevolutionary history. Unlike smaller islands, New Zealand has a mosaic of such communities – a Cretaceous polar Gondwanan *Nothofagus*-podocarp assemblage, a northern Zealandia kauri temperate rain forest and a specialised subalpine/alpine flora and fauna. There are no sharp geographic or compositional boundaries to these communities now because they share both space and species and have common evolutionary histories. But their existence has important implications for conservation planning.

The most important of these implications is that the biota is structured and there isn't *a* New Zealand biota. West Gondwanan elements found in the Subantarctic Islands and the alpine and subalpine habitats of the South Island are of special importance because this biota has been extirpated from West Antarctica and is now found only in these geographically restricted areas of southern New Zealand and southern South America. It will come under increasing pressure in New Zealand as climate change decreases the extent of alpine and subalpine habitats and pest species invade to higher altitudes. The rain forests of Northland also represent the remnant of another once more widespread biota – a southern Norfolk Ridge flora and fauna – now otherwise restricted to small islands such as Norfolk Island or the Three Kings Group. This biota should also have high conservation status because of its remnant status. The major environmental issue facing Northland kauri forest is its fragmentation, which has reduced its area, restricted movement of poorly dispersing taxa between fragments, increased all manner of edge effects and, from a Maori perspective (mātaurangi māori), reduced the mauri of the forest making it sick (e.g. spread of kauri die-back disease caused by *Phytophthora agathidicida*).

A second critical factor that has implications for conservation planning is the age of these assemblages. Mixed *Nothofagus*-podocarp forest is New Zealand's autochthonous element and can be traced back to pre-rift times. While 100 million years of evolution and extinction have changed its composition, there has nevertheless been a protracted period of coevolution between its constituent parts. Empirical studies have clearly established that isolated biotas, such as those found on islands or lakes, are prone to collapse when perturbed by the introduction of invasive species (e.g. Steadman 1995; Fritts and Rodda 1998). The connection between ecosystem stability and species diversity was reviewed in an influential paper by Kevin McCann (2000), who linked the greater stability of more diverse systems to a greater number of weakly interacting species' associations, which has a dampening effect and reduces the probability that invasive species will induce an extinction cascade. While different measures of stability and resilience yield different stability-diversity outcomes (McCann 2000; Ives and Carpenter 2007), and other factors such as response diversity (Mori et al. 2013) and mode of interaction (Ives and Carpenter 2007) have been implicated as causal factors, the central conclusion that decreased biodiversity increases average interaction strengths resulting in a decrease of ecosystem stability appears to be sound.

Because of their long history of coevolution, one would expect that many New Zealand species' interactions were strong, and that a combination of isolation from new immigrant sources and constant background extinction rates have reduced biodiversity over time, making these ecosystems particularly vulnerable to introduced predators. Overall, 58 New Zealand species of birds (26% of the avifauna) have become extinct since humans arrived some 800 years ago, with extinction rates for endemic forest birds reaching 41% (Tennyson 2006). Factors such as prey specialisation (e.g. Haast's eagle, *Harpagornis moorei*), flightlessness, ground nesting and mammalian predator naiveté are all important contributing factors to avian extinctions and demonstrate just how strong species' interactions had become. Alpine and subalpine environments are a little different because they are younger and have a relatively high diversity (in a

New Zealand context) due to accelerated speciation rates in response to the uplift of the Southern Alps. Areas where immigrant species from the northern kauri forests have mixed with the autochthonous Nothofagus-podocarp forests are also younger and should be more biodiverse.

There is evidence that avian biodiversity is greater in both alpine and subalpine habitats and in the composite forests of the lower North Island. For example, distribution maps for numbers of all bird taxa in native forest habitats (Robertson et al. 2007:392) and numbers of endemic taxa (Robertson et al. 2007:398) are both higher in these habitats. There is also evidence from bird distributions that these environments are more stable, as ecologists have predicted for environments with higher diversity. The numbers of introduced avian species are lower in these habitats, indicating lower invasion rates (Robertson et al. 2007:402) and endemic species such as whio (*Hymenolaimus malacorhynchos*), kakariki (*Cyanoramphus* spp), karearea (*Falco novaeseelandiae*), weka (*Gallirallus australis*), tititipounamu (*Acanthisitta chloris*), popokotea (*Mohoua albicilla*), toutouwai (*Petroica longipes*), korimako (*Anthornis melanura*) and kokako (*Callaeas wilsoni*) still survive there but have been extirpated (or have only small remnant populations) further north in Auckland and Northland (Robertson et al. 2007). Understanding the biological history of a place isn't just of academic interest. It has real life implications for how we view the world and manage environmental issues.

# References


Abbott L.D., Silver E.A., Galewsky J. 1994. Structural evolution of a modern arc-continent collision in Papua New Guinea. *Tectonics* 13: 1007–1034.

Adams C.G., Lee D.E., Rosen B.R. 1990. Conflicting isotopic and biotic evidence for tropical sea-surface temperatures during the Tertiary. *Palaeogeography, Palaeoclimatology, Palaeoecology* 77: 289–313.

Adams C.J. 1981. Migration of late Cenozoic volcanism in the South Island of New Zealand and the Campbell Plateau. *Nature* 294: 153–155.

Adams C.J., Campbell H.J., Griffin W.J. 2008. Age and provenance of basement rocks of the Chathem Islands: An outpost of Zealandia. *New Zealand Journal of Geology and Geophysics* 51: 245–259.

Adams C.J., Campbell H.J., Mortimer N., Griffin W.L. 2017. Perspectives on Cretaceous Gondwana break-up from detrital zircon provenance of southern Zealandia sandstones. *Geological Magazine* 154: 661–682.

Adams C.J., Morris P.A., Beggs J.M. 1979. Age and correlation of volcanic rocks of Campbell Island and Metamorphic basement of the Campbell Plateau, South-west Pacific. *New Zealand Journal of Geology and Geophysics* 22: 679–691.

Adams C.J., Seward D., Weaver S.D. 1995. Geochronology of Cretaceous granites and metasedimentary basement on Edward VII Peninsula, Marie Byrd Land, West Antarctica. *Antarctic Science* 7: 265–277.

Adams G.J., Gullen D.J. 1978. Potassium-argon ages of the granites and metasediments from the Bounty Islands area, southwest Pacific Ocean. *Journal of the Royal Society of New Zealand* 8: 127–132.

Ali J.R., Hall R. 1995. Evolution of the boundary between the Philippine Sea Plate and Australia: Palaeomagnetic evidence from eastern Indonesia. *Tectonophysics* 251: 251–275.

Amos J. 2017. Plate tectonics: When we discovered how the Earth really works. *Guardian newspaper,* 2nd October 2017.

Andrews P.B., Campbell H.J., Watters W.A. 1978. The Forty Fours: The most easterly outcrop of Mesozoic basement in the New Zealand region (Note). *New Zealand Journal of Geology and Geophysics* 21: 649–652.

Armstrong P.H., Martin G. 2015. *Geographers: Bibliographic Studies,* Volume 20. Bloomsbury, London, UK, 152 pp.

Audley-Charles M.G. 2004. Ocean trench blocked and obliterated by Banda forearc collision with Australian proximal continental slope. *Tectonophysics* 389: 65–79.

Austin C.C. 2000. Molecular phylogeny and historical biogeography of Pacific Island Boas (Candoia). *Copeia* 2000: 341–352.

Austin C.C., Rittmeyer E.N., Richards S.J., Zug G.R. 2010. Phylogeny, historical biogeography and body size evolution in Pacific Island Crocodile skinks Tribolonotus (Squamata: Scincidae). *Molecular Phylogenetics and Evolution* 57: 227–236.

Bagils R.Z., Ung V., Grand A., Vignes-Lebbe R., Cao N., Ducasse J. 2012. LisBeth: New cladistics for phylogenetics and biogeography. *Comptes Rendus Palevol* 11: 563–566.

Bailey A.M., Sorensen J.H. 1962. *Subantarctic Campbell Island. Denver Museum of Natural History, Proceedings Number 10.* A.W. & A.H. Reed, Auckland, New Zealand.

Balaguru A., Hall R. 2009. Tectonic evolution and sedimentation of Sabah, north Borneo, Malaysia. *Search and Discovery* Article 30084, 15 pp.

Baldwin S.L., Fitzgerald P.G., Webb L.E. 2012. Tectonics of the New Guinea region. *Annual Review of Earth and Planetary Sciences* 40: 495–520.

Baldwin S.L., Monteleone B.D., Webb L.E., Fitzgerald P.G., Grove M., Hill E.J. 2004. Pliocene eclogite exhumation at plate tectonic rates in eastern Papua New Guinea. *Nature* 431: 263–267.

Ball I.R. 1975. Nature and formulation of biogeographical hypotheses. *Systematic Zoology* 24: 407–430.

Ball I.R. 1990. The framing of biogeographical hypotheses. In: Garbary D.J., South G.R. (eds) *Evolutionary Biogeography of the Marine Algae of the North Atlantic.* Springer, Berlin, Germany, pp. 1–7.

Ballance P.F., Spörli K.B. 1979. Northland allochthon. *Journal of the Royal Society of New Zealand* 9: 259–275.

Barluenga M., Stölting K.N., Salzburger W., Muschick M., Meyer A. 2006. Sympatric speciation in Nicaraguan crater lake cichlid fish. *Nature* 439: 719–723.

Beggs J.Mac. 1978. Geology of the metamorphic basement and late Cretaceous Oligocene sedimentary sequence of Campbell Island, Southwest Pacific Ocean. *Journal of the Royal Society of New Zealand* 8: 161–177.

Beggs J.M., Challis G.A., Cook R.A. 1990. Basement geology of the Campbell Plateau: Implications for correlation of the Campbell Magnetic Anomaly System. *New Zealand Journal of Geology and Geophysics* 33: 401–404.

Berger J. 1973. *Ways of Seeing.* Penguin, London, UK.

Besnard G., Bertrand J.A.M., Delahaie B., Bourgeois Y.X.C., Lhuillier E., Thébaud C. 2016. Valuing museum specimens: High-throughput DNA sequencing on historical collections of New Guinea crowned pigeons (Goura). *Biological Journal of the Linnean Society* 117: 71–82.

Beu A.G., Maxwell P.A. 1990. Cenozoic Mollusca of New Zealand. New Zealand Geological Survey Paleontological Bulletin 58, 520 pp.

Biffin E., Hill R.S., Lowe A.J. 2010. Did Kauri (*Agathis*: Araucariaceae) really survive the Oligocene drowning of New Zealand? *Systematic Biology* 59: 594–602.

Boger, S.D., Spelbrink, L.G., Lee, R.I., Sandiford, M., Maas, R., Woodhead, J.D. 2017. Isotopic (U-Pb, Nd) and geochemical constraints on the origins of the Aileu and Gondwana sequences of Timor. *Journal of Asian Earth Sciences* 134: 330–351.

Bolnick D.I., Fitzpatrick B.M. 2007. Sympatric Speciation: Models and Empirical evidence. *Annual Review of Ecology, Evolution, and Systematics* 38: 459–487.

Bowring S.A., Williams I.S. 1991. Priscoan (4.00–4.03) orthogneiss from northwestern Canada. *Contributions to Mineralogy and Petrology* 134: 3–16.

Bronstein O., Kroh A., Tautscher B., Liggins L., Haring E. 2017. Cryptic speciation in pan-tropical sea urchins: A case study of an edge-of-range population of Tripneustes from the Kermadec Islands. *Scientific Reports* 7: 5948.

Brown G.K., Nelson G., Ladiges P.Y. 2006. Historical biogeography of *Rhododendron* section *Vireya* and the Melanesian Archipelago. *Journal of Biogeography* 33: 1929–1944.

Brune S., Autin J. 2013. The rift to break-up evolution of the Gulf of Aden: Insights from 3D numerical lithospheric-scale modelling. *Tectonophysics* 607: 65–79.

Bruxaux J., Gabrielli M., Ashari H., Prŷs-Jones R., Joseph L., Milá B., Besnard G., Thébaud C. 2018. Recovering the evolutionary history of crowned pigeons (Columbidae: Goura): Implications for the biogeography and conservation of New Guinean lowland birds. *Molecular Phylogenetics and Evolution* 120: 248–258.

Buckley T.R., Leschen R.A.B. 2013. Comparative phylogenetic analyses reveals long-term isolation of lineages on the Three Kings Islands, New Zealand. *Biological Journal of the Linnean Society* 108: 361–377.

Buckley T.R., Simon C. 2007. Evolutionary radiation of the cicada genus *Maoricicada* Dugdale (Memiptera: Cicadoidea) and the origins of the New Zealand alpine biota. *Biological Journal of the Linnean Society* 91: 419–435.

Bunce M., Worthy T.H., Phillips M.J., Holdaway R.N., Willerslev E., Haile J., Shapiro B., Scofield R.P., Drummond A., Kamp P.J., Cooper A. 2009. The evolutionary history of the extinct ratite moa and New Zealand Neogene paleogeography. *Proceedings of the National Academy of Sciences of the United States of America* 106: 20646–20651.

Calvalcanti M.J. 2009. Croizat: A software package for quantitative analysis in panbiogeography. *Biogeografía* 4: 5–7.

Campbell J.D. 2002. Angiosperm fruit and leaf fossils from Miocene silcrete, Landslip Hill, northern Southland, New Zealand. *Journal of the Royal Society of New Zealand* 32: 149–154.

Carpenter R.J., Jordan G.J., Macphail M.K., Hill R.S. 2012. Near-tropical Early Eocene terrestrial temperatures at the Australo-Antarctic margin, western Tasmania. *Geology* 40(3): 267–270.

Carr E.H. 1970. *What Is History?* Vintage Books, New York, NY.

Carter R.M. 1988. Post-breakup stratigraphy of the Kaikoura Synthem (Cretaceous-Cenozoic), continental margin, southeastern New Zealand. *New Zealand Journal of Geology and Geophysics* 31: 405–429.

Chappell P.R. 2013. *The Climate and Weather of Northland* (3rd Edition). NIWA Science and Technology Series Number 59, NIWA, Christchurch, New Zealand. 40 pp.

Charlton T.R. 2001. Permo-Triassic evolution of Gondwanan Eastern Indonesia, and the final Mesozoic separation of SE Asia from Australia. *Journal of Asian Earth Sciences* 19: 595–617.

Cibois A., Thibault J.C., Bonillo C., Filardi C.E., Pasquet E. 2017. Phylogeny and biogeography of the imperial pigeons (Aves: Columbidae) in the Pacific Ocean. *Molecular Phylogenetics and Evolution* 110: 19–26.

Coates B.J., Bishop K.D. 1997. *A Guide to the Birds of Wallacea.* Dove Publications, Alderley, Queensland.

Cochran J.R. 1983. Effects of finite rifting times on the development of sedimentary basins. *Earth and Planetary Science Letters* 66: 289–302.

Cockayne L. 1928. *The Vegetation of New Zealand*. Engelmann, Leipzig, Germany.

Conran J.G., Mildenhall D.C., Raine J.I., Kennedy E.M., Lee D.E. 2015. The monocot fossil pollen record of New Zealand and its implications for palaeoclimates and environments. *Botanical Journal of the Linnean Society* 178: 421–440.

Contreras L., Pross J., Bijl P.K., Koutsodendris A., Raine J.I., van de Schootbrugge B., Brinkhuis H. 2013. Early to Middle Eocene vegetation dynamics at the Wilkes Land Margin (Antarctica). *Review of Palaeobotany and Palynology* 197: 119–142.

Cook R.A., Sutherland R., Zhu H. 1999. Cretaceous-Cenozoic geology and petroleum systems of the Great South Basin, New Zealand. Institute of Geological & Nuclear Sciences Monograph 20, 190 pp.

Coope G.R. 1979. Late Cenozoic fossil Coleoptera: Evolution, biogeography, and ecology. *Annual Review of Ecology and Systematics* 10: 247–267.

Cooper A., Lalueza-Fox C., Anderson S., Rambaut A., Austin J., Ward R. 2001. Complete mitochondrial genome sequences of two extinct moas clarify ratite evolution. *Nature* 409: 704–707.

Crampton J., Laird M., Nicol A., Townsend D., van Dissen R. 2003. Palinspastic reconstructions of southeastern Marlborough, New Zealand, for mid-Cretaceous-Eocene times. *New Zealand Journal of Geology and Geophysics* 46: 153–175.

Craw R.C. 1978. Two biogeographical frameworks: Implications for the biogeography of New Zealand. A review. *Tuatar* 23: 81–114.

Craw R.C. 1988. Continuing the synthesis between panbiogeography, phylogenetic systematics and geology as illustrated by empirical studies on the biogeography of New Zealand and the Chatham Islands. *Systematic Biology* 37: 291–310.

Craw R.C., Weston P. 1984. Panbiogeography: A progressive research program? *Systematic Biology* 33: 1–13.

Crisp M.D., Trewick S.A., Cook L.G. 2011. Hypothesis testing in biogeography. *Trends in Ecology and Evolution* 26: 66–72.

Croizat L. 1964. Space, time, form: The biological synthesis. Published by the author, Caracas.

Davey F.J., Christoffel D.A. 1978. Magnetic anomalies across Campbell Plateau, New Zealand. *Earth and Planetary Science Letters* 41: 14–20.

Davies H.L. 2012. The geology of New Guinea – The cordilleran margin of the Australian continent. *Episodes* 35: 1–16.

Davies H.L., Perembo R.C.B., Winn R.D., KenGemar P. 1997. Terranes of the New Guinea Orogen. Geology, Exploration and Mining Conference, Madang, 10–12 October, 1997.

Davy B. 2014. Rotation and offset of the Gondwana convergent margin in the New Zealand region following Cretaceous jamming of Hikurangi Plateau large igneous province subduction. *Tectonics* 33: 1577–1595.

Decker J., Ferdian F., Morton A., Fanning M., White L.T. 2017. New geochronology data from eastern Indonesia – An aid to understanding sedimentary provenance in a frontier region. Proceedings, Indonesian Petroleum Association, Forty-First Annual Convention and Exhibition, Jakarta, Indonesia, May 2017.

Dietz R.S. 1961. Continent and ocean basin evolution by spreading of the sea floor. *Nature* 190: 854–857.

Dell R.K. 2000. *Fleming, Charles Alexander, Dictionary of New Zealand Biography*. Te Ara – The Encyclopedia of New Zealand, https://teara.govt.nz/en/biographies/5f9/fleming-charles-alexander (accessed 13 October 2018).

DeMets C., Gordon R.G., Argus D.F. 2010.Geologically current plate motions. *Geophysical Journal International* 181: 1–80.

DeMets C., Gordon R.G., Argus D.F., Stein S. 1990. Current plate motions. *Geophysical Journal International* 101: 425–478.

DeMets C., Gordon R.G., Argus D.F., Stein S. 1994. Effect of recent revisions to the geomagnetic reversal timescale. *Geophysical Research Letters* 21: 2191–2194.

Dickerson R.E., Merrill E.D., McGregor R.C., Schultze W., Taylor E.H., Herre A.W. 1923. Distribution of life in the Philippines. Manila Bureau of Science Monograph 21: 1–322.

DiVenere V.J., Kent D.V., Dalziel I.W.D. 1994. Mid-Cretaceous paleomagnetic results from Marie Byrd Land, West Antarctica: A test of post-100 Ma relative motion between East and West Antarctica. *Journal of Geophysical Research* 99: 15115–15139.

Driskell A.C., Norman J.A., Pruett-Jones S., Mangall E., Sonsthagen S., Christidis L. 2011. A multigene phylogeny examining evolutionary and ecological relationships in the Australo-papuan wrens of the subfamily Malurinae (Aves). *Molecular Phylogenetics and Evolution* 60: 480–485.

Dumbacher J.P., Fleischer R.C. 2001. Phylogenetic evidence for colour pattern convergence in toxic pitohuis: Müllerian mimicry in birds? *Proceedings of the Royal Society of London B* 268: 1971–1976.

Dumbacher J.P., Pratt T.K., Fleischer R.C. 2003. Phylogeny of the owlet-nightjars (Aves: Aegothelidae) based on mitochondrial DNA sequences. *Molecular Phylogenetics and Evolution* 29: 540–549.

Eagles G., Gohl K., Larter R.D. 2004. High resolution animated tectonic reconstruction of the South Pacific and West Antarctic margin. *Geochemistry, Geophysics, Geosystems* 5. doi:10.1029/2003GC000657.

Eaton J.A., van Balen B., Brickle N.W., Rheindt F.E. 2016. *Birds of the Indonesian Archipelago, Greater Sundas and Wallacea*. Lynx Editions, Barcelona, Spain.

Ebach M.C. 2017. *Reinvention of Australasian Biogeography*. CSIRO Publishing, Clayton, Australia.

Ebach M.C., Humphries C.J., Newman R.A., Williams D.M., Walsh S.A. 2005. Assumption 2: Opaque to intuition? *Journal of Biogeography* 32: 781–787.

Ebach M.C., Humphries C.J., Williams D.M. 2003. Phylogenetic biogeography deconstructed. *Journal of Biogeography* 30: 1285–1296.

Ebach M.C., Michaux B. 2017. Establishing a framework for a natural area taxonomy. *Acta Biotheoretica* 65: 167–177.

Elburg M.A., van Bergen M.J., Foden J.D. 2004. Subducted upper and lower continental crust contributes to magmatism in the collision sector of the Sunda-Banda arc, Indonesia. *Geology* 32: 41–44.

Eldredge N. 1979. Phenomenological levels and evolutionary rates. *Systematic Biology* 31: 338–347.

Eldredge N., Gould S.J. 1972. Punctuated equilibria. An alternative to phyletic gradualism. In: Schopf T.J.M. (ed.) *Models in Paleobiology*. Freeman, Cooper & Co., San Francisco, CA, pp. 82–115.

Ely K.S., Sandiford M., Hawke M.L., Phillips D., Quigley M., dos Rios J.E. 2011. Evolution of Atauro Island: Temporal constraints on subduction processes beneath the Wetar zone, Banda Arc. *Journal of Asian Earth Sciences* 41: 477–493.

Emberson R.M. 2002. The beetle fauna (Coleoptera) of the Chatham Islands, with additional species, corrections, and updated taxonomic, biological and distributional information to end 2002. Lincoln Research Archive, http://researcharchive.lincoln.ac.nz/handle/10182/4177 (accessed 14 October 2018).

Escapa I.H., Catalano S.A. 2013. Phylogenetic analysis of Araucariaceae: Integrating molecules, morphology, and fossils. *International Journal of Plant Sciences* 174: 1153–1170.

Evans M.R. 2012. Modelling ecological systems in a changing world. *Philosophical Transactions of the Royal Society B* 367: 181–190.

Filardi C.E., Smith C.E. 2005. Molecular phylogenetics of monarch flycatchers (genus *Monarcha*) with emphasis on Solomon Island endemics. *Molecular Phylogenetics and Evolution* 37: 776–788.

Fitzgerald P. 2002. Tectonics and landscape evolution of the Antarctic plate since the breakup of Gondwana, with emphasis on the West Antarctic Rift System and the Transantarctic Mountains. *Royal Society of New Zealand Bulletin* 35: 453–469.

Fleming C.A. 1949. The geological history of New Zealand. *Tuatara* 2: 72–90.

Fleming C.A. 1963. Age of the alpine biota. *Proceedings of the New Zealand Ecological Society* 10: 15–18.

Fleming C.A. 1965. The description of the New Zealand Cenozoic Mollusca: A historical survey. *New Zealand Journal of Geology and Geophysics* 8: 1149–1174.

Fleming C.A. 1966. Cenozoic history of Indo-Pacific and other warm-water elements in the marine Mollusca of New Zealand. *Venus* 25: 105–117.

Fleming C.A. 1971. A new species of cicada from rock fans in southern Wellington, with a review of three species with similar songs and habitat. *New Zealand and Journal of Science* 14: 443–479.

Fleming C.A. 1979. *The Geological History of New Zealand and Its Life*. Auckland University Press, Auckland, New Zealand.

Foden J.D. 1979. The petrology of some young volcanic rocks from Lombok and Sumbawa, Lesser Sunda Islands. Unpublished PhD thesis, University of Tasmania, Hobart.

Fritts T.H., Rodda G.H. 1998. The role of introduced species in the degradation of island ecosystems: A case history of Guam. *Annual review of Ecology and Systematics* 29: 113–40.

Georges A., Zhang X., Unmack P., Reid B.N., Le M., McCord W.P. 2014. Contemporary genetic structure of an endemic freshwater turtle reflects Miocene orogenesis of New Guinea. *Biological Journal of the Linnean Society* 111: 192–208.

Gilliard E.T. 1969. *Birds of Paradise and Bower Birds*. Weidenfeld and Nicolson, London, UK.

Glen R.A. 2005. The Tasmanides of eastern Australia. In: Vaughn A.P.M., Leat P.T., Pankhurst R.J. (eds) *Terrane Processes at the Margins of Gondwana*. Geological Society, London, UK, Special Publication 246, pp. 23–96.

Gohl K., Denk A., Eagles G., Wobbe F. 2013. Deciphering tectonic phases of the Amundsen Sea Embayment shelf, West Antarctica, from a magnetic anomaly grid. *Tectonophysics* 585: 113–123.

Goldberg J., Trewick S.A., Paterson A.M. 2008. Evolution of New Zealand's terrestrial fauna: A review of the molecular evidence. *Philosophical Transactions of the Royal Society B* 363: 3319–3334.

Goldstone J.A. 1998. Initial conditions, general laws, path dependence, and explanation in historical sociology. *American Journal of Sociology* 104: 829–845.
Goloboff P.A., Farris J.S., Nixon K.C. 2008. TNT, a free program for phylogenetic analysis. *Cladistics* 24: 774–786.
Golonka J. 2004. Plate tectonic evolution of the southern margin of Eurasia in the Mesozoic and Cenozoic. *Tectonophysics* 381: 235–273.
Gould S.J., Eldredge N. 1977. Punctuated equilibria: The tempo and mode of evolution reconsidered. *Paleobiology* 3: 115–151.
Grobys J.W.G., Gohl K., Eagles G. 2008. Quantitative tectonic reconstructions of Zealandia based on crustal thickness estimates. *Geochemistry, Geophysics, Geosystems* 9: Q01005.
Haddrath O., Baker A.J. 2012. Multiple nuclear genes and retroposons support vicariance and dispersal of the palaeognaths, and an Early Cretaceous origin of modern birds. *Proceedings of the Royal Society Series B: Biological Sciences* 279: 4617–4625.
Hakim A.S., Hall R. 1991. Tertiary volcanic rocks from the Halmahera Arc, Eastern Indonesia. *Journal of Southeast Asian Earth Sciences* 6: 271–287.
Hall R. 2011. Australia–SE Asia collision: Plate tectonics and crustal flow. In: Hall R., Cottam M.A., Wilson M.E.J. (eds) *The SE Asian Gateway: History and Tectonics of the Australia–Asia Collision*. Geological Society, London, UK, Special Publications 355, pp. 75–109.
Hall R. 2012. Late Jurassic-Cenozoic reconstructions of the Indonesian region and the Indian Ocean. *Tectonophysics* 570–571: 1–41.
Hall R., Cloke I.R., Nur'aini S., Puspita S.D., Calvert S.J., Elders C.F. 2009. The North Makassar Straits: What lies beneath? *Petroleum Geoscience* 15: 147–158.
Hall R., Patria A., Adhitama R., Pownall J.M., White L.T. 2017. Seram, the Seram Trough, the Aru trough, the Tanimbar trough and the Weber Deep: A new look at major structures in the eastern Banda Arc. Proceedings, Indonesian Petroleum Association Forty-First Annual Convention and Exhibition, Jakarta, Indonesia, May 2017, pp 1–19.
Hansford J.P., Turvey S.T. 2018. Unexpected diversity within the extinct elephant birds (Aves: Aepyornithidae) and a new identity for the world's largest bird. *Royal Society Open Science* 5: 181295.
Harold A.S. 2005. Peixes do Rio Negro. *Fishes of the Rio Negro. Copeia* 1: 212–214.
Harris R. 2006. Rise and fall of the Eastern Great Indonesian arc recorded by the assembly, dispersion and accretion of the Banda Terrane, Timor. *Gondwana Research* 10: 207–231.
Harshman J., Braun E.L., Braun M.J., Huddleston C.J., Bowie R.C.K., Chojnowski J.L., Hackett S.J., Han K.L., Kimball R.T., Marks B.D., Miglia K.J. 2008. Phylogenomic evidence for multiple losses of flight in ratite birds. *Proceedings of the National Academy of Sciences of the United States of America* 105: 13462–13467.
Hartwich S.J., Conran J.G., Bannister J.M., Lindqvist J.K., Lee D.E. 2010. Calamoid fossil palm leaves and fruits (Arecaceae: Calamoideae) from Late Eocene Southland, New Zealand. *Australian Systematic Botany* 23: 131–140.
Hayward B.W., Moore P.R., Gibson G. 1990. How warm was the Late Oligocene in New Zealand? Coconuts, reef corals and larger foraminifera. *Geological Society of New Zealand Newsletter* 90: 39–41.
Heads M. 2008. Panbiogeography of New Caledonia, south-west Pacific: Basal angiosperms on basement terranes, Ultramafic endemics inherited from

volcanic island arcs and old taxa endemic to young islands. *Journal of Biogeography* 35: 2153–2175.
Heads M. 2014. *Biogeography of Australasia*. Cambridge University Press, Cambridge, UK.
Heads M. 2017. *Biogeography and Evolution in New Zealand*. CRC Press, Florida, USA.
Heenan P.B., McGlone M.S. 2013. Evolution of New Zealand alpine and open-habitat plant species during the late Cenozoic. *New Zealand Journal of Ecology* 37: 105–113.
Heenan P.B., Mitchell A.D., de Lange P.J., Keeling J., Paterson A.M. 2010. Late-Cenozoic origin and diversification of Chatham Islands endemic plant species revealed by analyses of DNA sequence data. *New Zealand Journal of Botany* 48: 83–136.
Henderson I.M. 1991. Biogeography without area? *Australian Systematic Botany* 4: 59–71.
Hill K.C., Hall R. 2003. Mesozoic-Cenozoic evolution of Australia's New Guinea margin in a west Pacific context. *Geological Society of America* Special Publication 22: 265–289.
Hill R.S. 2004. Origins of the southeastern Australian vegetation. *Philosophical Transactions of the Royal Society of London B* 359: 1537–1549.
Hipsley C.A., Müller J. 2014. Beyond fossil calibrations: Realities of molecular clock practices in evolutionary biology. *Frontiers in Genetics* 5: 138.
Hoernle K., White J.D.L., van den Bogaard P., Hauff F., Coombs D.S., Werner R., Timm C., Garbe-Schönberg R.A., Cooper A.F. 2006. Cenozoic intraplate volcanism on New Zealand: Upwelling induced by lithospheric removal. *Earth and Planetary Science Letters* 248: 335–352.
Hollis C.J., Beu A.G., Raine J.I., Strong C.P., Turnbull I.M., Waghorn D.B., Wilson G.J. 1997. Integrated biostratigraphy of Cretaceous-Paleogene strata on Campbell Island, Southwest Pacific. Institute of Geological and Nuclear Sciences science report 97/25, 56 pp.
Hollis C.J., Taylor K.W.R., Handley L., Pancost R.D., Huber M., Creech J.B., Hines B.R., Crouch E.M., Morgans H.E.G., Crampton J.S., Gibbs S., Pearson P.N., Zachos J.C. 2012. Early Paleogene temperature history of the Southwest Pacific Ocean: Reconciling proxies and models. *Earth and Planetary Science Letters* 349–350: 53–66.
Holmes A. 1931. Radioactivity and earth movements. *Transactions of the Geological Society of Glasgow* 18: 559–606.
Hopper J.R., Funck T., Tucholke B.E., Larsen H.C., Holbrook W.S., Louden K.E., Shillington D., Lau H. 2004. Continental breakup and the onset of ultraslow seafloor spreading off Flemish Cap on the Newfoundland rifted margin. *Geology* 32: 93–96.
Irestedt M., Jønsson K.A., Fjeldså J., Christidis L., Ericson P.G.P. 2009. An unexpectedly long history of sexual selection in birds-of-paradise. *BMC Evolutionary Biology* 9: 235.
Ives A.R., Carpenter S.R. 2007. Stability and diversity of ecosystems. *Science* 317: 58–62.
Jackson J.A., Carroll E.L., Smith T.D., Zerbini A.N., Patenaude N.J., Baker C.S. 2016. An integrated approach to historical population assessment of the great whales: Case of the New Zealand southern right whale. *Royal Society Open Science* 3: 150669.

Johnson M.A., Clark J.R., Wagner W.L., McDade L.A. 2017. A molecular phylogeny of the Pacific clade of *Cyrtandra* (Gesneriaceae) reveals a Fijian origin, recent diversification, and the importance of founder events. *Molecular Phylogenetics and Evolution* 116: 30–48.

Johnston P. 2011. New morphological evidence supports congruent phylogenies and Gondwana vicariance for palaeognathous birds. *Zoological Journal of the Linnean Society* 163: 959–982.

Jones B.G. 1970. Paleontology and paleocology of Pakaurangi point, Kaipara, New Zealand. *Transactions of the Royal Society of New Zealand* 7: 137–176.

Jønnson K.A., Blom M.P.K., Päckert M., Ericson P.G.P., Irestedt M. 2018. Relicts of the lost arc: High-throughput sequencing of the *Eutrichomyias rowleyi* (Aves: Passeriformes) holotype uncovers an ancient biogeographic link between the Philippines and Fiji. *Molecular Phylogenetics and Evolution* 120: 28–32.

Jordan G.J., Carpenter R.J., Bannister J.M., Lee D.E., Mildenhall D.C., Hill R.S. 2011. High conifer diversity in Oligo-Miocene New Zealand. *Australian Systematic Botany* 24: 121–136.

Kaiser H., Carvalho V.L., Ceballos J., Freed P., Heacox S., Lester B., Richards S.J., Trainor C.R., Sanchez C., O'Shea M. 2011. The herpetofauna of Timor-Leste: A first report. *ZooKeys* 109: 19–86.

Kearns A.M., Joseph L., Cook L.G. 2013. A multilocus coalescent analysis of the speciational history of the Australo-Papuan butcherbirds and their allies. *Molecular Phylogenetics and Evolution* 66: 941–952.

Keey R.B. 2004. Some plants of Campbell and Enderby Islands. *Canterbury Botanical Society Journal* 38: 40–45.

Kennedy E.M. 2003. Late Cretaceous and Paleocene terrestrial climates of New Zealand: Leaf fossil evidence from South Island assemblages. *New Zealand Journal of Geology and Geophysics* 46: 295–306.

Kennedy E.M., Spicer R.A., Rees P.M. 2002. Quantitative palaeoclimate estimates from Late Cretaceous and Paleocene leaf floras in the northwest of the South Island, New Zealand. *Palaeogeography, Palaeoclimatology, Palaeoecology* 184: 321–345.

King P.R. 2000. Tectonic reconstructions of New Zealand: 40 Ma to the Present. *New Zealand Journal of Geology and Geophysics* 43: 611–638.

Kipf A., Hauff F., Werner R., Gohl K., van den Bogaard P., Hoernle K., Maicher D., Klügel A. 2014. Seamounts off the West Antarctic margin: A case for non-hotspot driven intraplate volcanism. *Gondwana Research* 25: 1660–1679.

Knapp M., Mudaliar R., Havell D., Wagstaff S.J., Lockhart P.J. 2007. The drowning of New Zealand and the problem of *Agathis*. *Systematic Biology* 56: 862–870.

Kranitz M.L., Biffin E., Clark A., Hollingsworth M.L., Ruhsam M., Gardner M.F., Thomas P., Mill R.R., Ennos R.A., Gaudeul M., Lowe A.J., Hollingsworth P.M. 2014. Evolutionary diversification of New Caledonian Araucaria. *PLOS ONE* 9: e110308.

Kuschel G. (ed.) 1975. *Biogeography and Ecology in New Zealand*. Dr. W. Junk Publisher, The Hague, the Netherlands.

Lamb S., Mortimer N., Smith E., Turner G. 2016. Focusing of relative plate motion at a continental transform fault: Cenozoic dextral displacement >700 km on New Zealand's Alpine Fault, reversing >225 km of Late Cretaceous sinistral motion. *Geochemistry, Geophysics, Geosystems* 17: 1197–1213.

Landis C.A., Campbell H.J., Begg J.G., Mildenhall D.C. 2008. The Waipounamu Erosion Surface: Questioning the antiquity of the New Zealand land surface and terrestrial fauna and flora. *Geological Magazine* 145: 173–197.

Lee D.E., Lee W.G., Jordan G.J., Barreda V.D. 2016. The Cenozoic history of New Zealand temperate rainforests: Comparisons with southern Australia and South America. *New Zealand Journal of Botany* 54: 1–28.

Le Péchon T., Dai Q., Zhang L.-B., Gao X.-F., Sauquet H. 2015. Diversification of Dombeyoideae (Malvaceae) in the Mascarenes: Old taxa on young islands? *International Journal of Plant Science* 176: 211–221.

Lewin R. 1985. Hawaiian *Drosophila*: Young Islands, Old Flies: Molecular evidence on Hawaiian drosophilines indicates that they colonized the 6-million-year-old islands more than 40 million years ago. *Science* 229: 1072–1074.

Linse K., Griffiths H.J., Barnes D.K.A., Clarke A. 2006. Biodiversity and biogeography of Antarctica and sub-Antarctic mollusca. *Deep Sea Research II* 53: 985–1008.

Lithoreau P., Ducrocq S. 2007. Family Anthracotheriidae. In: Prothero L.D.R., Foss S.E. (eds) *The Evolution of the Artiodactyls*. Johns Hopkins, Baltimore, MD, pp. 89–105.

Liu N., Zhu Y., Wei Z., Chen J., Wang Q., Jian S., Zhou D., Shi J., Yang Y., Zhong Y. 2009. Phylogenetic relationships and divergence times of the family Araucariaceae based on the DNA sequences of eight genes. *Chinese Science Bulletin* 54: 2648–2655.

Livermore R., Nankivell A., Eagles G., Morris P. 2005. Paleogene opening of Drake Passage. *Earth and Planetary Science Letters* 236: 459–470.

Lohman D.J., de Bruyn M., Page T., von Rintelen K., Hall R., Ng P.K.L., Shih H.-T., Carvalho G.R., von Rintelen T. 2011. Biogeography of the Indo-Australian archipelago. *Annual Review of Ecology, Evolution, and Systematics* 42: 205–226.

Luyendyk B.P. 1995. Hypothesis for Cretaceous rifting of east Gondwana caused by subducted slab capture. *Geology* 23: 373– 376.

Macqueen P., Seddon J.M., Austin J.J., Hamilton S., Goldizen A.W. 2010. Phylogenetics of the pademelons (Macropodidae: *Thylogale*) and historical biogeography of the Australo-Papuan region. *Molecular Phylogenetics and Evolution* 57: 1134–1148.

Mason R.G., Raff A.D. 1961. Magnetic Survey off the West Coast of North America, 32° N. Latitude to 42° N. Latitude. *Geological Society of America* 72: 1259–1266.

Matthews K.J., Seton M., Müller R.D. 2012. A global-scale plate reorganisation event at 105–100 Ma. *Earth and Planetary Science Letters* 355–356: 283–298.

Matthews K.J., Williams S.E., Whittaker J.M., Müller R.D., Seton M., Clarke G.L. 2015. Geologic and kinematic constraints on Late Cretaceous to mid Eocene plate boundaries in the southwest Pacific. *Earth-Science Reviews* 140: 72–107.

McAdoo D., Laxon S. 1997. Antarctic tectonics: Constraints from an ERS-1 satellite marine gravity field. *Science* 276: 556–561.

McCann K.S. 2000. The diversity-stability debate. *Nature* 405: 228–233.

McClelland P. 2011. Campbell Island – Pushing the boundaries of rat eradications. In: Veitch C.R., Clout M.N., Towns D.R. (eds) *Island Invasives: Eradication and Management*. IUCN, Gland, Switzerland, pp. 204–207.

McClelland P., Tyree P. 2002. Eradication: The Clearance of Campbell Island. *New Zealand Geographic* 58: 86–94.

McEwen M. 2005. *Charles Fleming, Environmental Patriot*. Craig Potton, Nelson, New Zealand.

McEwen M. 2006. *Charles Fleming's Cape Expedition Diary*. McEwen Associates, Wellington, New Zealand.

McGlone M.S. 1985. Plant biogeography and the late Cenozoic history of New Zealand. *New Zealand Journal of Botany* 23: 723–749.

McKenzie D.P., Parker R.L. 1967. The North Pacific: An example of tectonics on a sphere. *Nature* 216: 1276–1280.

McKewen M. 2005. *Charles Fleming, Environmental Patriot*. Craig Potton, Nelson, New Zealand.

Meredith R.W., Mendoza M.A., Roberts K.K., Westerman M., Springer M.S. 2010. A phylogeny and timescale for the evolution of Pseudocheiridae (Marsupialia: Diprotodontia) in Australia and New Guinea. *Journal of Mammalian Evolution* 17: 75–99.

Metcalfe I. 2006. Palaeozoic and Mesozoic tectonic evolution and palaeogeography of East Asian crustal fragments: The Korean Peninsula in context. *Gondwana Research* 9: 24–46.

Metcalfe I. 2009. Palaeozoic and Mesozoic tectonic evolution and palaeogeography of East Asian crustal fragments: The Korean Peninsula in context. *Gondwana Research* 9: 24–46.

Metcalfe I. 2011. Tectonic framework and Phanerozoic evolution of Sundaland. *Gondwana Research* 19: 3–21.

Metcalfe I. 2013. Gondwanan dispersion and Asian accretion: Tectonic and palaeogeographic evolution of eastern Tethys. *Journal of Asian Earth Sciences* 66: 1–33.

Meudt H.M., Simpson B.B. 2006. The biogeography of the austral, subalpine genus *Ourisia* (Plantaginaceae) based on molecular phylogenetic evidence: South American origin and dispersal to New Zealand and Tasmania. *Biological Journal of the Linnean Society* 87: 479–513.

Meurk C.D., Foggo M.N., Wilson J.B. 1994. The vegetation of subantarctic Campbell Island. *New Zealand Journal of Ecology* 18: 123–168.

Michaux B. 1988. Organotaxism: An alternative way of seeing the fossil record. *Journal of Theoretical Biology* 133: 397–408.

Michaux B. 1989. Morphological variation of species through time. *Biological Journal of the Linnean Society* 38: 239–255.

Michaux B. 1991. The evolution of the Ancillinae with special reference to New Zealand tertiary and Recent *Amalda* and *Alocospira* (Gastropoda: Olividae: Ancillinae). *Venus* 50: 130–149.

Michaux B. 1994. Land movements and animal distributions in east Wallacea (eastern Indonesia, Papua New Guinea and Melanesia). *Palaeogeography, Palaeoclimatology, Palaeoecology* 112: 323–343.

Michaux, B. 1996. The origin of southwest Sulawesi and other Indonesian terranes: a biological view. *Palaeogeography, Palaeoclimate and Palaeogeography* 122: 167–183.

Michaux, B. 2008. Alfred Russel Wallace - Biogeographer. In: *Beyond Natural selection. The intellectual Legacy of Alfred Russel Wallace*. Oxford University Press, Oxford.

Michaux, B. 2009. Reciprocality between biology and geology: Reconstructing polar Gondwana. *Gondwana Research* 16: 655–668.

Michaux, B. 2010. Biogeology of Wallacea: geotectonic models, areas of endemism, and natural biogeographical units. *Biological Journal of the Linnean Society* 101: 193–212.

Michaux B. 2014. *Tewkesbury Walks: Essays in Biogeography and Evolution*. Springer, New York, NY.

Michaux, B., Leschen, R. A. B. 2005. East meets west: biogeology of the Campbell Plateau. *Biological Journal of the Linnean Society* 86:95–115.

Mitchell K.J., Llamas B., Soubrier J., Rawlence N.J., Worthy T.H., Wood J., Lee M.S.Y., Cooper A. 2014. Ancient DNA reveals elephant birds and kiwi are sister taxa and clarifies ratite bird evolution. *Science* 344: 898–900.

Moore P.R., Wallace R. 2000. Petrified wood from the Miocene volcanic sequence of Coromandel Peninsula, northern New Zealand. *Journal of the Royal Society of New Zealand* 30: 115–130.

Mori A.S., Furukawa T., Sasaki T. 2013. Response diversity determines the resilience of ecosystems to environmental change. *Biological Reviews of the Cambridge Philosophical Society* 88: 349–364.

Morrone J.J. 2015. Track analysis beyond panbiogeography. *Journal of Biogeography* 42: 413–425.

Morrone J.J., Carpenter J.M. 1994. In search of a method for cladistic biogeography: An empirical comparison of component analysis, Brooks Parsimony Analysis, and three-area statements. *Cladistics* 10: 99–153.

Morrone J.J., Crisci J.V. 1995. Historical biogeography: Introduction to methods. *Annual Review of Ecology and Systematics* 26: 373–401.

Mortimer N. 2018. Evidence for a pre-Eocene proto-Alpine fault through Zealandia. *New Zealand Journal of Geology and Geophysics* 61: 251–259.

Mortimer N., Campbell H.J., Tulloch A.J., King P.R., Stagpoole V.M., Wood R.A., Rattenbury M.S., Sutherland R., Adams C.J., Collot J., Seton M. 2017. Zealandia: Earth's hidden continent. *Geological Society of America Today* 27: 27–35.

Mortimer N., Davey F.J., Melhuish A., Yu J., Godfrey N.J. 2002. Geological interpretation of a deep seismic reflection profile across the Eastern Province and Median Batholith, New Zealand: Crustal architecture of an extended Phanerozoic convergent orogeny. *New Zealand Journal of Geology and Geophysics* 45: 349–363.

Mortimer N., Rattenbury M.S., King P.R., Bland K.J., Barrell D.J.A., Bache F., Begg J.G., Campbell H.J., Cox S.C., Crampton J.S., Edbrooke S.W., Forsyth P.J., Johnston M.R., Jongens R., Lee J.M., Leonard G.S., Raine J.I., Skinner D.N.B., Timm C., Townsend D.B., Tulloch A.J., Turnbull I.M., Turnbull R.E. 2014. High-level stratigraphic scheme for New Zealand Rocks. *New Zealand Journal of Geology and Geophysics* 57: 402–419.

Müller-Wille S. 2013. Systems and how Linnaeus looked at them in retrospect. *Annals of Science* 70: 305–317.

Musser G.G. 1981. The giant rat of Flores and its relatives east of Borneo and Bali. *Bulletin of the American Museum of Natural History* 169: 71–175.

Nelson G., Ladiges P.Y. 1991. Three-area statements: Standard assumptions for biogeographic analysis. *Systematic Biology* 40: 470–485.

Nelson G., Ladiges P.Y. 1996. Paralogy in cladistic biogeography and analysis of paralogy-free subtrees. *American Museum Novitates* 3167: 1–58.

Nelson G., Platnick N.I. 1981. *Systematics and Biogeography: Cladistics and Vicariance*. Columbia University Press, New York, NY.

Norvick M.S. 1979. The tectonic history of the Banda Arcs, eastern Indonesia: A review. *Journal of the Geological Society* 136: 519–526.

Norvick M.S., Smith M.A. 2001. Mapping the plate tectonic reconstruction of southern and southeastern Australia and implications for petroleum systems. *APPEA Journal* 41: 15–35.

Nugraha A.M.S., Hall R. 2018. Late Cenozoic palaeogeography of Sulawesi, Indonesia. *Palaeogeography, Palaeoclimatology, Palaeoecology* 490: 191–209.

Nyári A.S., Peterson A.T., Rice N.H., Moyle R.G. 2009. Phylogenetic relationships of flowerpeckers (Aves: Dicaeidae): Novel insights into the evolution of a tropical passerine clade. *Molecular Phylogenetics and Evolution* 53: 613–619.

Ober J. 2002. Not by a nose. In: Cowley R. (ed.) *What If?* 2. Putnam's Sons, New York, NY.

O'Connor T. 1999. *New Zealand's Subantarctic Islands*. Reed Books, Auckland, New Zealand.

Oliver L.A., Rittmeyer E.N., Kraus F., Richards S.J., Austin C.C. 2013. Phylogeny and phylogeography of *Mantophryne* (Anura: Microhylidae) reveals cryptic diversity in New Guinea. *Molecular Phylogenetics and Evolution* 67: 600–607.

Ota T., Kaneko Y. 2010. Blueschists, eclogites, and subduction zone tectonics: Insights from a review of Late Miocene blueschists and eclogites, and related young high-pressure metamorphic rocks. *Gondwana Research* 18: 167–188.

Otake M., Takahashi H., Koseki T., Yoshiyama H. 2002. Geology, geochemistry and geochronology of the Bajawa area, central Flores, Indonesia: Geologic structure and evolution of the Bajawa depression. *Bulletin of the Geological Survey of Japan* 53: 161–173.

Oxburgh E.R. 1972. Flake tectonics and continental collision. *Nature* 239: 202–204.

Page R.D.M. 1987. Graphs and generalized tracks: Quantifying Croizat's panbiogeography. *Systematic Zoology* 36: 1–17.

Page R.D.M. 1990. Tracks and trees in the antipodes: A reply to Humphries and Seberg. *Systematic Zoology* 39: 288–299.

Page R.D.M. 1994. Maps between trees and cladistic analysis of historical associations among genes, organisms, and areas. *Systematic Biology* 43: 58–77.

Parenti L.R., Ebach M.C. 2009. *Comparative Biogeography: Discovering and Classifying Biogeographical Patterns of a Dynamic Earth*. University of California Press, Berkeley, CA.

Payn K.G., Dvorak W.S., Myburg A.A. 2007. Chloroplast DNA phylogeography reveals the island colonization route of *Eucalyptus urophylla* (Myrtaceae). *Australian Journal of Botany* 55: 673–683.

Phillips M.J., Gibb G.C., Crimp E.A., Penny D. 2010. Tinamous and moa flock together: Mitochondrial genome sequence analysis reveals independent losses of flight among ratites. *Systematic Biology* 59: 90–107.

Pieters P.E., Ryburn R.J., Trail D.S. 1979. Geological Reconnaissance in Irian Jaya 1976 and 1977. Bureau of Mineral Resources, Geology and Geophysics, pp. 95.

Pigram C.J., Davies H.L. 1987. Terranes and the accretion history of the New Guinea orogeny. *BMR Journal of Australian Geology and Geophysics* 10: 193–211.

Pizzey G., Knight F. 2012. *A Field Guide to the Birds of Australia* (9th Edition). Harper Collins, Sydney, Australia.

Platnick N.I., Nelson G. 1978. A method of analysis for historical biogeography. *Systematic Zoology* 27: 1–16.

Pocknall D.T. 1989. Late Eocene to Early Miocene vegetation and climate history of New Zealand. *Journal of the Royal Society of New Zealand* 19: 1–18.

Pole M. 1994. The New Zealand flora – Entirely long-distance dispersal? *Journal of Biogeography* 21: 625–635.

Pole M. 2008. The record of Araucariaceae macrofossils in New Zealand. *Alcheringa* 32: 405–426.

Pole M. 2014. The Miocene climate in New Zealand: Estimates from paleobotanical data. *Palaeontologia Electronica* 17(2,27A): 79 pp.

Pole M., Douglas B., Mason G. 2003. The terrestrial Miocene biota of southern New Zealand. *Journal of the Royal Society of New Zealand* 33: 415–426.

Pole M., Vajda V. 2009. A new terrestrial Cretaceous-Paleogene site in New Zealand – Turnover in macroflora confirmed by palynology. *Cretaceous Research* 30: 917–938.

Polhemus D.A., Polhemus J.T. 1998. Assembling New Guinea: 40 million years of island arc accretion as indicated by the distributions of aquatic Heteroptera (Isecta). In: Hall R., Holloway J.D. (eds) *Biogeography and Geological Evolution of SE Asia.* Backhuys, Leiden, the Netherlands, pp. 327–340.

Poole I., Cantrill D.J. 2006. Cretaceous and Cenozoic vegetation of Antarctica integrating the fossil wood record. In: Pirrie F.J.E., Crame J.A. (eds) *Cretaceous-Tertiary High-Latitude Palaeoenvironments, James Ross Basin, Antarctica.* Geological Society Special Publication, London. 258: 63–81.

Popper K. 1957. *The Poverty of Historicism.* Routledge & Paul, London, UK.

Powell C.McA., Roots S.R., Veevers J.J. 1988. Pre-breakup continental extension in East Gondwanaland and the early opening of the eastern Indian Ocean. *Tectonophysics* 155: 261–283.

Raby P. 2001. *Alfred Russel Wallace: A Life.* Princeton University Press, Princeton, NJ.

Rangin C., Pubellier M., Azéma J., Briais A., Chotin P., Fontaine H., Huchon P., Joliet L., Maury R., Müller C., Rampnoux J.-P., Stephan J.-F., Tournon J., Cottereau N., Dercourt J., Ricou L.-E. 1990. The quest for Tethys in the western Pacific. 8 Paleogeodynamic maps for Cenozoic time. *Bulletin of the Geological Society of France* 8: 907–913.

Rasmussen M., et al. 2011. An aboriginal Australian genome reveals separate human dispersals into Asia. *Science* 334: 94–98.

Raven P.H. 1973. Evolution of the subalpine and alpine plant groups in New Zealand. *New Zealand Journal of Botany* 11: 177–200.

Rayment W., Davidson A., Dawson S., Slooten E., Webster T. 2012. Distribution of southern right whales on the Auckland Islands calving grounds. *New Zealand Journal of Marine and Freshwater Research* 46: 431–436.

Read J., Francis J. 1992. Responses of some Southern Hemisphere tree species to a prolonged dark period and their implications for high-latitude Cretaceous and Tertiary floras. *Palaeogeography, Palaeoclimatology, Palaeoecology* 99: 271–290.

Reichgelt T., Kennedy E.M., Conran J.G., Mildenhall D.C., Lee D.E. 2015. The early Miocene paleolake Manuherikia: Vegetation heterogeneity and warm-temperate to subtropical climate in southern New Zealand. *Journal of Paleolimnology* 53: 349–365.

Rigg J.W.D., Hall R. 2011. Structural and stratigraphic evolution of the Savu Basin, Indonesia. In: Hall R., Cottam M.A., Wilson M.E.J. (eds) *The SE Asian Gateway: History and Tectonics of the Australia–Asia Collision.* Geological Society Special Publication, London. 355: 225–240.

Ritchie D.D., Turnbull I.M. 1985. Cenozoic sedimentary rocks at Carnley Harbour, Auckland Islands, Campbell Plateau. *New Zealand Journal of Geology and Geophysics* 28: 23–41.

Robertson A.H.F. 2007. Overview of tectonic settings related to the rifting and opening of Mesozoic ocean basins in the eastern Tethys: Oman, Himalayas and eastern Mediterranean regions. In: Karner G.D., Manatschal G., Pinheiro L.M. (eds.) *Imaging, Mapping, and Modelling Continental Lithosphere Extension and Breakup*. Geological Society, London, UK, Special Publications 282, pp. 325–388.

Robertson C.J.R., Hyvönen P., Fraser M.J., Pickard C.R. 2007. *Atlas of Bird Distribution in New Zealand 1999–2004*. The Ornithological Society of New Zealand, Wellington, New Zealand.

Roosmawati N., Harris R. 2009. Surface uplift history of the incipient Banda arc-continent collision: Geology and synorogenic foraminifera of Rote and Savu Islands, Indonesia. *Tectonophysics* 479: 95–110.

Rosen D.E. 1978. Vicariant Patterns and historical explanation in biogeography. *Systematic Zoology* 27: 159–188.

Rosenbaum G., Lister G.S., Duboz C. 2002. Reconstruction of the tectonic evolution of the western Mediterranean since the Oligocene. *Journal of the Virtual Explorer* 8: 107–130.

Ruiz G.M., Fofonoff P.W., Carlton J.T., Wonham M.J., Hines A.H. 2000. Invasion of coastal marine communities in North America: Apparent patterns, processes, and biases. *Annual Review of Ecology and Systematics* 31: 481–531.

Russ R. 2007. History, exploration, settlement and past use of the sub-Antarctic. *Papers and Proceedings of the Royal Society of Tasmania* 141: 169–172.

Sahlins M. 1985. *Islands of History*. The University of Chicago Press, Chicago, IL and London, UK.

Sandiford M. 2008. Seismic moment release during slab rupture beneath the Banda Sea. *Geophysical Journal International* 174: 659–671.

Satyana A.H., Armandita C., Tarigan R.L. 2008. Collision and post-collision tectonics in Indonesia: Roles for basin formation and petroleum systems. Proceedings, Indonesian Petroleum Association, Thirty-Second Annual Convention & Exhibition, Jakarta, Indonesia, May 2008.

Satyana A.H., Purwaningsih M.E.M. 2011. Collision of micro-continents with eastern Sulawesi: Records from uplifted reef terraces and proven-potential petroleum plays. Proceedings, Indonesian Petroleum Association, Thirty-Fifth Annual Convention & Exhibition, Jakarta, Indonesia, May 2011.

Schellart W.P., Lister G.S., Toy V.G. 2006. A Late Cretaceous and Cenozoic reconstruction of the Southwest Pacific region: Tectonics controlled by subduction and slab rollback processes. *Earth-Science Reviews* 76: 191–233.

Schulte I.I.J.A., Melville J., Larson A. 2002. Molecular evidence for ancient divergence of lizard taxa on either side of Wallace's Line. *Proceedings of the Royal Society of London Series B* 270: 597–603.

Schweizer M., Wright T.F., Peñalba J.V., Schirtzinger E.E., Joseph L. 2015. Molecular phylogenetics suggests a New Guinean origin and frequent episodes of founder-event speciation in the nectariverous lories and lorikeets (Aves: Psittaciformes). *Molecular Phylogenetics and Evolution* 90: 34–48.

Scott, J.M., Turnbull, I.M., Auer, A., Palin, J.M. 2013. The sub-Antarctic Antipodes Volcano: a <0.5 Ma HIMU-like Surtseyan volcanic outpost on the edge of the Campbell Plateau, New Zealand. *New Zealand Journal of Geology and Geophysics* 56:134–153.

Scott J.M., Turnbull I.M., Sagar M.W., Tulloch A.J., Waight T.E., Palin J.M. 2015. Geology and geochronology of the sub-Antarctic Islands/Tini Heke, New Zealand. *New Zealand Journal of Geology and Geophysics* 58: 202–212.

Sdrolias M., Müller R.D., Gaina C. 2001. Plate tectonic evolution of eastern Australian marginal ocean basins. Eastern Australian Basin Symposium 2001, Melbourne, Australia. pp. 227–237.

Severin T. 1998. *The Spice Island Voyage*. Abacus, London.

Sewell D.M., Wheatley C.J.V. 1994. The Lerokis and Kali Kuning submarine exhalative gold-silver-barite deposits, Wetar Island, Maluku, Indonesia. *Journal of Geochemical Exploration* 50: 351–370.

Shermer M. 2002. *In Darwin's Shadow*. Oxford University Press, Oxford, UK.

Siddoway C.S. 2008. Tectonics of the West Antarctic Rift System: New light on the history and dynamics of distributed intracontinental extension. In: Cooper A.K., Barrett, P., Stagg, H., Storey, B., Stump, E., and Wise, W. (eds) *Antarctica: A Keystone in a Changing World*. The National Academies Press, Washington, DC, pp. 91–110.

Sloan P.R. 1972. John Locke, John Ray, and the problem of the natural system. *Journal of the History of Biology* 5: 1–53.

Smith C.H. 2018. The Alfred Russel Wallace Page Website, http://people.wku.edu/charles.smith/index1.htm (accessed 12 October 2018).

Smith I.W.G. 2002. *The New Zealand Sealing Industry*. New Zealand Department of Conservation, Wellington, New Zealand.

Smith J.V., Braun E.L., Kimball R.T. 2012. Ratite nonmonophyly: Independent evidence from 40 novel loci. *Systematic Biology* 62: 35–49.

Smith J.V., Braun E.L., Kimball R.T. 2013. Ratite Nonmonophyly: Independent Evidence from 40 Novel Loci. *Systematic Biology* 62:35–49.

Smyth H.R., Hall R., Nichols G.J. 2008. Cenozoic volcanic arc history of East Java, Indonesia: The stratagraphic record of eruptions on an active continental margin. *The Geological Society of America Special Paper* 436: 199–222.

Smyth H.R., Hamilton P.J., Hall R., Kinny P.D. 2007. The deep crust beneath island arcs: Inherited zircons reveal a Gondwana continental fragment beneath East Java, Indonesia. *Earth and Planetary Science Letters* 258: 269–282.

Somers M.R. 1994. The narrative constitution of identity: A relational network approach. *Theory and Society* 23: 605–649.

Spakman W., Hall R. 2010. Surface deformation and slab-mantle interaction during Banda arc subduction rollback. *Nature Geoscience* 3: 562–566.

Stampfli G.M., Borel G.D. 2004. The TRANSMED transects in space and time: Constraints on the paleotectonic evolution of the Mediterranean domain. In: Cavazza, W., Roure, F.M., Spakman, W., Stampfli, G.M., Ziegler, P.A. (eds) *The TRANSMED Atlas. The Mediterranean Region from Crust to Mantle*. Springer-Verlag, Heidelberg, Germany.

Standley C.E., Harris R. 2009. Tectonic evolution of forearc nappes of the active Banda arc-continent collision: Origin, age, metamorphic history and structure of the Lolotoi Complex, East Timor. *Tectonophysics* 479: 66–94.

Steadman D.W. 1995. Prehistoric Extinctions of Pacific Island Birds: Biodiversity Meets Zooarchaeology. *Science* 267: 1123–1131.

Stöckler K., Daniel I.L., Lockhart P.J. 2002. New Zealand Kauri (*Agathis australis* (D. Don) Lindl., Araucariaceae) survives oligocene drowning. *Systematic Biology* 51: 827–832.

Stokes A., Dobbs K., Mantel P., Pierce S. 2005. *Fauna and Flora of the Great Barrier Reef World Heritage Area*. Great Barrier Reef Marine Park Authority, Townsville, Australia.

Strogen D.P., Seebeck H., Nicol A., King P.R. 2017. Two-phase Cretaceous-Paleocene rifting in the Taranaki Basin region, New Zealand; implications for Gondwana break-up. *Journal of the Geological Society* 174: 929–946.

Subarsyah S., Rahardiawan R. 2016. Geological structures appearances and its relation to mechanism of arc-continent collision Northern Alor-Wetar Islands. *Bulletin of the Marine Geology* 31: 55–66.

Suggate R.P. 1963. The alpine fault. *Transactions of the Royal Society of New Zealand* 2: 105–129.

Sutherland J.I. 2003. Miocene petrified wood and associated borings and termite faecal pellets from Hkatere Peninsula, Kaipara Harbour, North Auckland, New Zealand. *Journal of the Royal Society of New Zealand* 33: 395–414.

Sutherland R. 1999. Cenozoic bending of New Zealand basement terranes and Alpine Fault displacement: A brief review. *New Zealand Journal of Geology and Geophysics* 42: 295–301.

Sutherland R., Davey F., Beavan J. 2000. Plate boundary deformation in South Island, New Zealand related to inherited lithospheric structure. *Earth and Planetary Science Letters* 177: 141–151.

Sutriyono E. 2003. Provenance Study and Tectonic Implications on Rock Sequences in the Lengguru fold belt of Western papua: Constraints from zircon fission track thermochronology. *Forum Teknik* 27: 121–131.

Tennyson A.J.D. 2006. *Extinct Birds of New Zealand*. Te Papa Press, Wellington, New Zealand.

Tennyson A.J.D., Worthy T.H., Jones C.M., Scofield R.P., Hand S.J. 2010. Moa's ark: Miocene fossils reveal the great antiquity of moa (Aves: Dinornithiformes) in Zealandia. *Records of the Australian Museum* 62: 105–114.

Toledo-Piza M. 2002. *Peixes do Rio Negro. Fishes of the Rio Negro.* Universidade de Sao Paulo, Sao Paulo, Brazil.

Toussaint E.F.A., Hall R., Monaghan M.T., Sagata K., Ibalim S., Shaverdo H.V., Vogler A.P., Pons J., Balke M. 2014. The towering orogeny of New Guinea as a trigger for arthropod megadiversity. *Nature Communications* 5: 4001.

Trainor C.R. 2005a. Species richness, habitat use and conservation of birds of Alor Island, Lesser Sundas, Indonesia. *Emu* 105: 127–135.

Trainor C.R. 2005b. Birds of Tapuaafu peninsula, Roti island, Lesser Sundas, Indonesia. *Forktail* 21: 121–131.

Trainor C.R., Andersen A.N. 2010. The ant fauna of Timor and neighbouring islands: Potential bridges between the disjunct faunas of South East Asia and Australia. *Australian Journal of Zoology* 58: 133–144.

Trainor C.R., Coates B.J., Bishop K.D. 2007. *Birds of Timor-Leste*. Bird Life International and Dove Publications, Desa Putera, Indonesia.

Trainor C.R., Soares T. 2004. Birds of Atauro Island, Timor-Leste (East Timor). *Forktail* 20: 41–48.

Trewick S.A., Gibb G.C. 2010. Vicars, tramps and assembly of the New Zealand avifauna: A review of molecular phylogenetic evidence. *Ibis* 152: 226–253.

Trewick S.A., Paterson A.M., Campbell H.J. 2007. Hello New Zealand. *Journal of Biogeography* 34: 1–6.

Ung V., Michaux B., Leschen R.A.B. 2016. A comprehensive vicariant model for Southwest Pacific biotas. *Australian Systematic Botany* 29: 424–439.

Unmack P.J., Allen G.R., Johnson J.B. 2013. Phylogeny and biogeography of rainbow fishes (Melanotaeniidae) from Australia and New Guinea. *Molecular Phylogenetics and Evolution* 67: 15–27.

Vajda V., Raine J.I. 2003. Pollen and spores in marine Cretaceous/Tertiary boundary sediments at mid-Waipara River, North Canterbury, New Zealand. *New Zealand Journal of Geology and Geophysics* 46: 255–273.

Vajda V., Raine J.I. 2010. A plynological investigation of plesiosaur-bearing rocks from the Upper Cretaceous Tahora formation, Mangahouanga, New Zealand. *Alcheringa* 34: 359–374.

Van Leeuwen T., Allen C.M., Kadarusman A., Elburg M., Michael Palin J., Muhardjo, Suwijanto 2007. Petrologic, isotopic, and radiometric age constraints on the origin and tectonic history of the Malino Metamorphic complex, NW Sulawesi, Indonesia. *Journal of Asian Earth Sciences* 29: 751–777.

van Whye J. 2015. *The Annotated Malay Archipelago.* National University of Singapore Press, Singapore.

Van Wyck N., Williams I.S. 2002. Age and provenance of basement metasediments from the Kubor and Bena Blocks, central Highlands, Papua New Guinea: Constraints on the tectonic evolution of the northern Australian cratonic margin. *Australian Journal of Earth Sciences* 49: 565–577.

Veevers J.J. 2012. Reconstructions before rifting and drifting reveal the geological connections between Antarctica and its conjugates in Gondwanaland. *Earth-Science Reviews* 111: 249–318.

Veevers J.J., Powell C.McA., Roots S.R. 1991. Review of seafloor spreading around Australia. I. Synthesis of the patterns of spreading. *Australian Journal of Earth Sciences* 38: 373–389.

Villeneuve M., Martini R., Bellon H., Réhault J.P., Cornée J.J., Bellier O., Burhannuddin S., Hinschberger F., Honthaas C., Monnier C. 2010. Deciphering of six blocks of Gondwanan origin within Eastern Indonesia (South East Asia). *Gondwana Research* 18: 420–437.

Wagstaff S.J., Breitwieser I., Swenson U. 2006. Origin and relationships of the austral genus *Abrotanella* (Asteraceae) inferred from DNA sequences. *Taxon* 55: 95–106.

Wallace A.R. 1869. *The Malay Archipelago.* Macmillan, London, UK.

Wallace L.M., Stevens C., Silver E., McCaffrey R., Loratung W., Hasiata S., Stanaway R., Curley R., Rosa R., Taugaloidi J. 2004. GPS and seismological constraints on active tectonics and arc-continent collision in Papua New Guinea: Implications for mechanics of microplate rotations in a plate boundary zone. *Journal of Geophysical Research* 109: B05404.

Wardle P. 1963. Evolution and distribution of the New Zealand flora, as affected by Quaternary climates. *New Zealand Journal of Botany* 1: 3–17.

Wardle P., Ezcurra C., Ramírez C., Wagstaff S. 2001. Comparison of the flora and vegetation of the southern Andes and New Zealand. *New Zealand Journal of Botany* 39: 69–108.

Waters J.M., Craw D. 2006. Goodbye Gondwana? New Zealand biogeography, geology, and the problem of circularity. *Systematic Biology* 55: 351–356.

Watkinson I.M., Hall R. 2017. Fault systems of the eastern Indonesian triple junction: Evaluation of Quaternary activity and implications for seismic hazards. In: Cummins P.R., Meilano I. (eds) *Geohazards in Indonesia: Earth Science for Disaster Risk Reduction*. Geological Society Special Publication 441, London, UK, pp. 71–120.

Watkinson I.M., Hall R., Ferdian F. 2011. Tectonic re-interpretation of the Banggai-Sula – Molucca Sea margin, Indonesia. In: Hall R., Cottam M., Wilson M.E.J. (eds.) *The SE Asian Gateway: History and Tectonics of the Australia – Asia Collision.* Geological Society, London, UK, Special Publication 355, pp. 203–224.

Weaver S.D., Storey B.C., Pankhurst R.J., Mukasa S.B., DiVenere V.J., Bradshaw J.D. 1994. Antarctica-New Zealand rifting and Marie Byrd Land lithospheric magmatism linked to ridge subduction and mantle plume activity. *Geology* 22: 811–814.

Weissel J.K., Watts A.B. 1979. Tectonic evolution of the Coral Sea Basin. *Journal of Geophysical Research* 84: 4572–4582.

White C.M.N., Bruce M.D. 1986. *The Birds of Wallacea.* British Ornithologists' Union, London, UK.

White H. 1984. The question of narrative in contemporary historical theory. *History and Theory* 23: 1–33.

Wiley E.O. 1981. *Phylogenetics. The Theory of Phylogenetic Systematics.*Wiley-Interscience, New York, NY.

Wiley E.O., Lieberman B.S. 2011. *Phylogenetics. Theory and Practice of Phylogenetic Systematics* (2nd Edition). Wiley-Blackwell, Hoboken, New Jersy.

Wilf P., Escapa I.H. 2014. Green web or megabiased clock? Plant fossils from Gondwanan Patagonia speak on evolutionary radiations. *New Phytologist* 207: 283–290.

Wilf P., Little S.A., Iglesias A., Del Carmen Zamaloa M., Gandolfo M.A., Cúneo N.R., Johnson K.R. 2009. Papuacedrus (Cupressaceae) in Eocene Patagonia: A new fossil link to Australasian rainforests. *American Journal of Botany* 96: 2031–2047.

Wilke T., Schultheiß R., Albrecht C. 2009. As time goes by: A simple fool's guide to molecular clock approaches in invertebrates. *American Malacological Bulletin* 27: 25–45.

Williams-Ellis A. 1966. *Darwin's Moon: A Biography of Alfred Russel Wallace.* Blackie, London, UK.

Wilson H. 2007. *Alpine Plants.* Te Ara – the Encyclopedia of New Zealand, http://www.TeAra.govt.nz/en/alpine-plants (accessed 3 November 2018).

Williamson P.G. 1981. Palaeontological documentation of speciation in Cenozoic molluscs from Turkana Basin. *Nature* 293: 437–443.

Winkworth R.C., Wagstaff S.J., Glenny D., Lockhart P.J. 2005. Evolution of New Zealand mountain flora: Origins, diversification and dispersal. *Organisms Diversity & Evolution* 5: 237–247.

Wobbe F. 2014. Crustal and Lithospheric Dynamics of the Southern Pacific and the West Antarctic Margin. Unpublished PhD thesis, Jacobs University, School of Engineering and Science.

Wobbe F., Goh K., Chambord A., Sutherland R. 2012. Structure and breakup history of the rifted margin of West Antarctica in relation to Cretaceous separation from Zealandia and Bellingshausen plate motion. *Geochemistry, Geophysics, Geosystems* 13: Q04W12, doi:10.1029/2011GC003742.

Wood P.L., Heinicke M.P., Jackman T.R., Bauer A.M. 2012. Phylogeny of bent-toed geckos (Cyrtodactylus) reveals a west to east pattern of diversification. *Molecular Phylogenetics and Evolution* 65: 992–1003.

Wood R.A., Stagpoole V.M. 2007. Validation of tectonic reconstructions by crustal volume balance: New Zealand through the Cenozoic. *Geological Society of America Bulletin* 119: 933–943.

Woodhead J., Hergt J., Sandiford M., Johnson W. 2010. The big crunch: Physical and chemical expressions of arc/continent collision in the Western Bismark arc. *Journal of Volcanology and Geothermal Research* 190: 11–24.

Zahirovic S., Seton M., Müller R.D. 2014. The Cretaceous and Cenozoic tectonic evolution of Southeast Asia. *Solid Earth* 5: 227–273.

Zimmermann S., Hall R. 2014. Provenance of Mesozoic sandstones in the Banda arc, Indonesia. Proceedings, Indonesian Petroleum Association Thirty-Eighth Annual Convention and Exhibition, May 2014, 13 pp.

Zwiers P.B., Borgia G., Fleischer R.C. 2008. Plumage based classification of the bowerbird genus *Sericulus* evaluated using a multi-gene, multi-genome analysis. *Molecular Phylogenetics and Evolution* 46: 923–931.

# Index